30 Years' Review of China's Science & Technology

(1949 — 1979)

30 Years' Review of China's Science & Technology
(1949 — 1979)

Orient.
Q
127
C5
A15
1981

WORLD SCIENTIFIC

World Scientific Publishing Co Pte Ltd
P.O. Box 128
Farrer Road
Singapore 9128

ISBN 9971-950-48-0

Copyright © 1981 by World Scientific Publishing Co Pte Ltd. All rights reserved. This book, or parts thereof, may not be reproduced in any form or by any means, electronic or mechanical, including photocopying, recording or any information storage and retrieval system now known or to be invented, without written permission from the publisher.

Printed by the Singapore National Printers (Pte) Ltd.

PREFACE

A review of China's science and technology was published recently. It attempts to reflect some of the scientific and technological advances made in China over the past 30 years. It also contains a historical record of important events that occurred during the period. Brief particulars of eminent scientists who have made significant scientific and technological accomplishments are also included.

In order to allow greater accessibility to non-Chinese readers, we have decided to publish this review in English. This volume is based on English translations of the articles provided by the Shanghai Scientific Press, with minor editorial modifications.

The Editorial,
World Scientific.

Foreword

The Nature Journal Yearbook 1979 is now out in print. This is the first yearbook of natural science in our country and we warmly greet the birth of such a book.

The publication of this Yearbook during this time is by no means accidental. The establishment of the People's Republic of China in 1949 enabled our country to march towards the building of a properous socialistic New China.

The realization of the "Four Modernizations" (Industry, Agriculture, National Defence, Science and Technology) of socialism has been the long-cherished common desire of our entire nation. Immediately after the suppression of the "Gang of Four" and in response to the Party's call, the whole nation started a new surge in marching towards new heights in science and technology, striving to achieve an early realization of socialistic modernization. The publication of such yearbooks should be able to serve this purpose.

The number of articles and treatises on various subjects of natural science in the whole world published every year exceeds 100,000. This is a clear indication of the rapid advancements in science and technology. Facing such voluminous and extensive literature, one will certainly have difficulties in reading and understanding major progresses made in each subject. Consequently, there is a necessity for us to compile a book containing such information for easy reference, and we believe this Yearbook serves the purpose. It took the editorial staff much pain and effort to compile this book whose primary objective is to meet the needs of the "Four Modernizations".

Now that many yearbooks of natural science have already been published elsewhere in the world, one will tend to raise such a question: "Is it necessary to compile a yearbook of our own?" The answer is in the affirmative. China has suffered a great deal during the ten years of disaster under the dictatorship of Lin Biao and the "Gang of Four". It has also caused a further widening of the gap in terms of scientific and technological advancement between China and the advanced countries. However, it should be noted that since the founding of the People's Republic of China, we have established a scientific-technical base on a definite scale and have built up a contingent of technicians and engineers. In scientific research, we have made some achievements peculiar to our country. Following the acceleration of the pace of the "Four Modernizations", the progress made in this respect is becoming increasingly distinguished. Therefore, it is absolutely necessary to review the achievements made, reflect the work done in scientific research and to accumulate our own data and materials for our reference. All these can never be done nor substituted by any foreign yearbook. Moreover, sooner or later, we have to compile and publish our own yearbook of natural science. During this time when much emphasis is placed on the realization of the "Four Modernizations", we have, therefore, decided to compile on trial, the first yearbook of natural science of New China.

As 1979 is the 30th anniversary of the founding of the Republic of China, the first part of the Yearbook on "Feature Articles" reflects the progress made on various subjects of natural science since the founding of New China. It is good to recall the progress of natural science in New China since its establishment and record the main achievements in the first Yearbook. Further, since we are carrying forward the cause pioneered by our predecessors and forging ahead into the future, such retrospection will urge us to strive for greater successes.

In the course of compiling this Yearbook, the editorial board of Nature Journal received warm support from the scientific circles of our country, especially from the scientists of the older generation, who ensured that the Yearbook appeared on time and guaranteed its quality. We hereby express our heartiest thanks to all of them for their valuable support and assistance.

We hope that a better Yearbook will be compiled each year and the contents of which, reflecting the science and technology of our country, will be richer and more varied each year. In conclusion, we like to use this opportunity to wish our scientific and technical enterprise an ever-increasing prosperity.

Ma Feihai
April 1, 1980

Contents

Feature Articles

References
Index

1	**DEVELOPMENT OF THE NATURAL SCIENCES IN CHINA OVER THE PAST 30 YEARS** *by Yan Jici* Academia Sinica.............. (1)	9	**A SURVEY ON THE DEVELOPMENTS OF MATHEMATICS IN NEW CHINA** *by Su Buchin and Sun Laixiang* Fudan University.............. (53)
2	**SCALING THE NEW HEIGHTS OF NUCLEAR SCIENCE AND TECHNOLOGY** *by Wang Ganchang* Institute of Atomic Energy, Academia Sinica.............. (10)	10	**A THEORY OF POLYMERIZATION OF SILICIC ACID IN AQUEOUS SOLUTION** *by Tai Anpang and Chen Yungsan* Coordination Chemistry Institute, Nanjing University.............. (64)
3	**REVIEW OF ACOUSTICS RESEARCH IN CHINA** *by Maa Dahyou* Institute of Acoustics, Academia Sinica.............. (14)	11	**RECENT DEVELOPMENT IN THE STUDY OF THEORETICAL ORGANIC CHEMISTRY IN CHINA** *by Sheng Huaiyu* Shanghai Institute of Organic Chemistry, Academia Sinica.............. (71)
4	**TWENTY YEARS OF THE INSTITUTE OF SEMICONDUCTORS — A SURVEY** *by Huang Kun, Wang Shouwu, Lin Lanying, Cheng Zhongzhi and Wang Shoujue* Institute of Semiconductors, Academia Sinica.............. (24)	12	**A SURVEY ON ASTRONOMY RESEARCH IN NEW CHINA** *by Yi Zhaohua and Qu Qinyue* Nanjing University.............. (84)
5	**CHINA'S FIRST LASER** *by Wang Zhijiang and Wang Nenghe* Shanghai Institute of Optics and Fine Mechanics, Academia Sinica.............. (30)	13	**ON THE ADVANCES AND DEVELOPMENTS OF WEATHER PREDICTION IN CHINA** *by Shu Jiaxin* Bureau of Meteorology, Shanghai..... (91)
6	**ADVANCES IN BIOCHEMISTRY AND MOLECULAR BIOLOGY IN CHINA** *by Wang Yinglai and Shi Jianping* Shanghai Institute of Biochemistry, Academia Sinica.............. (33)	14	**NEW FEATURES OF THE EARTHQUAKE SCIENCE IN CHINA** *by Mei Shirong* State Seismological Bureau........ (100)
7	**RECENT ADVANCES OF CHINESE PALAEOANTHROPOLOGY** *by Wu Ruking (Woo Jukang) and Wu Xinzhi* Institute of Vertebrate Palaeontology and Paleoanthropology, Academia Sinica.............. (41)	15	**A SUMMARY OF MARINE RESEARCH IN CHINA** *by Shen Zhendong* National Bureau of Oceanology .... (112)
		16	**WINDING ROADS AND A BRIGHT FUTURE — 30 YEARS OF CHINESE PSYCHOLOGY** *by Xu Liancang* Institute of Psychology, Academia Sinica.............. (117)
8	**BRILLIANT ACHIEVEMENTS OF PALAEONTOLOGICAL RESEARCH IN CHINA** *by Lu Yanhao, Zhou Mingzhen, Hao Yichun and Li Zingxue* Manjing Institute of Geology and Palaeontology Academia Sinica.............. (46)	17	**A BRIEF INTRODUCTION TO TRADITIONAL CHINESE MEDICINE** *by Tsai Chingfong* Academy of Traditional Chinese Medicine................ (125)
		18	**COMMEMORATING THE CENTENARY OF THE BIRTH OF THE GREAT SCIENTIST ALBERT EINSTEIN** *by Zhou Peiyuan* Beijing University .... (139)

CHAPTER 1

DEVELOPMENT OF THE NATURAL SCIENCES IN CHINA OVER THE PAST 30 YEARS

by Yan Jici

1979 is the thirtieth anniversary of the founding of the People's Republic of China: it is also the thirtieth anniversary of the establishment of the Chinese Academy of Sciences. The past thirty years have been years in which many important changes took place in China's history and in which we surmounted many difficulties and forged ahead in our scientific undertakings. Today, in the new historical period of striving for the realization of the 'Four Modernizations', it is absolutely necessary to look back over what has been achieved in the last thirty years of development of natural sciences so that we may draw useful lessons for reference in our future work.

I

Ever since the May 4th Movement (1919), the Chinese people have regarded "science and democracy" as part of the anti-imperialist, anti-feudal New Culture Movement. Taking science as the weapon to fight for national progress and to bring about a rich, strong and prosperous country, many patriotic people, with lofty ideals in mind, tried to find a way to "save the country through science" But under the reactionary rule at that time, there was no possibility of scientific development, let alone the road of "national salvation through science". In the pre-liberation days, besides universities and colleges, there were in China very few scientific research institutions, with personnel numbering only a little over six hundred. Among those institutions were the predecessors of the Chinese Academy of Sciences, namely, the Central Research Institute and the Beijing Research Institute. They covered more than twenty scientific research units and had a personnel of over 200. What those research workers were asked to do was merely to maintain an outward show. They struggled hard under extremely difficult conditions. There was nothing of modern science and technology to speak of except regional research work in the fields of geology, biology, meteorology, etc. and some other research that they could carry out without modern laboratory equipment. Only with the victory of the people's revolution did our scientific undertakings begin to develop on a large scale.

In November 1949, the Chinese Academy of Sciences was established under the care of the Central Party Committee and the People's Government headed by Chairman Mao Zedong, marking the beginning of a new historical period for China's scientific undertakings. Led by Premier Zhou, representatives of natural science workers throughout the country met in 1950. In June 1950, the Government Administration Council of the Central People's Government gave important instructions on the basic tasks of the Chinese Academy of Sciences, pointing out that the stipulations concerning scientific work laid down in the common programme of the Chinese People's Political Consultative Conference would be taken as the general guiding principle for the science workers of our country. According to that principle, the basic tasks of the Chinese Academy of Sciences were: to establish the concept of science for the people, to strive for an integration of scientific research with local conditions, to absorb the world's advanced scientific experience, to engage in basic and applied researches in a planned way and to strengthen the common ties between different branches of science. After that, various specialized scientific research institutes were set up one after another by government departments, for instance, the Chinese Academy of Agricultural Science, the Chinese Academy of Medical Science and other research institutes in the fields of geology, metallurgy, machinery, railway, communications, hydroelectric power, petroleum, coal, chemical industry, electronics, post and telecommunications, architecture, textile industry, forestry and national defence. Scientific experiment institutes were also established by various provinces and cities as well as by factories, mines and other enterprises. In the meantime, the teaching quality of institutions of higher learning was considerably improved, their level was correspondingly raised and their scope expanded. Thus, foundations were laid for a scientific and technological system formed from the five quarters — the Chinese Academy of Sciences, institutions of higher learning, industrial enterprises, national defence departments and local scientific research institutes.

With the rehabilitation of the national economy and the all-round fulfillment of the First Five-Year Plan for Economic Construction (1953–1957) ahead of time, initial foundations were built up for the industrialization of our country. The development of socialist construction required the scientific undertakings to quicken their pace. With a view to strengthening the academic leadership of the Chinese Academy of Sciences, in June 1955 we set up under the Academy: the Department of Mathematics, Physics and Chemistry, the Department of Biology and Earth Science, the Department of Science and Technology and the Department of Philosophy and Social Sciences (the predecessor of the Chinese Academy of Social Sciences), and 230 outstanding scientists were elected to be members of the department commissions. The set up for professional work brought into fuller play the collective strength of scientists in guiding our scientific undertakings. In 1965, the Party Central Committee convened a conference on the problems of intellectuals and Chairman Mao issued a call to march in science, assigning the task of coming up to advanced world standards economically, scientifically and culturally in a few decades. Premier Zhou made an important speech on the questions of how to give full play to the role of intellectuals, how to improve their working and living conditions and so on. Then, under the joint sponsorship of Premier Zhou, Chen Yi, Li Fuchun, Guo Moruo and other comrades and through the vigorous efforts of specialized science workers, the first long-range plan of our country for developing science and technology was worked out, namely, the Twelve-Year Plan (1956–1967). This magnificent programme drew a great number of scientific workers into the socialist construction of our country.

Under the organized leadership of the State Science Council and the National Defence Science Council headed by Comrade Nie Rongzhen and technical personnel, the main tasks of the Twelve-Year Plan were completed in 1962, five years ahead of schedule. The completion of this long-term programme filled in some of the major gaps in the fields of science and technology and strengthened certain branches of basic science. For example, the building and development of new technology such as atomic energy, radioelectronics, transistors, automation, computing techniques, air jetting and rocketry, optics and precision instruments, and the strengthening and improving of basic sciences like mathematics, astronomy, physics, chemistry, biology and earth science and of some technological sciences — all this exercised a profound and vital influence in developing science and technology of our country and in speeding up our socialist construction. To lose no time in continuing the rapid expansion of our scientific undertakings as well as to offset the bad influence exerted on them by certain "Left" errors which had emerged in a past period of time, we worked out in 1961, "Suggestions on the Present Work of Research Institutes" (i.e., A Fourteen-Article Proposal concerning Scientific Research Work). In 1962 a conference on scientific and technological work was held in Guangzhou. Chaired by Comrade Nie Rongzhen, the conference had two important speeches delivered by Premier Zhou and

Comrade Chen Yi respectively and then saw the formulation of the second long-term programme, namely, the Ten-Year Plan (1963–1972). While hearing reports on the Ten-Year Plan, Chairman Mao clearly pointed out that it was imperative to take the strongholds of science and technology, without which productivity would never go up. In 1964 China successfully exploded an atom bomb, smashed the nuclear monopoly of imperialism and social-imperialism and became another of the countries in the world that possessed nuclear weapons. By 1965 the scientific research institutes throughout the country had grown to over 1,600 with a personnel of more than one hundred thousand. Of those institutes, one hundred and six belonged to the Chinese Academy of Sciences, having a personnel of 22,000; they have been expanded by nearly a hundred fold in the past 17 years; institutions of higher learning have increased by more than four hundred times. Over fifty national specialized institutes were set up under the Chinese Association of Science and Technology. Mass participation in scientific experiments and popular science was in full swing. The gap between our science and technology including some new fields and the world's advanced levels narrowed; in some fields we had approached or surpassed the advanced international levels.

The course of science, however, was tortuous. Just when our scientific undertakings were flourishing, Lin Biao, Chen Boda, the "Gang of Four" and their underlings in the scientific and technological circles did their best, during the cultural revolution, to sabotage scientific research, violently destroying our scientific research institutes. They absurdly took as an inexorable law the fallacy that "while the satellite is going up, the red flag is falling to the ground". They opposed and undermined the revolutionary movement for scientific experiments. Completely blind to the ideological progress of the vast numbers of intellectuals, they denied that the great majority of intellectuals were people to rely on, villifying numerous intellectuals as "Stinking Number Nine" and "targets of dictatorship", persecuting and tormenting scientific and technical workers, of whom quite a lot were persecuted to death. They entirely rejected basic theory and experimental research by disintegrating scientific research institutions and disbanding scientific and technical personnel. By 1973, of the scientific research institutions under the Chinese Academy of Sciences, only fifty-three remained and there were merely a little over 13,000 left of the research personnel. The Chinese Academy of Agricultural Science was almost completely broken up with its personnel sent away to various places. Schools of higher learning were subjected to great sabotage with the result that the growth of intellectuals of the younger generation was adversely retarded. All scientific and technological organizations under the Chinese Association of Science and Technology were suspended from work; publication of academic periodicals was stopped one after another; scientific undertakings were hit by an unheard-of catastrophe. The gap between our science and technology and the world's advanced levels widened again.

In October 1976, to execute the will of the people, the Party Central Committee headed by Chairman Hua smashed the "Gang of Four" who had played havoc with the nation and the people, thus saving China and its people a second time and the cause of science as well. As Guo Moruo, President of the Chinese Academy of Sciences, put it, "We have won liberation for the second time .... The springtime of science is here!" At the National Science Conference convened by the Central Committee of our Party in March 1978, Comrade Hua Guofeng issued the great call to "raise the scientific and cultural level of the entire Chinese nation". Taking a broad and long-term view in expounding the gigantic role played by scientific research endeavours in the modernizing socialist construction, Chairman Hua gave tremendous encouragement to the scientists and technicians of the whole country. In his report Comrade Deng Xiaoping clarifies the rights and wrongs of the two-line struggle on the scientific and technological fronts by profoundly elucidating the point that science and technology are part of the productive forces and that the overwhelming majority of the intellectuals have already become a part of the working class. He also expounded on questions concerning the system of leadership of scientific research institutes, thus pointing the way to the formulation of our national policy on science. At the conference, Comrade Fang Yi gave a report on the programme and measures in the development of science and technology. In the report he put forward the chief goals and focal points of the National Programme for Scientific and Technological Development from 1978 to 1985, demanding that stress be laid on the eight comprehensive fields of science and technology which directly affect the overall situation (namely, agriculture, energy sources, materials science, electronic computer technology, laser, space physics, high-energy physics and genetic engineering), and on important new branches of science and technology as well. He also demanded that remarkable successes be achieved so as to push forward the high-speed development of science and technology and the national economy.

Due to the inspiration of the science conference, the whole Party is being mobilized to bring about a new upsurge in scientific activities in a big way. Many disintegrated scientific research units have been restored. According to statistics, by the first half of 1979, scientific research institutes all over the country had amounted to more than 2,400, with a scientific research personnel of 310,000 in all. Among them the scientific research institutes under the Chinese Academy of Sciences had reached 120 with their personnel increased to over 23,000; institutions of higher learning had grown to over 600; the national specialized institutes attached to the Chinese Association of Science and Technology had increased to over 80 — these had topped all previous records. Full of confidence, vast numbers of scientists and technicians are striving to fulfil the various tasks put forward at the science conference.

II

In the entire history of development of science and technology, thirty years is a short span. But in this very short period of time, world science and technology has advanced by leaps and bounds. With excessively weak foundations and the ten years of great disaster brought on by Lin Biao and the "Gang of Four", our science and technology are still behind advanced world levels. Generally speaking, there are quite a few fields where either our foundations are weak or we have gaps to fill. Even in some of the established branches of science, we are still ten to twenty years behind. However, it should be pointed out that we have initially built up a scientific and technological research system of our own and that we have trained quite a big contingent of scientists and technicians who have reached fairly high levels and produced a good number of important results in scientific research. Our scientific and technological work has made significant contributions towards the strengthening of national defence capability and the development of national economy, and has created favourable conditions for the further advancement of science and technology. The successful experiments on atom bombs, hydrogen bombs and ballistic missiles, the launch of man-made satellites and their successful return to the earth and the synthesis of bovine insulin — all these are manifestations of our achievements in science and technology.

We have an age-old tradition in mathematics. Since the founding of our People's Republic, pure mathematics has been developing considerably and the contributions of our research, especially in additive theory of numbers, Sieve method and Goldbach's conjecture, have won recognition both at home and abroad. We have made important contributions to the development of topology and achieved outstanding results in the theory of functions, the theory of groups, differential equations, functional analysis, etc. In applied mathematics we have developed our own finite element method and its theory; our research in information theory, mathematical theory of computer science and other theories has grown out of nothing. Good progress has been made in the development of mathematical statistics, mathematical logic, the theory of probability and operational research; success has been achieved in the popularization of the optimum seeking method and the overall planning method; good results have been obtained in control theory.

In the last thirty years, we have built the Beijing, Shaanxi, Yunnan and other observatories and astronomical instrument plants and man-made satellite observation stations. To meet the urgent needs of our national defence and economy, we have also developed time and polar motion services, independent compilation and calculation of astronomical almanacs and prediction of solar motion. Since 1963 the accuracy of our world time service has been on a level with the world's advanced ranks. Research is being gradually carried out in the fields of the solar and stellar systems, extragalactic physics and cosmic science. In recent years some significant work has been done in theoretical study.

Our theoretical research and experimental work is crowned with gratifying results in the fields of solid-state physics, nuclear physics, high-energy physics, optics, acoustics, etc. Large numbers of physicists have participated, either directly or indirectly, in the manufacture of atomic bombs, hydrogen bombs, ballistic missiles and man-made earth satellites. Still more people have made contributions towards the development of our machinery, electronics, metallurgy and instrument-making industry and towards the building from scratch of new industries such as atomic energy and semi-conductors as well as other fields of our national economy and national defence construction. Years of effort have now made it possible for us to set up large installations like high-energy accelerators, the production of which has gotten underway. Our physicists discovered in 1959, the Σ negative hyperon, advanced in 1966 the theory of the straton model and discovered in 1972 at the Yunnan Cosmic Ray Station, a heavy particle which is probably more than ten times heavier than a proton. Quite recently, the Ding Zhaozhong Group, which some of our high-energy physics scholars have joined, has obtained experimental evidence concerning the existence of the gluon; the Mo Wei Group, in which our other high-energy physics scholars have worked, has made significant progress concerning the unity of weak interaction and electromagnetic interaction in its experiments on μ — neutrino and electron elastic scattering. These are some of the instances which show our success in carrying out international cooperation in science and technology.

Over thirty years we have rapidly established and developed many branches of chemistry which used to be nonexistent, for example, theoretical chemistry, structural chemistry, high polymer chemistry and physics, organic synthesis, catalytic and ultimate organic chemistry and colloid chemistry. We have made some achievements which have distinctive features of their own. The synthesis of insulin and analysis of crystal structure have come up to advanced world standards. Primary catalysts for producing aviation kerosene, petroleum and the chemical industry, synthetic butadiene rubber and rare-earth for isoprene rubber, synthetic mica, silicon-flouride materials, etc. have made positive contributions to the modernization of our industry, especially to the development of the synthetic rubber industry and oil industry. At the same time, they have provided the "A-bomb and H-bomb" and the artificial satellite with a lot of new materials of crucial importance such as solid rocket propellants, fuel cells and solar energy cells, temperature control coating, altitude control hydrazine decomposition catalysts, synthetic rubber, plastics and lubricating materials, the last three of which can resist high and low temperature heating and strong oxidizers. They have thus made a contribution towards the modernization of national defence construction. In recent years, remarkable progress has been made and miraculous results produced in the fields of synthetic chemistry, structural chemistry, quantum chemistry, catalysis, rare-earth chemistry, analysis and separation, such as linear patterns of organic homologues, advancing of the active centre model of nitrogen-fixing enzyme, extraction and separation of rare earth elements, high-effect deoxidant catalysts, materials resisting high-temperature heating and erosion, and high-efficiency miniature chromatogram technique.

In the field of biology we have changed the pre-liberation practice of laying particular stress on descriptive research. Systematic and experimental biological research have developed rapidly. Several major gaps have been filled and weak branches of the science strengthened. Significant results have been obtained in the field of basic theory research like nucleoplasm relations. Meanwhile, a large amount of research work has been done in the light of the needs of industry, agriculture, medicine and national defence. For instance, use of hybrid vigour and genetic breeding, locust control, biological control of plant diseases and insect pests, artificial stimulation of fish oestrus, cultivation of kelp and laver, new-type microbial fermentation, principles of acupuncture analgesia, use of Chinese herbal medicine, biological soil and resources surveying, making of topographic maps, variety introduction and domestication of animals and plants like castor silkworms – all of which have contributed towards the socialist construction of our country.

In the field of earth science, we have either filled gaps or strengthened foundations in geophysics, atmospheric physics, marine physics, marine chemistry, engineering geology and hydrogeology, and we are conducting a comprehensive marine investigation and a coastline survey throughout the country. In recent years, we have made fairly good progress in our investigation of the continental shelf areas along the West Pacific Ocean, the South China Sea and the East China Sea and in our investigation of marine pollution and others. We have built a country-wide network of meteorological observatories. New progress has been made in the basic research of dynamic meteorology. Geotectonic theory has provided some theoretical guidance for the development of geology and detection of mineralization. We have basically completed a nation-wide network of seismograph stations and have, on the basis of mass participation, conducted fairly good earthquake prediction at Haicheng, Longling and Songpan. We have made surveys of the Heilungjiang River valley, Sinjiang, Qinghai, Ningmeng and the middle reaches of the Hunghe River and worked out natural atlases of our People's Republic. Systematic research has been directed towards glaciers, desert control, frozen earth of the Qinghai-Xizang area, mud-rock flow of the Southwest area. Fairly good results have been achieved in our investigation of Mount Qomolangma and in our inspection into the course and causes of the appearance of the Qinghai-Xizang Plateau and its influence on natural environment.

In the field of technological science, with the ever

increasing expansion of the national economy and national defence construction, we have greatly strengthened our knowledge in mining and ore dressing, metallurgy, machinery, electrical engineering, civil and architectural, hydraulic and electric power engineering, chemical engineering, engineering heat physics, material science, applied mechanics, applied optics, etc. We have also filled the gaps in the new fields of science and technology such as electronics, computers, semiconductors, automatic control, laser and infrared rays, space science and technology, and anticorrosive science. The successful building of the Nanjing-Yangtse River Bridge, the comprehensive utilization of vanadium and titanium, magnetite and rare-earth ores, the successes achieved in developing the production of alloy steel and non-ferrous alloys in the light of our resources, the successful manufacture of 12,000-ton hydraulic presses, giant turbogenerators with inner water-cooled stators and rotors, large-size electronic computers, large-scale integrated circuit, high-resolving power electron microscopes and a number of large optical installations for precise measurements. The results obtained in the triple flow theory of calculating vane wheel machinery, in the multiple-shock theory of calculating the strength of mechanical parts and in the theory of metal breaking — all these show that our technological and scientific research work has achieved high levels and made important contributions to socialist construction.

In the field of agricultural science, our scientific workers have contributed greatly towards the development of agricultural production and agronomy, for example, the improvement and utilization of alkaline and red soil low-yield land, research on the development of improved species of rice, wheat, cotton and other staple crops by selection and on laws of their high yield, successful breeding of hybrid rice and allocotoploid triticale, law of the migration of locusts and their prevention and control, comprehensive prevention and control of wheat rust and midges, research on the regular pattern of the winter migration of armyworms, rabbit-attenuated vaccine against swine fever and rinderpest, diagnosis of infectious horse anaemia and acquisition of its immunity, and utilization of rural methane.

In the field of medicine and sanitation, certain technical barriers have been broken down. Of the many severe epidemic diseases, parasitic diseases and endemic diseases, some have been basically eliminated and some have had their incidence and death rate reduced considerably (for instance, the plague, kala-azar, Keshan disease and so on). The successful manufacture of "cotton phenol", a male contraceptive, has opened a new way for family planning. After an overall investigation, we have basically found out about the nation-wide distribution of major malignant tumour occurrences and worked out a collection of charts to show the distribution of those tumour occurrences throughout the country. The treatment of coronary heart disease, fracture, acute abdominal and other diseases through the combination of traditional Chinese and Western medicine has heightened curative effects and promoted theoretical research in medicine. The study of Chinese herbal medicine has greatly enriched pharmaceutical resources. The virus of trachoma was first successfully separated and cultured by our country. Acupuncture anaesthesia, replantation of severed limbs, treatment of extensive burns, early diagnosis of liver cancer and surgical treatment of oesophagus cancer — all these achievements have come up to the world's advanced levels.

III

Looking back over the course of 30 years' development of natural science in China, we have experience to sum up and lessons to bear in mind.

In order to bring about the smooth development of our scientific undertakings, it is imperative, first of all, to have a stable and unified political situation.

The founding of New China put an end to her state of political turmoil and repeated foreign aggression during the previous hundred years, thus paving the way for economic development as well as scientific and cultural prosperity. At an important moment in 1955 when our socialist revolution had achieved tremendous success and our socialist construction had reached a new upsurge, Chairman Mao correctly pointed out: "We have entered a period, a new period in our history, in which what we have set ourselves to do, think about and dig into, in socialist industrialization, socialist transformation and the modernization of our national defence, and we are beginning to do the same thing with atomic energy." This is a very wise thesis. The main tasks for the new historical period are to put an end to the backwardness of our country in economy as well as in science and technology, to reform the relations between production and the infrastructure that are not suited to the economic construction, and the development of science and technology and to protect and expand production under the new relations of production. The fundamental aim of scientific research institutes is to produce positive results in research and train talented people so as to push production forward. Because of its continuity, scientific research requires a politically stable environment in which research workers can be wholly absorbed in their work and make unremitting efforts to achieve high and steady yields. Let there be no endless political movements like "Take class struggle as the key link", "Politics pounds at everything" and what not. This is a bitter lesson to be learnt from the ten years of disaster created by Lin Biao and the "Gang of Four" and this bitter truth has been repeatedly shown by thirty years of experience, both positive and negative.

Secondly, we should make a correct class assessment of the contingent of intellectuals and fully rely on the scientific and technical personnel.

Whether we can make a correct class assessment of the intellectual contingent forms the important basis of whether we can correctly carry out the policy towards intellectuals. Instead of causing an outflow of intellectuals, the victory of the Chinese revolution made large numbers of scientists return from abroad to take part in socialist construction. Together with the scientists at home, they formed the backbone of the scientific and technological contingents. This was a victory of the Party's policy towards intellectuals. At the same time, we also took care to foster and train young scientific and technological workers. Through fostering graduate students, sending students abroad and through expansion in higher education, we brought up a great number of well-trained people. Numerous inventors and crackajacks at technical innovation emerged from among the mass of workers and peasants. Together they became the backbone of the vast contingents of scientists and technicians. Practice has proved that the great majority of our intellectuals, scientists and technicians, who have long since been a part of the working class are a reliable force to be counted on. However, for a long time past, such progress and fundamental change of our intellectuals failed to win deserved recognition, as it was wrongly thought that all intellectuals were bourgeois. What is more, Lin Biao and the "Gang of Four" took advantage of this wrong thinking to persecute and torment intellectuals by labelling them as "Stinking Number Nine", "base of capitalist restoration" and "targets of dictatorship". This is an extremely bitter lesson. Since the smashing of the "Gang of Four", it has been quite clear that the overwhelming majority of the intellectuals are already a part of the working class and that they are a reliable force in our Party and country. A series of measures has been taken to give scope to the leading role in scientific and technological work. Those important measures have brought order out of chaos and will exert their far-reaching influence on the development of scientific and technological undertakings.

Thirdly, we should develop academic democracy and carry out the policy of letting a hundred schools of thought contend.

The leadership of science and technology depends not on taking administrative measures but chiefly on strengthening academic leadership in accordance with the laws of scientific and technological development. The chief method of exercising academic leadership is to encourage democracy in academic research and allow a hundred schools of thought contend.

Natural science takes nature as its object of study; different branches of science study the specific laws of motion of different kinds of matter. Experimentation should be the sole basis for judging the theories of natural science. Differing academic viewpoints must not be labelled in a sweeping way as such and such or of such and such classes, still less must administrative means be employed to support one school of thought and suppress another. Experience both at home and abroad has testified that things like that will hinder the development of science and are doomed to failure in the end.

After summing up its experience in directing scientific research work, our Party put forward the policy of letting a hundred flowers blossom and a hundred schools of thought contend. We corrected in our practical work the errors which had been committed by advocating under foreign influence one academic point of view and keeping down another in genetic engineering and other branches of research during the initial period of the founding of our country. We are now pushing forward the cause of science and making it flourish through the adoption of correct policies: advocating free exchange of views, free debate and free competition among different schools of thought and at various symposiums, and encouraging scientists to strengthen their practice in science and respect facts with a scientific approach so that they may learn from each other and to make up for each other's deficiencies. We will also pay great attention to the development of international academic exchange, absorb new academic thinking and take an active part in international cooperation in science and technology. During the cultural revolution, the "Gang of Four" set up forbidden zones here and there, strangling democracy and stifling science. This is a lesson always to be remembered.

Fourthly, we should do a good job in coordinating and organizing forces from different organizations.

Our scientific and technological base comprises five main divisions, namely, the Chinese Academy of Sciences, institutions of higher learning, industrial branches, national defence departments and local scientific research institutes. We should organize these forces well, share out the work and act in harmonious cooperation in the eight major fields laid down in the National Programme for Scientific and Technological Development as well as in the key projects of various branches of science and specialities. First of all, we must get the cream of our scientific and technical forces and their core members organized in a practical and effective manner, so that we may avoid unnecessary overlaps and the practice of non-cooperation with each going his own way at work. Meanwhile, we should sum up past experience and lessons, absorb good methods at home and abroad, earnestly reform our scientific and technological management, arouse the enthusiasm of forces from various sources and develop the range and quality of science and technology. Our past plans have all achieved good results. We firmly believe that the fulfilment of the Eight-Year Plan formulated in 1978 will make our science and technology undergo further great changes and will lay a solid foundation for the realization of their modernization.

Fifthly, we should establish a system of "basic research — applied research — developmental research".

In old China, science developed lopsidedly. It was, on the whole, divorced from production. After liberation, the Chinese Academy of Sciences and various schools of higher learning, while building up and developing basic science, put in a lot of effort to integrate scientific research with industrial and agricultural production and with national defence construction; on the other hand, production departments were gradually building up their own contingents of scientists and technicians. Now, we have instituted a preliminary system of practising division of labour on a cooperative basis with each paying particular attention to his own job.

Basic research tends to be neglected because it is often considered to be divorced from practical work. What is more, Lin Biao and the "Gang of Four" slandered basic research as an "evil theoretical trend", but our Party and state have persistently given their full attention to basic theoretical research. Chairman Mao and Premier Zhou gave repeated instructions on the strengthening of research work in basic theory. The main task of basic research is to discover the basic laws of nature and then accumulate basic knowledge in every field of natural science. Not only is this of great significance to the long-term development of science and technology but also plays an important role in developing new techniques, in training scientists and technicians and in absorbing the strong points of other countries better and faster. It is therefore, necessary to correct the mistaken idea that basic research is divorced from politics, divorced from practical work and divorced from the masses. We have to take a broad and long-term view of basic research, back it up steadily and try hard to turn its results into a part of applied techniques.

Applied research, covering a wide range of fields, can directly solve various scientific and technical problems existing in national economy and national defence construction. At present we will make further efforts to develop applied research and strengthen developmental research which is still quite weak now (including intermediate experiment, popularization experiment and production experiment). Proceeding from economic management and its system of organization, we can best encourage industrial departments to energetically unfold scientific research and to outdo each other in applying the results of scientific research so as to turn science and technology into part of the productive forces as quickly as possible. Our stress on the modernization of science and technology is the key to realizing the Four Modernizations. Scientific research must go ahead of production and construction. This is an objective reflection of the importance of science and technology and points out the law of objective reality. Recognition of this law has yet to be followed up by due attention and support from all involved. Many intermediate links between science and production need to be constructed in a down-to-earth way; if not, scientific development may not necessarily bring about an advance in production.

Sixthly, we must strengthen the building of a contingent of scientific research workers and improve both their working and living conditions.

Our ranks of scientists and technicians should not only be of considerable size but have a fairly high level as well. Over a long period of time, owing to the lack of strict training and the failure to sum up experience in scientific research in an indepth way, there still exists quite a big gap between the level of many of our scientific and technical workers and the requirements of modern scientific research work. Every effective measure must be taken to raise the level of the present scientific and technical personnel. Major efforts should be devoted to developing elementary, secondary and higher education and to launching mass activities in scientific experiments so that scientific research institutions may be infused constantly with new blood. In most scientific research institutions and schools of higher learning, scientific instruments, laboratory apparatus and other equipment are frightfully obsolete. Living conditions are difficult for the scientific research personnel, especially for the young and middle-aged group research workers. All this calls for immediate improvement. If we have a mammoth force of high-level scientific and technical personnel and advanced experimental computer equipment, we can certainly realize the modernization of science and technology.

The course of 30 years' development of natural science in China has provided us with much experience and many lessons, both positive and negative. With the flourishing of our scientific cause in mind, we should tackle these problems conscientiously, be receptive to new ideas and study new problems. In the history of our country, science and technology has blossomed in radiant splendour. Through we have now fallen behind in modern science and technology, we can still accomplish much provided we know how to learn and how to sum up experience and lessons, and provided we are doubly diligent in our work. There are bright prospects for our scientific undertakings. Academy President Guo Moruo once said, "Only socialism can liberate science, and the building of socialism is possible only when it is done on the basis of science. Socialism is indispensable to science; science is all the more indispensable to socialism". Today, the great socialist construction of our motherland is moving at an unprecedented rate and scale. The whole nation is working hard to fulfil the grand programme for the Four Modernizations. The modernization of science and technology, being the key to carrying out the Four Modernizations, is the most pressing need of the Party and the people. Our comrades on the scientific and technological fronts should go all out to make the country strong and prosperous and do a good job in their scientific and technological endeavours, thus making due contributions to the building of a modern and powerful socialist state.

CHAPTER 2

SCALING THE NEW HEIGHTS OF NUCLEAR SCIENCE AND TECHNOLOGY

by Wang Ganchang

On the occasion of the celebration of the thirtieth anniversary of the founding of our great motherland, I, as an aged scientist, cannot but feel excited with a variety of thoughts flooding my mind while recalling the numerous events which occurred in the course of the development of nuclear science in our country, seeing today's flourishing scenes and looking forward to the prospects of the 'New Long March.'

Before liberation, China was a poor and backward country and the people lived in dire poverty. The foundation of scientific research was then quite poor, and it was unimaginable to develop the new branch of science, the nuclear science, and the use of nuclear energy.

After liberation, the Communist Party of China and the people's government showed great concern for the development of scientific research. Under the very difficult conditions in the early period after the founding of our People's Republic, the Institute of Modern Physics of the Academy of Sciences of China was established. (It was the predecessor of the Institute of Atomic Energy). Their scientists and technicians engaged in nuclear research and some students who returned from abroad were also gathered at this institute. Research work on nuclear physics, cosmic ray physics and radiochemistry was carried out. Facing the economic and technical blockade by the imperialists, we started our work from the very beginning. We were determined to develop the devices and equipment by ourselves ranging from nuclear emulsions, the different kinds of counters, scintillators, cloud chambers to high tension multipliers, electrostatic accelerators and electron linear accelerators. I remember at that time, some of my colleagues had to take the risk of radiation hazards and contamination to repair a radium source for a hospital or to extract radon from it in order to construct a neutron source. We also established a cosmic ray experimental station on the Luoxue mountain at an altitude higher than 3000 m in Yunnan province. The work done in that period laid a preliminary technical foundation for further development of nuclear science and technology. Since then a group of young scientists has grown up gradually, and most of them have now become the backbone of the scientific and technical contingents in different institutions and factories. At the same time, we popularized the general knowledge of nuclear science and technology among the broad masses of the people and cadres under the call and support of the Party.

In the 1950's, guided by the twelve-year national

programme for the development of science and technology laid down under the direction of the late Premier Zhou Enlai, a relatively complete and comprehensive programme for nuclear science and technology research was established, and different institutions were set up in succession. Research on nuclear physics and engineering, preparation and separation of radioactive and stable isotopes, controlled thermonuclear reaction, radiogeology, particle accelerator technology, detection of nuclear radiation and radiation protection and dosimetry, etc., was carried out systematically at these institutions. The application of radioactive isotopes to medical science and agriculture was becoming widespread. In many universities and colleges, numerous research institutions in related specializations were also set up and thus a great number of scientific and technical personnel in nuclear science and technology and related fields were trained for the country. In addition, we imported some necessary and more advanced experimental facilities and techniques from abroad. A number of nuclear industry bases were set up one after another.

In the sixties, having surmounted numerous difficulties caused by the Soviet Union, we achieved with the common efforts of scientists, workers and cadres, a series of high-level results in many important fields of nuclear science and technology. The successful explosion of the atomic bomb and the hydrogen bomb served to symbolize the fact that development of science and technology in China had reached a new stage and it was an important step towards catching up with the advanced world standards.

In the seventies, the great majority of scientists and technicians still persevered in scientific research inspite of the sabotage and interference of Lin Biao and the 'Gang of Four'. Since the smashing of the 'Gang of Four', the bright sunny spring has come again for the research in nuclear science and technology. We have to work hard to make up for the lost time and decrease the distance we lag behind, compared with the world standards. In the recent two years, encouraging progress has been achieved in the fields of nuclear physics, radiochemistry and its applications, and the application of nuclear techniques in inter-displinary sciences.

This year the Party Central Committee put forward the call to transfer the key point of our work to the socialist modernization in our country. The valiant, industrious and ingenious Chinese people will certainly achieve their goal steadily and improve on the present backward situation.

In this 'New Long March' of Modernizations of industry, agriculture, national defence and science and technology, scientists and engineers engaged in nuclear science and technology have their own important obligations. In order to promote the development of nuclear science and technology and draw a lesson from the past negative and positive experiences, I would like to make some suggestions which are of importance in my point of view.

First of all, scientific research personnel should be equipped with modern experimental facilities. Since the 1960's, experimental facilities in developed countries had undergone a technical transformation characterized by the increasing use of the computer, automation and the adoption of new achievements made in other branches of science. In our country, the experimental facilities still remain at the level of the early or middle fifties. Let me take the fundamental research in nuclear physics for example. We are short of advanced accelerators with the suitable energy range, which are able to accelerate different ions of a wide mass range. As for the detection instruments, they are not only small in quantity and variety but also poor in quality. The use of the computer in nuclear physics experiments has not become popular yet, and data acquisition and processing with online computers is only at the beginning stage. One cannot achieve first-class or excellent experimental results without advanced experimental techniques and equipment, even when the idea or proposal is a good one. In 1942, I proposed an experiment to directly verify the existence of the neutrino by observing the nuclear recoil occuring in the K-capture process of very light nuclei. In old China it was absolutely impossible to carry out such an experiment, and I could do nothing more than the publication of my proposal. Half a year later, an American physicist experimentally verified the existence of the neutrino according to my proposal. Again, in the mid-fifties, a 10 GeV proton synchrotron was completed in the Joint Institute for Nuclear Research in Dubna, which was the accelerator with the highest energy in the world at that time. But no outstanding result was achieved on this machine for several years, for its beam intensity was rather low and there was also a lack of the necessary equipment for analysis and detection. In 1959 my group started to work on a high energy π-meson beam, using a relatively large propane bubble chamber by the standards at that time. In 1960 we discovered an antisigma negative hyperon ($\overline{\Sigma}^-$) and found some anti-lambda hyperon ($\overline{\Lambda}^\circ$) events. At that time we did have an idea of working with an antiproton beam, and I was sure that in the process of proton-antiproton annihilation, we could discover different kinds of anti-hyperons more easily and could accumulate more events of anti-hyperons to investigate their properties. But owing to the limitations of our experimental facilities, this idea could not be realized. On the other hand, although we did accumulate a great amount of information about the strong interactions of hadrons, we would not find any new elementary particles in the form of very short-lived resonance states because of the limitations of the computer and measuring equipment.

It is obviously necessary to import some advanced experimental facilities in order for us to catch up with the modern standards. Nevertheless, the purpose of importing these facilities is to allow us to learn and have a good grasp of their properties so that we can improve on them and make creative changes to meet the practical requirements, thereby making progress in our own way. We would never be able to catch up with the modern standards if we should only rely on the imported facilities. We should instead establish our own bases of research, design and manufacturing. We have to set up a programme for development so that we can provide modern and advanced experimental facilities for the researchers in a step by step manner by combining the two ways — to import from abroad and to manufacture by ourselves.

It should be pointed out that it is difficult to completely modernize our facilities and equipment at present. The financial situation of our country is still rather weak, so our alternative is to tap the potentials of the existing facilities and to make innovations on them. While concentrating our efforts on building new important facilities, we should also place emphasis on improving and making full use of the existing ones. Their performance should be improved as far as possible and, if conditions permit, research work of even higher levels should be carried out with them. The foreign advanced equipment should be imported according to their relative importance and urgency. For fast effects with our limited investment, we should begin with the improvement of the speed and accuracy of the detecting and measuring systems, as well as with the automatic processing of experimental data. In some engineering projects, instead of obtaining complete sets of equipment from abroad, we should import only those key parts which cannot be designed and fabricated in China at present.

Secondly, we should be innovative in our approach. In fundamental research, we should be bold to bring forth new ideas, adopt new techniques and strive for new results. In order to catch up with the world standards, it is necessary to learn and master the advanced knowledge and sometimes even copy or repeat what others have done before. Of course this stage is inevitable, but the ultimate purpose of research is to create. It is most unfortunate that there is at present more copying than creation in some of our research work. Even worse, some research workers are contented with copying or checking.

When determining the direction of our research work, we should grasp the frontier problems and try our best to make outstanding achievements. In the fundamental research of nuclear physics, we should pay attention to the various research aspects, such as the fundamental interactions, new forms of nuclei and new modes of nuclear motion, etc. In the research on energy sources, apart from trying to make full use of fission energy, we should also strengthen the research work on the use of controlled thermonuclear fusion energy and explore them in different ways and means. This is one of the major problems in research on energy sources. In addition, the application of nuclear techniques in material science, condensed matter physics, superconductivity, atomic physics, radiochemistry, biophysics and the study of chemical processes should be developed.

The third point is that we should devote much attention to the development and uses of nuclear energy. The purpose of the basic research of natural science is to explore and understand the laws of nature in order that we may better the world we live in. Nuclear techniques have been widely applied to various branches of science, and their use in industry, agriculture, medicine and other fields have infinitely broad prospects. An even more important aspect, however, is the use of nuclear energy. In many developed countries in the world, nuclear power plants are now being quickly developed because the demands for electricity are increasing rapidly. From the long term point of view, nuclear energy will certainly become the main energy source. In China the geographical distribution of hydro-power and coal resources is not uniform, though she is rich in these energy resources. Consequently we have to develop nuclear energy as the complementary means to the hydro-power and coal resources. We have to consider this problem with a view to the long term interests and work out our nuclear energy policy. The programme for the development of nuclear energy will have to be set up, taking into consideration the experience in the development of energy resources in foreign countries, and in accordance with the practical conditions of the resources and economic ability of our country. We must develop and construct nuclear power plants in China and make them benefit the people, and thereby strive towards the realization of the 'Four Modernizations'.

The fourth point is that we should make persistent efforts to carry out the research work once the direction is determined. Any hasty decision to develop or abandon a research project would be a terrible waste of our limited resources and also damaging to the training of research personnel. If any possibility of new development in research should arise, we must grasp and pursue it until we understand it thoroughly. Only in this way can progress be made and breakthroughs achieved. In the history of the development of science, there are many well-known examples from which we can get a good deal of enlightenment. The English physicist Chadwick discovered the neutron by analysing the discrepancy in the theoretical explanations of previous experiments and carrying out a series of careful experimental measurements and calculations. He was thus able to reject the wrong inference, which held that the hitherto unknown particle was a kind of high energy gamma-ray, and to come to the correct conclusion. The discovery or parity non-conservation in weak interactions was similarly a result of careful analysis of experimental results and theoretical explanations.

As a final point, I would like to stress that we have to carry out the principle of extensive cooperation and work in good collaboration and solidarity. The achievements we have made in the past twenty years are the results of extensive cooperation and we should continue to work in this spirit. We need to have an overall organization of all our efforts — the import and construction of experimental facilities, establishment of research programmes, and the training of research personnel. Only in this way can we bring our limited financial funds, material resources and manpower to full use and concentrate our efforts on making progress; otherwise there will be unnecessary repetition and waste in our research work, thereby delaying our development.

On the occasion of the celebration of the thirtieth anniversary of our National Day, we deeply cherish the memory of our late Chairman Mao and Premier Zhou. We shall never forget their great concern for the development of nuclear science and technology in our country. The present favourable conditions for scientific research greatly inspires me, and I shall work conscientiously for the country and the people all my life and make my contributions to the cause of the 'New Long March', though I am already in my seventies. At the same time, I expect much of the middle-aged and young scientists and engineers in this country. They are the main force in the realization of the 'Four Modernizations' and the development of nuclear science and technology in our country. Let us make common efforts to scale the new heights of nuclear science and technology.

CHAPTER 3

REVIEW OF ACOUSTICS RESEARCH IN CHINA

by Maa Dahyou

We have a long history of acoustical investigation in China. In this century, however, under the feudal-reactionary rule, acoustics, just like other branches of science, was left far behind in its development. Only after the national liberation in 1949 was research in acoustics revived. A relatively strong research force gradually formed, and some important contributions were made in support of the socialistic reconstruction in China.

I. Ancient Acoustics

China is one of the few nations which made outstanding contributions to acoustics in ancient times. Before 6000 B. C., water wave designs had already appeared on potteries of the Yangshao culture. Wang Chong (27–96 A. D.) in his essays "Lun Heng," further discussed the similarities between sound and water waves and clearly stated the wave nature of sound. Later on, towards the end of the 12th century, Zhang Zai of the Sung dynasty observed that "sound is produced by the interaction of solids and gases". He further explained that the sound of thunder was due to gas acting on gas, the sound of the drums from solid acting on solid, the sound of an arrow from solid acting on gas, and the sound of the flute from gas acting on solid — that was the ancient theory of sound production. It is clear from the way words were used that the ancient Chinese people had much understanding of the nature of sound. As defined in "Shi Xu" (Introduction of Poems), "sheng" (声 , sound) with rigour makes "yin" (音 , tone); "yin" in harmony makes "yue" (乐 , music). So the three words, "sheng", "yin" and "yue" were clearly differentiated, and as a general term, "shengyin" (声音 , sound) was used. The word "xiang" (响 , sound sensation), though derived from "sheng" and "yin", is different from both: ' "xiang" is attached to "sheng" just as shadow to a body'. Specifically, "xiang" is used to describe the effect or "sheng", particularly the sensation produced by sound on people. There is another character "zao" (噪, noise), which means annoyance produced by yelling or just disturbance. The exact usage of different characters describing the phenomena of sound shows that our ancestors had made keen observations and achieved understanding of the physical world. The facts that the size and shape of the vibrating body affect the tone it emits, the damping of vibrations, and the distance sound is transmitted, were all discussed in "Kaogong Ji" (On Engineering) written in the late "Chun Qiu" period (fifth century

B.C.). At that time, the relationship between the tone of a string and its length and thickness was known quantitatively. Resonance phenomena and the application of resonators were recorded by Mozi (468–376 B. C.). The use of pottery below stages or on the walls as resonators to amplify or absorb sound in theaters had long been known as a sound control technique. The Wall of Echoes (a smooth circular wall with a 65 m diameter, along which any faith voice will propagate for two hundred metres, the Three-Tone Stone (the stone at the centre of the Wall of Echoes, on which one can hear a series of echoes), and the circular stone altar with a 23 m diameter on which one can hear the echoes from the surrounding stone fence as if from a well underneath, all in the Temple of Heaven, Beijing (Fig. 1), were built more than five hundred years ago. Also of interest are the ancient wash basins, from which

Fig. 2. Sprinkling washbasin.

water springs by rubbing on the side wall, dating back to the Han dynasty, two thousand years ago.

The most systematic investigation in China, however, was in the field of musical acoustics. It was recorded in "Lu Shi Chun Qiu" that Huangdi (c. 2550 B. C.) ordered Linglun to make a standard flute out of bamboo and to vary its length to obtain 12 tones; Fuxi, even earlier, had made the "qin" (string instruments), lengthened or shortened by one third to get 13 tones. In fact, ever since the Zhou dynasty (11 century B. C.) music had always had a special status in the feudal culture of China. The musical scales and instruments were studied officially for 3000 years. Harmonious tones were obtained by increasing or decreasing the length of the pipe (flute, falgeolet, etc.) by one third. This was one of the earliest theories in acoustics and the standard tone was decided in terms of the length of a pipe and written-recorded in "Gunazi", 6th century B. C. A musical scale was formed by this one-third rule at about the same time as the legendary Pythagorean scale was established, but the latter was based on strings. Using the one-third rule, one could get a pentatonic scale and also 12 tones in approximately an octave, and it was known that tones would sound in unison when the interval was an octave. Therefore, every possible method to achieve this was explored and in most of the last two thousand years of Chinese history, the research work on musical scales concentrated on devising a procedure by which one could get an exact octave based on the one-third increase-decrease rule. Zhu Zaiyu of the Ming dynasty actually proposed in 1584 A.D., an equi-tempered scale with intervals of $\sqrt[12]{2}$, exactly like that found today on the keyboard of a modern musical instrument. This was the first time such a scale was proposed.

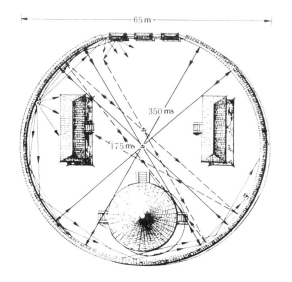

Fig. 1. The Wall of Echoes, the Three-tone Stone and the Circular stone altar in the Temple of Heaven.

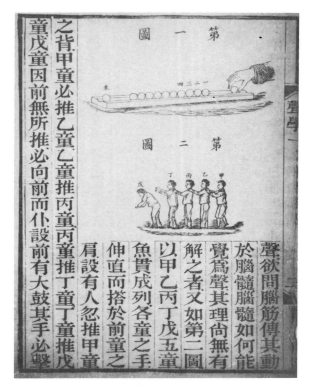

Fig. 3. A page of Tyndall's *Acoustics*.

II. Acoustic Work before Liberation

Until the national liberation, research work in acoustics in China was very limited. This was not due to the lack of scholarship on the part of the Chinese people. More than a hundred years ago, some classical works such as "Tyndall's Acoustics" (Fig 3), were translated into Chinese. In the twenties, Chao Yuanren studied the four tones in the Chinese language. Ye Chisun, Shi Ruwei, and others of the Qinghua University studied the acoustics of their auditorium and measured the sound-absorbing power of Chinese garments in 1929. Zhou Tongqing measured the depth of the Yangzijiang by ultrasonic reflection in 1941. The numerous works of Zhou Peiyuan on the theory of turbulence also concerned acoustics. These were the pioneering works in China.

Qian Xuesen's investigations of sound propagation in rarefied gases and most of the original investigations of Yan Jici on the piezoelectric quartz between 1927 and 1937 were completed outside China. So were some of the theoretical investigations on normal modes by the author. There was essentially no organized research work in China at that time, and practically no equipment. Quite a few scientific workers made some attempts but gradually gave up, because of the lack of support. The only work that survived in the severe conditions was that started and directed by the late Liu Fu in the twenties on an ancient musical scale and speech acoustics (or experimental linguistics). Fig 4 shows one of his inventions, the Lingraph, an instrument for measuring and plotting pitch variation from a speech wave before the electronic era, and a most advanced instrument at that time. Later the group was led by Luo Changpei and continued working on speech waves of the Chinese language, making important contributions.

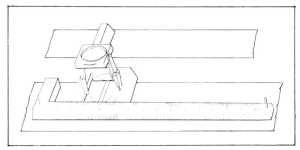

Fig. 4. The Lingraph.

III. Acoustics in the past thirty years

Before Liberation, only a few acousticians could persist in their work under the difficult conditions because of personal interests and a feeling of responsibility in science. Many branches of acoustics were ignored. In 1956, Premier Zhou Enlai organized and helped to draw up the 12-year plan for national science and technology development, and the science of acoustics entered a period of planned developments in China. Upon the completion of the 12-year plan was a 10-year plan of development drawn in 1962, and acoustics had its share as a technical science. During the Cultural Revolution, research work on acoustics like those in other disciplines was ill-treated and suffered great losses. After the "Gang of Four" was smashed, another 10-year plan was drawn in 1977, listing acoustics as a basic science. The aim was to serve the economic reconstruction and the national defence by stressing on fundamentals and advancements. The acousticians of the whole nation are marching towards the advanced levels of the world confidently.

During the last thirty years, research work on most of the branches of acoustics was started and established. Now we have a solid foundation on physical acoustics, mechanical vibration, noise, ultrasonics, speech communication, physiological and psychological acoustics, architectural acoustics, electro-acoustics, acoustical measurements, underwater sound, atmospheric acoustics, geoacoustics and musical acoustics with efforts also being directed towards bioacoustics, astroacoustics and nonlinear acoustics. Many professional organisations were set up by acoustics workers. A technical committee on acoustics was formed within the Chinese Physical Society in 1964, which organized the First National Congress on Acoustics. There were 118 delegates, with a total of 483 people from some 150 institutions taking part in its activities, and 134 papers were presented on physical acoustics, underwater sound, ultrasonics, electro-acoustics and acoustical measurements, noise and vibration, architectural acoustics, physiological acoustics and psychological acoustics. The Acoustical Society of China was inaugurated in 1977, and the Second National Congress on Acoustics was organized with contributions from more branches of the science (musical acoustics, atmospheric acoustics and geoacoustics in addition to those branches that had appeared in the First Congress) in which 200 delegates participated and 340 papers were presented. The growth was obvious. Scientific work met tremendous difficulties during the Cultural Revolution, but quite a few workers still persisted with good accomplishments. The acousticians, in response to the call for fundamental research by the late Premier Zhou Enlai organized the national acoustical symposium in 1973, with more than a hundred participants presenting 85 papers. In addition to the national meetings mentioned above, there had also been many professional meetings on architectural acoustics, ultrasonics, flaw detection, sound insulation, underwater sound, medical ultrasonics environmental acoustics, etc. on a smaller scale, which served to encourage the growth of ideas and experiences as well as to promote research and development in acoustics.

IV. Main Contributions

In the last thirty years, the acoustical community which has grown steadily, helped the industry to build up and strengthen related manufacturing and engineering units (for example in ultrasonic equipment, underwater equipment, broadcasting equipment, electroacoustic instruments, mufflers and absorbing materials, musical instruments, etc.) so that their products could satisfy the basic requirements of the country; and considerable educated personnel for industrial design and engineering. They were also very successful in research work. Besides completing tasks for the State, they have published some three hundred papers and brief notes. Only a small part of the most important could be enumerated in the following, and omissions were unavoidable.

1. Architectural Acoustics

This is one branch of acoustics that we started to study earlier than the others and managed better. There were many accomplishments of architectural acoustics in ancient China, and some are still useful today. After the Liberation, scientific workers repeated these investigations using contemporary methodology and obtained some understanding of the underlying scientific principles. For instance, the Wall of Echoes, the Three-Tone Stone and the stone altar in the Temple of Heaven were investigated and explained. A study was carried out on the ancient wash-basins, waves on the water surface analyzed and a model study made. It was found that sidewall vibrations not only produced water springs in the basin, but also produced subharmonic waves on the water surface under certain conditions (Fig. 5).

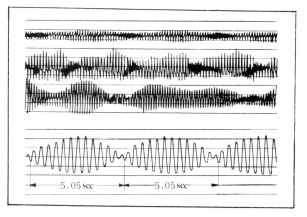

Fig. 5. Subharmonic water waves.

Theoretical investigations on normal modes yielded the directional and spatial distributions under specific excitations (e.g. pure-tone excitation, which gave a Raleigh distribution, narrow band white noise excitation, etc). A series of investigations on the second criterion besides reverberation was carried out, and model experiments with spark response and impulse response inside a room were made. A proposal was made to study the acoustics of auditoriums by the method of directional diffuseness in a plane, and a related acoustic lens with directivity in a plane was developed. Directional distributions were also investigated with directional microphones. Based on this work, the acoustical institutions in Beijing collaborated in 1959 in the design work for the

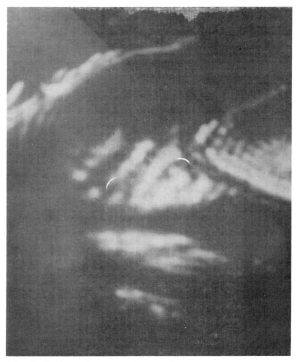

(a) Soft tissues between the thumb and the forefinger.

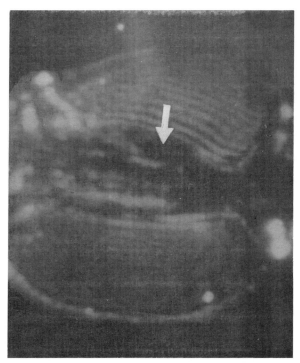

(b) A mouse tumour (the black spot indicated by the arrowhead).

Great Hall of the People, as well as in the sound field measurement and articulation tests upon the completion of the building. In this hall of 90,000 m^3, appropriate sound-absorbing materials and diffusion means were used, and together with distributed and semi-distributed sound systems, an audience of 10,000 could listen to lectures or appreciate music very satisfactorily. This is an achievement even by today's standards, and provided good experience for similar tasks later on. With these successes in acoustical designs, a number of broadcasting studios, auditoriums, theatres, recording halls, with good acoustics were built in China and in foreign countries. The capital gymnasium in Beijing, the Hangchow Theatre and the Hangchow gymnasium are all suitable for musical performances besides their primary functions.

In the area of acoustical materials, a large amount of work was done in the investigation of porous absorbing materials, emphasizing impedance characteristics. Artificial materials were developed. Much work was also done on sound insulation structures and insulation measurements. Special attention was given to the insulation of light structures, and appropriate measurement methods devised.

2. Ultrasonics

Ultrasonics is a branch of acoustics that has attracted much interest in China since the early days. A lot of work was done before 1952 on ultrasonic defect detection. After the 12-year perspective plan was drawn, organized research work was started, especially on the application of ultrasonics (in defect detection, processing, seed treatment, visualization, medical applications etc.). The successful experience in the use of the liquid whistle in the fields of dye material pulverization, emulsification, seed treatment, etc, was spectacular. Unfortunately, some leaders of certain institutions exaggerated the power of ultrasonics and created an "ultrasonic wave" movement in the whole country in 1961. Great difficulties were encountered in serious research and development work in ultrasonics as a result of the over-exaggeration which undermined the people's confidence in the subject. The science stagnated and regular research resumed only after 2 years. At about the same time, the industry reorganized and built up their applied research teams and ultrasonic equipment enterprises.

In basic research, ultrasonic vibrations of a bar, ultrasonic emulsification and ultrasonic absorption of bubbles in water were studied. A laboratory of molecular acoustics was established and a series of investigations on relaxation, absorption and sound absorption of aqueous suspensions were carried out. Instrumentation for ultrasonic attenuation study in solids was set up and some work carried out. Careful studies were made on the sound velocity and attenuation in visco-elastic and compressible fluids, carried out in the fifties and first part of the sixties. In 1965, studies on surface wave transducers were started, and high frequency surface acoustic wave (SAW) studies were initiated in 1970. The first SAW pulse-compressing filter was successfully fabricated in 1977. At present, SAW devices are in production and function satisfactorily in a range of equipment. Research work on acoustical holography was started in 1972 and later successfully applied to the visualization of under-water subjects (Fig. 6).

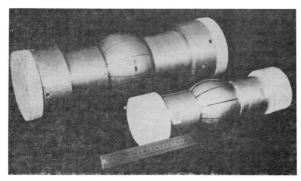

Fig. 6. Acoustical holography.

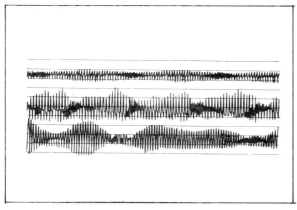

(a) When the amplitude has the critical value, ordinary waves and half-frequency waves appear in turn on the water surface.

Ultrasonic flaw detection and the treatment of herb seeds were very successful. Measurements of sound velocity in fluids attained rather high

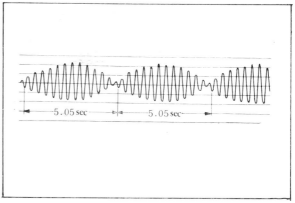

(b) When the amplitude is too large, there are only half-frequency waves.

accuracy (10-5), and consequently became applicable to the determination of fluid composition and velocity inside pipelines. In medical applications, the technique of ultrasonic diagnosis gradually gained sophistication and was applied to a variety of problems such as the inspection of blood vessel connections during the rejoining of severed limbs. Ultrasonic therapy was also successful in the treatment of some important diseases. Meanwhile ultrasonic diagnosis by Doppler shift developed in recent years was very powerful, and a special symposium on the subject was held in 1978. Piezoelectric crystals and ceramics are transducing materials that interest all acousticians. Research work on piezoquartz was continued after liberation. Slightly later, Rochell salt started to be produced, and that facilitated the manufacture of microphones, pickups and hydrophones. From 1963 onwards, barium titanate materials were developed and successfully produced. Single crystal drawing techniques based on buoyant force were developed in recent years, facilitating the preparation of single crystals in slab or tubing forms directly. This was useful in the development of high frequency piezoelectric materials (such as $LiNbO_3$). Work was also started on ultrasonic microscopy and hypersonics.

3. Underwater sound

Research work on underwater sound was non-existent in China before liberation and was organized only after the creation of the 12-Year Plan. A lot of research work was completed in these years. We have accumulated quite a collection of data for different types of hydrographical and geological conditions in shallow seas, on sound propagation, absorption, reverberations, scattering, fluctuations, correlations, wave distortions, noise in nature, etc., as well as in deep sea sound channels and convergent regions, after large scale under-water acoustical surveying on the continental shelf of China, in the South China Sea and in the Pacific. The experimental work became more systematic and a fair amount of theoretical work was completed. Original contributions were made on the theory of normal modes between 1960 — 1962, when the acousticians obtained a formula for the coefficient of attenuation in the extreme normal modes in terms of the sea floor reflection coefficient and span, and the approximate solution of sound field for arbitrary sound velocity distributions was obtained by phase integration, yielding a general formula and a clearer picture of the characteristics of the normal modes. Similar results were published overseas in 1964. Some developments were obtained in the study of approximate calculations. With the concept of the "three-parameter model" of sea floor reflection, the difficulties of linking the near field with the far field was avoided, and the theoretical results were also valid for the "transition distance". We have suggested the method for deriving the sea floor reflection coefficient from the results of sound field measurements, and good results were obtained. Measurements of back-scattering from the sea floor were also made at large angles. Theoretical investigations were carried out on the sound fluctuations and noise in shallow sea, introducing the concepts of local spectrum and the spectrum scale of noise field, yielding important results which indicated the possibility of the existence of a low noise channel. The numerical results obtained agreed well with the experimental measurements. As for noise emission from ships, we obtained the average power and dynamic spectra, and analyses of line spectra were carried out. Model analysis was used to determine the sound field in stratified media due to directional plane radiations. In the investigation on the deep sea sound field, the smoothed average field as well as the fine structure in the sound field in the inversion point convergent region in deep sea were computed, and the result compared with experimental results. In addition, work in the laboratories was carried out on sound absorption in sea water, local scattering of sea water and sound propagation in pure Kaolin suspension. Theoretical and experimental investigations on nonlinear acoustics were also started and experimental parametric arrays were built. Computer programs were written for shallow sea forecast. Large scale development work on under-water transducers was very fruitful. Many kinds of sound projectors were developed, among which the flexuotensile type of piezo-ceramic transducers has the advantages of a wider frequency band, higher power output and being lightweight (Fig. 7).

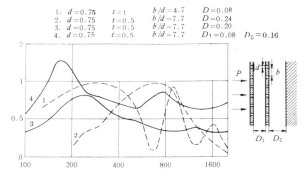

Fig. 7. Double-layer microperforated panel structure and its absorption characteristics.

Projectors of high power and of good frequency response in the range 1 — 10 kHz were developed. Receiving transducers were also developed, including the Type BS—I as a standard hydrophone, which was very efficient in the frequency range 20 Hz — 100 kHz, with good temperature stability. Also developed were many types of underwater sound instruments and equipment for fishery and navigation. At the present time, we are manufacturing enough transducers, instruments and equipment of good quality for home use.

4. Noise and Noise Control

This practice started in China in 1958 after being strongly advocated by some authorities, but at that time our industry was still very weak, and the issue was given little attention. The problem of noise received its recognition only after the late sixties when its control as a part of the environmental program appeared on the agenda. The office of Environment Protection was formed under the state council and also in the provinces and main cities. Control stations and groups were established. Significant research and development work on noise and noise control have been carried out. Now, the ordinary means of noise control, e.g. absorption, sound insulation, damping, vibration isolation, enclosures and ear protectors, have been mastered and put into use. Different kinds of factories and shops have also been systematically surveyed, and on this basis, the classification of sound sources according to their noise emission characteristics was proposed. A survey was also made on the effects of noise on people, including hearing and non-hearing physiological effects, speech-hearing, etc. and quite a lot of data were accumulated. Based on these data and experience overseas, guidelines on noise control standards were proposed: for hearing and health conversation, 70 – 90 dBA (average for an 8-hour working day); indoors for conversation, mental work, 50 – 70 dBA and for rest (including sleep), 30 – 50 dBA. The environmental law was passed and published by the Standing Committee of the National Assembly of People's Deputies. Standards for industrial enterprises were published by the hygiene and labour protection authorities, ruling that the highest permissible level is 90 dBA for an 8-hour working day. Standards for impulsive noise and for buildings are under consideration. Measurements and analyses of traffic noise were made and it is considered that 60 – 80 dBA is acceptable on the side of a main road. The main problem in noise arises mainly from the sounds made by horns, which raise the average sound level peak by some 10 dB, making Chinese cities among the noisiest in the world. The most urgent measure that has to be taken in order to reduce traffic noise is to strengthen strict traffic control, prohibiting indiscriminate honking particularly at night. Noise emissions of different vehicles are also being studied. Sound absorbing materials were studied theoretically and experimentally, and mineral blocks were developed. The microperforated panel, developed in 1966, with sub-millimeter perforations for absorption instead of porous absorbing materials (Fig. 8) is being further developed and studied theoretically. It was not only successfully used in sound absorbing structures in severe situations, but was used in mufflers very satisfactorily. This is because it is less affected by air current than ordinary porous materials. As a single-value representation of sound insulation, proposals were made to use the sound insulation at 500 Hz or the difference between the C-level of incident sound and the A-level of transmitted sound; further discussion is planned. Proposals to use the standard tapping machine as a standard sound source in sound insulation and reverberation measurements aroused great interest among acousticians. The method seems to be simple, dependable and easy to use in any laboratory, as well as in practical situations. Flow noise attracted widespread interests because it exists practically everywhere (from rocket and jet aircrafts to boiler blow-down and pipeline leakage). It has been proposed that the sound field of turbulent noise should be expressed in A-weighted sound levels in order to estimate its detrimental effect on man. Both the over-all sound pressure levels and the A-weighted sound levels were found to be functions of plenum chamber pressure and explicit formulae were obtained for these relations. This provided a better understanding of turbulent noise. One outcome of this study is that for the same efflux permit area under the same plenum chamber pressure, the noise (measured in dBA) is found to be proportional to the third power of the diameter. The micropore diffuser thus designed from this theory has a predictable and excellent performance. Study was also made on the shock-cell noise produced in choked air jets and some interesting phenomena were found. Investigations were also made on porous diffuser-mufflers and open-hole diffusers. Mufflers have been properly installed on vehicles with good results. On the other hand, vibration isolation and enclosures were found very effective for the control of impact noise, and 30 dB reductions were obtained in the manufacturing of concrete assembly elements and for the cement ball mill. Sound sources in textile machines are also investigated, but this kind of work is only beginning. In addition, some work has been done on noise environment testing; a 168 dB travelling wave tube and a 152 dB reverberation room were built and acoustic fatigue studied, with some tests on animals. At high intensity, the wave shape changed into a saw-tooth and saturation was observed and analyzed in a long travelling tube, and it was found that modulation and the total harmonic content of the sound wave were higher than predicted from saw-tooth wave theory. This is an interesting phenomenon in nonlinear acoustics.

5. Speech

The acoustical study of speech started in the twenties was continued after liberation and has grown in scale. Research and development in speech communications were taken up by the industry in addition to the universities and research institutes. The frequency analysis of vowels, its formants and intonation measurements were taken up by many laboratories. The investigations led to the determination of the standard spectra of vowels, the average spectrum of Putonghua (colloquial Chinese), and articulation test methods. "Speech-noise" was proposed independantly and experiments carried out. A speech analysis-synthesis system (vocoder) with special emphasis on Chinese tone language was developed. Based on the statistics of seven hundred thousand Putonghua words (more than a million characters), the frequency of occurrence was determined for vowels, consonants, tones, syllables (characters) and words. The frequencies of occurrence of letters in the alphabet and of characters were determined when the language was transcribed into spelling form according to the scheme for the Chinese phonetic alphabet. Logatoms (nonsense syllable lists) were composed for the Putonghua, balanced in both phonemes and tones. The relations between the articulation scores for different phonetic units were obtained together with their dependence on speech transmission parameters (such as loudness, frequency band, signal to noise ratio, etc). From these we obtained the general rules of speech articulation. The articulation score of tones is nearly independent of transmission conditions and almost always perfect. Owing to this, the variation of tones and some properties of consonants were emphasized in the design of vocoders and high intelligibility and naturalness were obtained for 2400 bits per second, making the channel vocoder practical. We have also determined the near-field (near the mouth) average spectrum of Putonghua, on the request of the C.C.I.T.T. It was found that the average spectrum depends slightly on the level of utterance (the spectrum shifts towards higher frequencies for loud voices), and it is suggested that speech levels must be indicated in discussing the average spectrum. In automatic speech recognition, we succeeded in recognising 10 vowels of Chinese in 1961 and since then great progress has been made. Using suitable spectrum sampling, time normalization and pattern matching on computers, rather high efficiency of recognition (nearly 99%) was obtained, and the recognition may be made perfect (100%) if some display or error-correction technique is used. This, incidentally, applies to Putonghua as well as other dialects or languages. The recognition is valid only for a finite vocabulary, but it is already almost practical for computer input, voice-controlled automatic systems, etc.

6. Electroacoustics and acoustical measurements

As stated above, we were rather successful in our study and development of piezoelectric crystals and ceramics. Microphones of all types are now in production. The condenser microphones with directly plated diaphragms developed by us possess the high stability and low temperature coefficient required for microphones. Electret microphones are also manufactured, and high quality measuring electret microphones, as well as good quality antinoise microphones and crystal louspeakers have been developed. The electro-pneumatic loudspeakers and flexuo-tensile hydrophones are also being developed, which made the radiation of a few kilowatt sound power in air and in water feasible. Wide frequency band standard hydrophones are also in production. Sound pressure standards for air and under-water sound were established and calibration methods standardised. Some progress was made in acoustical measuring instruments, but difficulties still exist and it is far from meeting the national demand. Statistical analysers and sound spectrographs were developed towards the end of the sixties. Later on, when the computers became available, computer programs (in BASIC and FORTRAN) were written for fast Fourier transforms (FFT), and software was developed and stored in a program library. Lately, real time FFT hardware was developed and put into production; twelve functions were available namely FFT, IFFT, ODFT, autospectra, cross-spectra, convolution spectra, auto-correlation, convolution digital filter, probability density and transient capture. An A/D converter was provided which allowed direct interfacing with the computer. Reverberation rooms, insulation laboratories, anechoic rooms and anechoic pools for water sound measurements were built at various institutions, facilitating acoustical measurement.

7. Miscellaneous

A series of investigations on physiological acoustics and the electrophysiology of hearing was carried out. Experiments on animals to study the effect of impulsive sound were performed. A survey was carried out on the effects of continued exposure to high intensity noise on man in order to draw up noise standards. Investigations were made in the fields of atmospheric acoustics and geoacoustics. It was found that the infrasound produced by typhoons had more or less a constant period of 6 seconds, and hence the localization of typhoon centers by infrasound was feasible. During earthquakes, it was found that sounds of all frequencies including infrasonic and ultrasonic were produced in the earth besides the seismic waves, and that acoustical detection proved to be more sensitive than seismic wave detection. Besides, low frequency sound was also observed in the air. Whether this was radiated from the epicenter or produced locally was still left unascertained. Infrasonic receivers and monitoring systems were installed with instruments produced in China. Work has also been done in musical acoustics, the study of non-electronic and electronic musical instruments.

V. Concluding Remarks

The acoustical method is one of three main physical methods (the others being the optical or electromagnetic method and the method of bombardment by particles) with which man observes the world. Sound waves propagate in all material media (solid, liquid, gas, plasma etc.) revealing their mechanical properties. For these reasons, acoustics is indispensible in improving the means of ideological intercourse between men, improving the living and working environment, quality control of products, non-destructive testing of products for safe utilization, inspection of the microstructure of materials, prospecting, medical applications, and even in the study of mental activity. In view of these, it is clear that fully exploring and exploiting acoustics will be symbolic of the Four Modernizations. The progress of acoustics in China followed a tortuous path in the last thirty years. Now with the reorganization and strengthening of the study programmes in acoustics, we are sure that we can work better and harder from now and advance the science and technology of acoustics in China to a high level that is to bring prosperity to our great country.

CHAPTER 4

TWENTY YEARS OF THE INSTITUTE OF SEMICONDUCTORS — A SURVEY

by Huang Kun, Wang Shouwu, Lin Lanying, Cheng Zhongzhi and Wang Shoujue

The earliest research work on semiconductors in this country was started in 1952 at the former Institute of Applied Physics of the Chinese Academy of Sciences. With the launching of the national 12-year plan for development of science and technology in 1956, the research group and their work were greatly expanded. The rapid progress that followed laid the basis for the present Institute of Semiconductors, which was established as an independent institute of the Chinese Academy of Sciences in 1960. The Institute has been expanding steadily, especially during the sixties, and is now a multidisciplinary research institute undertaking a wide range of research and development work. The subjects include semiconductor materials, integrated circuits, microwave devices, optoelectronics, advanced processing technology, electronic instrumentation, physical and chemical testing and analysis, and semiconductor physics. In the following, we shall present a brief historical survey of the various achievements of the Institute during its twenty years' development.

I

The materials research at this Institute has, over the years, covered a fairly wide range of materials, including Ge, Si, GaAs, GaP, InP, Cu_2O, InSb, InAs, ZnSb, PbTe, Bi_2Te_3 families of thermoelectric materials, chalcogenide and oxide families of amorphous semiconductors. The materials research has always been closely integrated with various device researches undertaken by the Institute. Thus, owing to their great practical importance, GaAs and Si were in a class by themselves; they have been subjects of long-term extensive studies.

The early research started in the late fifties with zone-refining, single-crystal growing, impurity doping, measurement and control of minority carrier lifetime of germanium. Early work on silicon was also started during this period and the first single crystal was produced in the laboratory in 1958. This was followed by several years' work spent on the design and construction of appropriate single crystal puller and zone-refining apparatus to pave the way for the growth of single crystal silicon of high purity and crystal perfection. With the crystal puller specially designed to suppress vibration and other causes of mechanical disturbance and stray nucleation, dislocation-free crystals were successfully grown in 1962. Subsequently, with zone-melting apparatus specially designed with a stationary induction heating coil to give stability of operation, and introducing wet hydrogen ambient for boron removal, high purity single crystal silicon was grown. N- and P-type crystals with respective carrier concentrations of $6 \times 10^{11}/cm^3$ and $2 \times 10^{12}/cm^3$ were obtained.

Systematic investigations were conducted on the effect of different ambients; namely, hydrogen, argon, vacuum, on zone-melt grown single crystal silicon. Thus it was discovered that hydrogen ambient invariably introduced defects in the grown crystal and only with the last zone-melt cycle conducted in vacuum or argon could one obtain reasonably good crystals. Work was also carried out to investigate how the growth condition affected the dislocation density and by what means could low dislocation density be achieved. The results of this investigation were presented at an international conference held in Prague in 1964 and apparently aroused much interest among the scientists then present. By 1965, zone-melt single crystal silicon could be grown with "O" dislocation density and high purity crystals were grown with resistivity of $10^4 \Omega$ cm and minority carrier lifetime of 1000 μsec. The extensive work on silicon single crystal during this period provided the basis for the very active and fast expanding silicon device research.

The first single crystal of GaAs was grown at the Institute by the horizontal Bridgeman method in 1962. This first work was succeeded by the growth of heavily doped crystals, development of the ECZ method and the growth of large diameter Cr-doped high resistivity crystals for certain specific applications.

During the late sixties, in the course of development of microwave devices, it was discovered that resistivity of the GaAs crystals had an anomalous negative temperature coefficient. Investigation in this problem eventually led to an effective method for reducing the concentration of the relevant deep-level centres by appropriate annealing. Thus GaAs material with positive temperature coefficient for resistivity and improved mobility was obtained. During the early seventies, an intensive investigation was undertaken to discover the cause of the growth defects in single crystal GaAs. This work succeeded in pinpointing the primary factors causing the production of dislocations in growing GaAs single crystals by the horizontal Bridgeman method. On this basis, by using seed crystals to control the crystal growth direction and with appropriate control of growth temperature, growth speed as well as stochiometry, low dislocation density ($10^2/cm^3$) Te-doped (subsequently also Si-doped) GaAs single crystals were successfuly grown in 1974. In the following years, the mechanism for supression of dislocations was intensively investigated and a theoretical model was proposed which could adequately account for all the experimentally observed facts. The growth of high purity GaAs was a difficult and very painstaking task, on which a great deal of work was done at the Institute. Recently the work made good progress and has successfully achieved the reproducible growth of epitaxial GaAs with low temperature mobility $\mu 77K$ above $1.5 \times 10^5 cm^2/Vs$. both by VPE and by LPE, the highest $\mu 77K$ attained being $2.1 \times 10^5 cm^2/Vs$.

Owing to the growing importance of the semiconducting materials that have to be grown under high pressures, a crystal pull which could withstand pressures above 100 bars was specially designed and constructed a few years ago. GaP and InP have since been grown and provided the basis for new subjects for research previously impossible.

Historically, the Institute also carried out at different times, research on materials designed for specific applications. Thus since the late fifties, a systematic investigation was carried out on the Bi_2Te_3 type compounds and their mixed crystals with a view of developing their thermoelectric application (power generation and refrigeration). The material properties and material preparation problems investigated included the dependence of thermal conductivity on the composition ratio of the mixed crystals, the effect of different doping impurities on thermoelectric properties, the relation between the preparation condition and the quality of the preferential crystal orientation (by the Bridgeman method and by zone-melting) etc. On the basis of such basic investigations, thermoelectric materials were obtained in 1965, which could produce a maximum temperature difference of 78°C. With these materials, various small refrigerating units for different applications were successfully made at the Institute. During the seventies, a great deal of work was done on the amorphous oxide and chalcogenide semiconducting materials in an attempt to explore the possibility of making switching and memory devices with these materials.

II

The device research at the Institute started with the fabrication of germanium power transistors, surface-barrier transistors and high frequency transistors with the alloy and alloy-diffusion methods in the fifties. After 1960, in preparation for the development of silicon transistors, work was started on processing equipment and fabrication techniques, such as making ohm contact to silicon, developing appropriate diffusion processes etc. During 1962–1963, the first power transistor, high-speed switching transistor and high-frequency power transistor were successively fabricated by the silicon planar process. At this time, a number of problems in connexion with these transistors were subject to careful research, such as the factors affecting the maximum oscillation frequency, the working of the transistor in the saturation region and the saturation voltage, the high-frequency oscillation problem caused by the carrier transit delay etc. These investigations led to transistors with improved performance. The development of the planar process and the fabrication of the first silicon planar transistors was awarded a first class prize in 1964 by the National Committee of Science and Technology.

The above work also paved the way for the development of the integrated circuits. In 1964, the first early prototype bipolar integrated circuit was made and vapour phase deposition of SiO_2 was already introduced in the fabrication of the integrated circuits. At about this time attempts to make silicon MOS devices were also started. This effort eventually led to the making of the first MOS transistor in the country.

Even in the early days in the development of integrated circuits, it was felt that making the pattern of the photolithography masks by hand was very laborious and unsatisfactory, and the idea of developing some automatic equipment for this was proposed. However actual work along this direction was not started until a few years later; then the work made rapid progress by adopting a building-block mode of generating the mask pattern, which greatly simplified the mechanical construction of the equipment. An automatic pattern generator was thus successfully made in 1972; this was the first automatic pattern generator for photolithography mask making in the country. Subsequently it was used successfully in making masks for 256-bit and 1024-bit MOS shift registers, which were among the first generation of LSI being developed in the country.

In recent years, we have emphasized that, as an institute of the Chinese Academy, we should particularly encourage original ideas and innovation of new types of circuits. As a result, some very interesting work has been achieved. Of significance was the development of a new type of high-speed logic circuit, which has been designated as DYL circuit. It introduced a new "linear AND-OR gate" as the basic logic element; it is characterized by high-speed operation and is readily adapted to simplified processing. Thus integrated circuits consisting exclusively of this type of gates can be implemented without epitaxy and isolation, which are indispensible for usual integrated circuits. Practical experiments have shown that, with the most conventional processing (10 μm line width and 5 μm alignment allowance), a medium scale integrated circuit of this kind of gate can be fabricated with a delay time per gate of about 1 nsec. Since 1978, a fairly wide variety of DYL circuits has been designed and made in the laboratory, including a 12-bit first carry generator, which has the function of an equivalent TTL circuit consisting of 130 gates. Our experience confirmed the advantage of the DYL circuits in speed, simple processing and high yield. As the circuit functions of the new circuits and their exploitation differ rather radically from usual types of logic circuits, a micro-computer was built in 1979, using only DYL circuits. Experience on the exploitation of the new circuits was gained from the construction of the computer and its successful working testified to the practicality of the DYL circuits. Besides the DYL another type of new circuit was also developed which is a hybrid of TTL and ECL circuits. During the last two years, a variety of this type of circuits was designed and fabricated; these circuits showed that the same logic function can be implemented with less circuit elements compared with usual circuits. They also have the advantage of low power dissipation and are particularly adaptable to large scale integration.

During the last few years, a special project was undertaken to raise the yield of MOS LSI processing. This work has the background of a number of years of research on MOS circuitry and fabrication technology. Despite all the difficulties owing to limitations in equipment, processing conditions etc, the work made good progress and achieved very good results. Thus typical MOS LSI such as 4K RAM with full peripheries can now be fabricated with reasonably high yield with very conventional equipment, all produced by the home industry.

III

Microwave semiconductor electronics represents another branch of the Institute's device research activities. Historically our microwave device research has covered a wide range of subjects, from parametric amplification to wide-band low-noise amplification and frequency multiplication, from individual components to whole systems, from decimeter to centimeter and millimeter wave bands.

Before 1970, an X-band GaAs beam lead Schottky diode hybrid integrated circuit was successfully developed. From this work was developed the technology of hybrid circuitry with passive elements on ceramic substrate. For the actual circuit, a 3db bridge was then implemented on a ceramic substrate with high aluminium content. The use of beam lead structure for the active element gave close electrical coupling to the micro-strip. The high carrier mobility of GaAs is conducive to superior device performance, but presented more fabrication problems. Thus an appropriate planar process had to be developed for the fabrication of the GaAs device. In connexion with the fabrication of the active device, certain new measures were introduced with success. Thus chromium was used instead of nickel for the metallic contact. It was found that it gave higher mechanical strength and could form a barrier no higher than that of nickel contact. Besides, a chemical etching process was introduced to effect the chip isolation in the beam lead structure.

Since 1971, work has been started in the millimeter wave band. The research work centered around two projects, which had for their aim the respective development of a 4mm band oscillator and a 4mm band mixer. The active element of the oscillator was a P^+NN^+ Si avalanche diode, developed specifically for the purpose. The research project produced a high Q cavity-stabilized continuous wave oscillator, which compared favourably with a klystron in output power, frequency spectrum, frequency stability and noise figure. The mixer was based on a GaAs Schottky diode. An electroplating process was developed for making the metallic contact, which served to reduce the contact capacitance. The Schottky diode and the whole device were subject to an integral circuit design, such that the diode was directly packed in a section of waveguide. With this design, the mixer achieved a frequency conversion loss of 6−7 dB and noise figure N_F of 12 dB. These devices have been successfully used in the receiving circuit of radio-astronomical telescopes.

Sufficiently high applied voltage can cause avalanche breakdown to occur in the high field domain of a Gunn diode and consequently a relaxation type of oscillation can be set up, with a time constant which is related to the recovery of the specimen after the avalanche generates excess electrons and holes. For a number of years systematic experimental and theoretical investigations have been carried out on these and related phenomena. The use of specially fabricated planar Gunn diodes with epitaxially grown N^+ GaAs electrodes made it possible to experiment with much higher applied voltages (up to 10 times the threshold voltage). With such specimens the phenomenon of relaxation oscillation was systematically investigated; oscillations at 100 MHz with an efficiency of 30% have been obtained. Experimenting with the ohm contact showed that good alloy ohm contact could raise the oscillation frequency to the L band. The phenomenon of super-radiation following on the avalanche breakdown of the high-field domain was experimentally observed. This suggested that with a suitable optical cavity it might be possible to realize high-frequency modulated laser emission. The detailed mechanism of the relaxation oscillation was investigated both experimentally and theoretically; in particular computer simulation calculations have been carried out for the whole process occurring. The computer simulation work also extended to the investigation of stationary domains and discovered that they could go over to travelling domains under certain conditions; this prediction was later experimentally verified.

IV

Work on junction lasers at the Institute was first started as early as 1962. During the next year, pulse-operated GaAs junction lasers working at 77K with a maximum power output of 100W and pulse-operated GaAs lasers with a room temperature power output of 8W were produced in the laboratory. Subsequently work was started on hetero-junction lasers and the devices developed gradually found applications. Early work on double heterojunction lasers started in 1973; continuous laser emission at room temperature was first observed in 1975. From that time, work turned to careful analysis and continuous improvement of the material and device processing; in recent years research has been directed towards the problem of the degradation of the device with operating time. At present the heterojunction lasers fabricated in the laboratory have operating lifetimes exceeding 10000 hours and have been in use with experimental optical fibre communication systems.

In the past few years a new device, "PNPN negative resistance laser", was also developed which effectively builds a heterojunction laser into a PNPN structure such that, when the latter is switched to the high current state, laser emission is turned on. The device is designed as a 6-layer hetero-structure $N_P P_{PN} P$ such that the right-side transistor structure, which does not involve the active laser region, has a value very close to 1. This design provided a favourable condition for the implementation of an effective laser in the structure. With this design, the device showed the fairly low threshold current density of 2500A/cm² for lasing. The device could generate self-excited oscillations with frequency reaching 15 MHz.

Historically a number of projects have been undertaken to develop various optoelectronic devices for special applications. Development of silicon solar cells was one of the earlier works. By 1965 the conversion efficiency had reached 13%. The result achieved by the project later served to provide the power source for the second satellite launched in this country. A more recent project developed a photodiode for detection of radiation in the wavelength range 0.8–0.9 microns in connexion with applications of the GaAs junction laser. Other applied works included laser telemetry, optical communications (through space in atmosphere), surface barrier α-particle counter etc.

V

Physical investigations and development of measurement methods followed closely upon the preparation of the single crystal materials. Early work in the fifties was mostly related to the development of single crystal germanium, such as the measurement and investigation of surface recombination and minority carrier lifetimes.

In 1960, owing to the rapid development of semiconductor technology in the country, the National Committee of Development of Science and Technology decided to establish at this Institute, a centre for standard measurements on semiconducting materials and devices. For this purpose, an intensive study on the development of methods of measurement of minority carrier lifetime was undertaken. The study covered transient measurements, steady state measurements as well as measurements with sinusoidal modulation. For transient measurement, a new method was developed, which modified the double-pulse method by using light injection. The method preserved the non-destructive character of the double pulse method, but avoided all the troubles associated with the electric contact. By this method, reliable measurements of lifetime down to the order of 1 μsec were achieved. Steady state measurements were studied mainly for the purpose of measuring very short lifetimes (μsec-nsec), in connexion with investigations on recombination centres and development of fast switching devices. With photomagneto-electric methods and photoconduction-compensation methods, minority carrier lifetimes down to 10^{-9} sec were measured with Ge and InSb materials. A further method based upon the photomagnetoelectric effect employed an alternating magnetic field; by using difference-frequency measurement, the photovoltaic interference arising from the non-uniform resistivity of Si materials was excluded and the method proved well adapted to short lifetime measurements. A method based on sinusoidal modulation was developed by utilizing the Kerr effect of nitrobenzene to produce sinusoidally modulated light up to hundreds of KHz; from the phase shift of the observed photoconduction, minority carrier lifetime in the range $10^{-7}-10^{-8}$ sec could be measured.

In the meantime, systematic measuring methods were established and corresponding equipment built for transistor measurements, including all the frequency parameters (f_α, f_β, f_{max}) and the transfer parameters.

Other physical investigations at the time included electric conductivity and Hall effect of Si (20 – 300K), characterisation of compensated silicon, paramagnetic spin resonance experiments with silicon and surface field effects of germanium etc. The work on chemical analysis was also making progress. Thus, in collaboration with the Atomic Energy Institute and Applied Chemistry Institute, the method of activation analysis (including activation by charged particles) was established and the oxygen content in GaAs as well as more than ten kinds of impurity elements in Si were analyzed. With spectro-analysis, impurities in high purity silicon, arsenic and gallium were analyzed. Besides, impurity contents in Si, As, Ga, GaAs and other compounds were analyzed by spectrophotometry, polarography and mass spectrography.

A project was initiated a few years ago to establish at the Institute, a centre for physical and chemical analysis, for the purpose of standard measurements, and analysis of basic parameters of semiconducting materials and devices, ultra high-purity analysis, structure analysis and failure analysis. Progress has been made over a wide range of subjects. Thus the method of X-ray topography was established to investigate dislocations introduced in silicon device fabrication. An intensive investigation by atomic absorption spectro-analysis has been conducted on impurity contamination introduced by wafer polishing, cleaning, chemical reagents and deionized water. Various spectroscopic investigations have been undertaken relating to impurity atoms in semiconducting materials. Thus infrared absorption spectra relating to oxygen in Si and Ge were investigated down to 6°K; photoluminescent spectra of doped and high-purity GaAs were investigated down to 4°K. The results of these researches found direct application in the material preparation.

Facilities were established for spectroscopic investigation of cathodoluminescence produced by a scanning electron beam and effective studies on ion-implanted impurities and impurity doping in microregions in the compound semiconductors GaAs, $Ga_{1-x}Al_xAs$, $GaAs_{1-x}P_x$ were carried out. Other methods established for the detection and study of impurities and defects included SEM study of defects in materials and devices, trace analysis for impurities in ultra high-purity hydrogen by gaseous chromatographic analysis, and various electric measurement techniques, such as spreading resistance measurement, thermally stimulated capacitance measurement etc. The laboratories have been able to acquire more modern instruments, so recently it became possible to carry out back-scattering experiments with 2.5 MeV H^+ and He^+ ions and surface analysis by Auger electron spectroscopy.

VI

Electronic instrumentation work has always occupied an important place in the work of the Institute. The development of devices and IC's, as well as physical investigations and applied research, often required measuring apparatus which were not available in the country and had to be designed and built by our scientists and engineers at the Institute. Thus during the development of the first generation of germanium transistors, the whole set of transistor measurement equipment for low frequency and high frequency parameters, noise figures and I-V characteristics tracing etc was built at the Institute During the early sixties, through about five years work, the basis was laid for nanosecond pulse electronics work. Thus, in this period, we built a number of instruments, which were first of their kind in the country, such as the 1000 MHz sampling oscilloscope, pulse generator with 1 nsec rise time, pulse generator with 100 MHz pulse generation frequency etc.

In recent years, owing to the important development of LSI and the complexity of their measurement problem, a great deal of work has gone into automatic measurement systems. Thus in collaboration with the local electronics industry, an automatic measurement system for IC dynamic parameters was designed and built for the testing of high speed TTL circuits and semiconductor memory testing equipment was built for the testing of MOS LSI memories. All this instrumentation work was closely integrated with the device and IC research at the Institute and at the same time contributed to the development of the electronics equipment industry in the country. Thus an instrument first built at the Institute was often handed over as prototype for industrial production. Another piece of modern digital electronics equipment of general significance was a logic analyzer, which was built a year ago and handed over to the industry.

Besides, from time to time, work was undertaken for certain specific applications. A typical example was the development of a solid-state positioning satellite transponder, which was undertaken in view of the satellite development.

The above survey can only be a sketchy account of some of the Institute's representative achievements, during its twenty years' development. In particular, it has not covered some of the more recent developments. Thus in the past few years, we have encouraged more basic research and long-term projects. Typically, we have organized a semiconductor physics group to foster basic research and an advanced technology laboratory to work on fabrication technologies which promise to pave the way for future VLSI development. All our efforts have been to work towards a high level of scientific achievement as appropriate for an institute of the Academy in future years.

CHAPTER 5

CHINA'S FIRST LASER

by Wang Zhijiang and Wang Nenghe

In September 1961, China's first laser — a ruby laser — was successfully constructed by the Changchun Institute of Optics and Fine Mechanics, Academia Sinica. This can be marked as a significant event in the history of optical science in China.

Optics is a science with a long history. Due to its intimate relation with the people's daily life, scientific records of optical phenomena were found in many ancient Chinese books such as the "Mo Jing". The developments of modern physics at the turn of the century, as well as the subsequent birth of relativity theory and quantum mechanics, were closely related to the progress in optics. However, by the 1950's, optical science was generally perceived as having reached its summit in its development, with few future breakthroughs possible.

The Changchun Institute of Optics and Fine Mechanics was the first optical research facility established after the birth of the new China. In its early days, research in classical optics was pursued. By the end of the 1950's, the Institute had acquired a capable research team with quite advanced experimental facilities. One of the authors of this article worked at the Institute at that time.

The Institute was immersed in an active academic atmosphere at the end of the 1950's. A number of the research workers felt that some of the laws of classical optics seriously restricted the application of optical techniques. For example, light intensity could only be decreased, wavelength could not be reduced, light energy could not be concentrated and the image contrast would inevitably be reduced by the scattering medium, etc.

The fact that brightness could only be reduced had long been one of the fundamental principles in optical instrument design. It can in fact be considered as a generalization of the laws of thermodynamics in optics. As traditional light sources offered insufficient intensity, applications in optical engineering, such as optical ranging and remote illumination etc., were often quite seriously limited.

The dispersion of optical energy had always been a problem in the design of Raman spectroscopes. Since stimulated emission has a definite opening angle, the Raman scattering is characterized not only by its low quantum efficiency but also by a distribution over the 4π solid angle. The resulting radiant flux on the detectors (such as photographic films) is usually too weak to be recorded.

Although some devices based on the trapped energy principle in solid state energy band theory (such as infrared image converters) were available, it was still necessary to stimulate the electrons to the "trap" by electrical or thermal means. However, direct means to appreciably reduce the wavelength seemed impossible.

It was indeed puzzling that techniques such as amplification, frequency conversion, etc. had been successfully applied to electromagnetic waves in the radio frequency spectrum, while they were considered impossible in the optical range. Many questions in this respect were raised by the research workers at the Institute.

Having noticed the different nature between a radio transmitter and a light emitter, serious work was begun at the Institute on microwaves, which lie in between radio waves and light in the electromagnetic spectrum. The operational principles of the magnetron were investigated and two of its characteristics were focused. First, the microwave transmitter, quite similar to the radio transmitter, depends on the free oscillations of electrons. The magnetron, in fact, is a device which forces the electrons to move in a periodic magnetic field. Based on these considerations, the possibility of coherent light emission using a slow wave system was investigated. This idea, to date, has not become a reality because no suitable technique is available. In spite of this, the idea of using low energy electrons to generate coherent light is still valuable, with good potentials. Indeed in recent years we saw the invention of the free-electron laser, exploiting the passage of a high energy electron beam through a periodically alternating magnetic field to extract coherent light. It can be considered as an "optical magnetron".

Another characteristic of the microwave transmitter is related to the concept of wave forms (or modes). As is well known, this concept had played an important role in Planck's theory of blackbody radiation; even microwave guides were named directly after the wave forms. The concept, however, was little used in radio techniques and in traditional optical instrument designs, although the wave form classification method and its generalization became quite popular in laser physics lately. This method is of course not the only one available: the laser researchers in China had also used Planck's classification, i.e. taking the wave form number to be equal to $(4\pi v^2 \Delta v/c^3) \cdot V$. Calculations of the relaxation oscillation process in a multimode laser had been obtained this way.

We had also noticed earlier that the atomic lifetime is closely correlated, not just with the atom itself as described in textbooks of quantum mechanics, but also with the cavity confining the radiation. Specifically, we analysed the atomic radiation lifetime in Fabry-Perot interferometers.

By then much attention was paid to the research on masers and the limitations in their applications were noted. These were the limitations that made it impossible to replace the microwave transmitters, such as the magnetron. We did, however, investigate the possibility of extending the operational domain to include the optical band.

With this background preparation, we were able to grasp the essence of the seminal paper of Townes and Schawlow when it was published. We started to develop a device for light amplification by stimulated emission of radiation (laser) before Maiman's pioneering experiment was known to the public. We named our device "Ji Guang Qi" at the Third National Conference on Quantum Electronics (China) in 1964.

It is one of the essential steps towards the construction of a laser to seek a suitable active medium. The use of the ruby crystal as the active medium in our first laser was suggested by the maser studies. The energy levels of the ruby crystal were known since the 1930's and were carefully analysed in the early studies of ruby masers. It was realized that there are two absorption bands (U and Y) and energy can be transferred to the excited energy states of the R-line with a very high quantum efficiency.

Although our first laser was developed a year later than those in other countries, it had many distinctive features. First, while Maiman used a helical xenon lamp and diffused illumination for excitation, we used a linear xenon lamp and a sphere to form the image illuminator. Taking into account the fundamental principles of optical systems, it is obvious that our excitation efficiency is higher than that of Maiman's. We believe we were the first to use a spherical illuminator in the world. Although cylindrical illuminators received some attention from foreign scientists after Maiman's work, we did not think that a higher efficiency could be achieved. Multi-lamp cylindrical illuminators were also popular at that time. However, considering the fundamental relationship between illumination and brightness, we again concluded that when the diameters of the active medium and the xenon lamp were the same, multi-lamp illumination could not be superior. Experiments confirmed our conclusions.

The second feature of our first laser was that we used an extra outer cavity. Before 1961, highly reflective films were coated directly onto the two end-surfaces of the active medium in lasers. However, we found, through careful optical experiments (such as the "star point" test and Foucault shadow method), that ruby rods were actually optically inhomogeneous. This inhomogeneity had to be corrected by optically processing one of the two end-surfaces. The end surface after processing was neither a plane nor a perfect sphere. After optical inspection, we found that we had to take into account the effects of the boundary between air and the medium. In order to ensure that the optical path lengths of the laser resonator were actually equivalent, we separated the reflector mirror from the excitation medium, thus creating a semi-outer cavity.

At the time of our laser development, the technology of fine machining in China was better developed than that of coating of multi-layered dielectric films by vacuum sputtering. To reduce the losses at the reflector ends and to increase reflectivity, we pioneered the use of prisms as components of the resonator. We found the stable polarization state through careful analyses of polarization changes after multiple reflection in the prisms.

It was found that multi-layered dielectric films were easily damaged when the ruby laser energy was increased to 10 J in 1962. By replacing some of the reflective films with several glass plates of different thickness and spacing, we obtained good results. Shortly after this, we discovered and overcame the problem of instability in the operation of the laser induced by parasitic oscillations of side reflection. Also, nonlinear influences of the gain length of the medium on laser efficiency were noticed.

Our first laser was developed without the support of advanced technology, so we encountered many difficulties. For example, to obtain a pulsed xenon lamp, we had to do everything from the very beginning — we worked on electrode materials, glass materials for sealing of quartz tubes and the tungsten-thorium electrode, etc.

The evolution was natural from the maser to the laser, and from Townes's creative concept to Maiman's pioneering experiment. It was not accidental that our first laser was constructed at the Changchun Institute of Optics and Fine Mechanics, where well-trained workers had done a lot of work on the fundamental technologies necessary for the laser development. Their work included optical and mechanical designs, electric and electronic designs, fine machining, optical film coating, and even fabrication of glass and crystal growth. All these laid good foundations for our success.

Looking back at the development of our first laser, we realized that physical laws do have definite limitations on possible technological applications, but most of them are valid only under certain conditions. While the limitations severely restrict technological progress under one set of conditions, they may be bypassed under another set of conditions, and new prospects for applications are then possible. At present, the techniques and phenomena of amplification, frequency conversion, heterodyne and self-focusing are widely used in the field of laser science. Techniques of stimulated scattering, imaging via diffusing medium, self-adaptive technique via turbulent medium have been investigated extensively.

Today, we are able to not only produce various kinds of lasers (solid state, gas, semiconductor, liquid, chemical, gas-dynamic, large pulsed, CW, mode-locked, ultra-short pulsed, tunable lasers and lasers in the ultraviolet, visible, infrared and far infrared spectra) but also to provide wide applications in industrial processing, range finding, laser communication, holography, as well as medical treatment and seed breeding in agriculture. But laser science and technology in China is still developing, and we look forward to further advances that will bring us to the frontiers of the field.

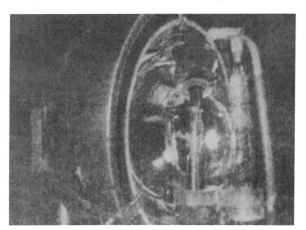

Physical structure of China's first ruby laser.

CHAPTER 6

ADVANCES IN BIOCHEMISTRY AND MOLECULAR BIOLOGY IN CHINA

by Wang Yinglai and Shi Jianping

The modernization of biological sciences is an integral part of the 'Four Modernizations'. Biochemistry and molecular biology have attracted their due attention among the various branches of modern natural science. The latter has been regarded as a key biological discipline and has been given special emphasis.

Before the liberation of the country from Kuomintang's rule, biochemistry was relatively very poorly developed. Among the few biochemists of that long period, Wu Hsien and his co-workers had done outstanding researches in protein denaturation, immunochemistry, clinical analysis and nutrition in the second and third decades of this century. But owing to the sad neglect of science in the old days, for a long period from the beginning of the Japanese invasion until liberation, scientific research was almost at a complete standstill. After the birth of new China, due to the close attention and strong support of the Communist Party and Government, biochemistry and molecular biology have taken root and have begun to flourish. In the fields of proteins, enzymes and nucleic acids, important results have been obtained, some of which have attained a very high level. The applications of biochemistry in agriculture, industry, medicine and national defence have also made important progress. In the following pages, we shall try to outline some of the work that has been done during the last thirty years in biochemistry and molecular biology.

I. Fundamental research

1. Structure and function of proteins and polypeptides

Research on the structures and functions of the important biomacromolecules especially proteins, enzymes and nuclei acids, form the basis of molecular biology. In these areas we have already gained solid ground and have made generally recognized contributions, the most outstanding of which being the total synthesis of crystalline bovine insulin in 1965. This is the first protein with full biological activity ever to be synthesized artificially and represents an important breakthrough in man's many efforts to gain a deeper insight into the mystery of biological phenomena. It holds an important place both in natural science and philosophy, and paves the way for the future synthesis of other proteins. In the early seventies, our scientific workers independently completed the X-ray crystallographic analysis of porcine insulin at 2.5 and 1.8A resolution.

The results agreed almost completely with those of Dorothy Hodgkin, who had been working on the same problem intermittently for over thirty years. This work has raised our X-ray crystallographic analysis to an advanced level, provided a basis for further work on the relationship between the structure and function of insulin, filled up the hitherto blank area of protein crystallography in our country and constituted an important contribution to the development of molecular biology in China.

Studies on the structure, function and mechanism of action of insulin and its analogues which have opened up on the basis of the above work, have become a traditional area of research in this country. Studies range from the replacement of parts of the C-terminal region of the B-chain by various peptide fragments and amino acid residues and observing the resulting functional change, to comparative studies on the structure and activity of insulin from different species leading to new important developments. Work has also been done on the linear polymerization of the B-chain of insulin.

Investigations on biologically active peptides which may be considered as being derived from the insulin work, have flourished and a good number of peptide hormones have been synthesized, including oxytocin, vasopressin, angiotensin II, luteinizing hormone-releasing factor, glucagon, thyroid hormone-releasing factor and others. Certain improvements have been introduced in the methodology of peptide synthesis, such as the use of the potassium salts of amino acids to react with the chloromethylated solid supports and the condensation of small peptide fragments on a special solid support, with satisfactory results. Work has also been conducted on morphine peptides. From porcine spinal cord and pituitary and human placenta, a number of biologically active peptides have been isolated, some of which may be hitherto unreported new peptides. Biologically active peptides have already become an active area of research in this country.

Systematic investigations on muscle proteins have been carried out in spite of long continued interruptions. From the results of comparative studies, it has been suggested that tropomyosin may be involved in the contracting and holding functions of the muscle. Studies have been made on the chemical structure, physico-chemical properties, and ultra-micro structure of tropomyosin and paramyosin from various sources after purification and crystallization. It has been found that tropomyosin may exist in the form of paracrystals. Trichosanthin has been purified and crystallized and systematic studies on its structure at various levels and its pharmacological action are underway. Work on snake venoms, toxins, virus proteins, blood serum proteins, hormones and their receptor proteins, has produced interesting results.

2. Structure and function of enzymes

Systematic and detailed studies on succinic dehydrogenase have been going on despite frequent interruptions over a long period of time. The enzyme has been successfully solubilized from mitochondrial membranes and purified over 200 fold, with an activity almost twice as high as that reported simultaneously in the United States. It has been found that the prosthetic group of this enzyme contains flavin adenine dinucleotide (FAD) and non-heme iron; the FAD is linked to the apo-enzyme by covalent bonding. Many interesting properties of the enzyme have been reported. This work has stimulated research on the enzyme system of the respiratory chain especially on the reconstitution of the system.

In work on the respiratory chain and oxidative phosphorylation, use has been made of the competitive action of two enzyme systems for a common linking factor to elucidate the oxidative pathways of some substrates through the cytochrome system, thus providing an effective method for the study of a complex enzyme system.

A series of investigations on the relation between chemical structure and enzyme activity led to the establishment of a quantitative relationship between the lowering of enzyme activity and the modification of the essential side chain groups. It has been possible, based on these studies to make a more accurate evaluation of experimental data, and the number of essential groups of certain enzymes has been estimated. Work has also been carried out on the kinetics of enzyme action and systematic studies have been made on the kinetics of the binding between an enzyme and its modifiers and the effect of substrate on the binding. Based on the kinetic equations obtained, it was shown that during the irreversible inhibition of an enzyme, the product formed from the substrate approaches a definite value with the increase of the reaction time; and from the determination of the product formation at definite time intervals, the rate constant for the enzyme-inhibitor system can be obtained and the type of inhibition ascertained. For a large number of enzymes, two ES intermediates are involved in their mechanism of catalysis. The kinetic behaviour of such systems towards inhibitors has been studied and a general equation obtained.

In the presence of NAD^+, carboxymethylated

D-glyceraldehyde-3-phosphate dehydrogenase gives rise to a new fluorophore at the active site of the enzyme upon irradiation with ultraviolet light. This is a new observation which may prove to be useful as a probe for the study on the interrelations of some important functional groups at the active site.

The nitrogenase system is very important both from the theoretical and practical point of view, and active research has been conducted on the purification, activity, and structure of the Mo-Fe-S-protein, Fe-S-protein and Fe-Mo-cofactor. Models for the active centre of nitrogenase have been proposed and have attracted international interest; they provide theoretical bases for chemical imitation and for research on model enzymes.

Hydrolases, including phosphoesterases, phytases, cholinesterases, nucleosidases and nucleotidases have been widely studied, including their purification and properties. Proteolytic enzymes from various animal and plant sources and protease inhibitors have also been purified and their mechanisms of action investigated. Studies on some oxide-reduction enzymes, enzymes connected with carbohydrate metabolism, urokinase and asparaginase have also produced interesting results.

3. Structure, function and synthesis of nucleic acids

The majority of works in this field has centred on tRNA. A mild preparative method has been devised enabling the large scale isolation of the nucleic acid. Treatment with phenol and subsequent fractionation with ammonium sulphate yielded a preparation which was homogeneous on electrophoresis and ultracentrifuge sedimentation and possessed high methionine-receptor activity.

Improvements have been made in the traditional two dimensional paper electrophoretic and chromatographic method to avoid the branching phenomenon and to raise the resolution. Using the modified method, investigations have been made on the soluble RNA from *Saccharomyces cerevisiae, Saccharomyces carlsbergensis,* the silk glands of *Bombyx mori* and *Attacus ricini*. It was found that the RNAs from the yeast are quite different in nucleotide composition from those of the silk glands.

Studies have been carried out to locate the combining sites on the tRNA molecules from *E. coli* by labelling with dinitrophenyl (DNP). The DNP was found to combine only with the 5-monophosphate terminals, thus providing a useful means for studying the primary structure of RNA. Using this procedure, the sequence of the nucleotides at the 5'-end of *E. coli* tRNA was elucidated.

By using enzymatic and chemical degradation, a simple paper chromatographic procedure has been devised for the determination of the distribution of nucleotides of tRNA. Employing Rushizky and Knight's paper on electrophoresis-chromatography mapping procedure, the products of pancreatic ribonuclease digestion of *E. coli* tRNA were separated and determined. It was found that the experimental values differed considerably from the values if the bases were arranged in a random sequence.

After alkaline hydrolysis, the alkali-resistant dinucleotides of yeast soluble ribonucleic acid was further treated with *E. coli* alkaline phosphatase. Four fractions were obtained and the composition was determined after digestion with a crude extract of *Naja Naja atra* snake venom.

Analyses of the enzymatic hydrolysates from tRNA and deaminated tRNA in the presence or absence of metallic ions revealed that Mg^{++} ions have a marked stabilizing effect on the secondary structure of tRNA. Thus tRNA in its secondary structural state is resistant to RNase degradation. Deamination of tRNA also led to a change in its secondary structure rendering it more susceptible to hydrolysis by RNase

Studies have been made on the structure-function relationship of tRNA using hydroxylamine. After the action of hydroxylamine on *E. coli* tRNA, the content of pyrimidine nucleotides was markedly decreased. But the extent of lowering of the cytidylic acid and uridylic acid was different depending on the pH of the reaction. The leucine-methionine accepting activity was lowered at different pH, but under the same conditions of treatment, the lowering of the accepting activity for the two amino acids was quite different.

Progress has been made in recent years in the synthesis of yeast $tRNA^{Ala}$. A number of nucleotide fragments have been synthesized, the longest of which was a hentetracontanucleotide fragment. At the same time, the two natural half fragments of yeast $tRNA^{Ala}$ have been successfully recombined by *E. coli*-T_4 RNA ligase.

The nucleotide sequence of the $tRNA^{Ala}$ from the posterior silk gland of *Attacus ricini* has been determined; it is the first work of its kind carried out in this country.

The effects of various physical and chemical factors on RNA and DNA have been carried out to supply information on the structure-function relationship of nucleic acids; it also provides theoretical background for research on radiation damage, its prevention and cure.

Changes in the nucleic acid compositions of hepatomas induced by 3-methyl-4-dimethylaminoazohenzene and those of various tissues after the use of antitumour drugs have been studied.

Studies on the nucleic acids of *Bombyx mori* and *Attacus ricini* are actively pursued. Progress has been made in the research on TMV and the mechanism of replication of the cytoplasmic polyhedrosis virus of the silkworm.

Methods for the preparation and analysis of nucleic acids have been established or improved. These included a spectrophotometric procedure for the determination of base composition, a two-wavelength spectrophotometric method for the estimation of nucleic acids, a colorimetric procedure for the analysis of DNA in the presence of salts, the use of a new fluorescent reagent for the microdetermination of the nucleotide sequence of oligonucleotides, polyamide thin-layer chromatography for the separation and identification of blocked nucleosides, nucleotides and oligonucleotides; sequence analysis of DNA fragments, improved gel electrophoresis for the isolation, purification and analysis of tRNA, the use of agarose-bead column chromatography for the isolation of the low molecular RNAs, etc. The stability of RNA preparations has also been investigated.

Polynucleotide phosphorylase, RNase N_1, RNA-ligase, snake venom phosphodiesterase and RNase T_1 have been isolated and purified. These have also been produced on a large scale and have thus facilitated research in the structure, function and synthesis of nucleic acids.

4. Metabolism

Systematic studies have been conducted on the metabolism of micro-organisms. It has been found that in the cells of *Streptomyces aureofaciens,* both the Embden-Meyerhof-Parnas (EMP) and hexose-monophosphate shunt systems coexist, that the fructose diphosphate (FDP) produced during the degradation of ribose-5-phosphate serves to link the two systems and that an increase in the concentration of phosphate shifts the metabolism of FDP towards the EMP pathway. A series of investigations on the relation between the physiology of *Streptomyces aureofaciens* and the production of aureamycin has been published. Studies also include the nitrogen nutrition of *Streptomyces griseus,* the growth of the *Streptomyces* in different sugars and the production of Streptomycin, and the changes in the composition of the culture medium and the influence of potassium ions. The ornithine cycle has been reported to exist in the *Streptomyces* cells; the action of the cycle continuously provides the guanido-group for the synthesis of Streptomycin. The formation of the enzyme system for acetolactate synthesis in *B. polymyxa* has been reported to be repressed by exogenous valine, isoleucine or leucine, all of which are the end products of the synthetic pathway catalyzed by enzymes involved in Acetolactate synthesis. α-Ketoglutaric acid rod-form bacterium (Peking 2990-6) can produce large amounts of glutamic acid in the culture medium. Systematic investigations have been made on the metabolic pathways of this organism, including the composition of the innoculation medium, the respiratory chain, the presence of aspartase simultaneously with glutamic acid accumulation, and the degradative pathways of glucose. These studies have helped to clarify the mechanism of glutamic acid fermentation.

Work on amino acid metabolism has centred mostly on tryptophan degradation and on the metabolism of amino acids in the silk glands of the silkworms. The former chiefly involved the process of degradation of tryptophan to nicotinic acid and quinolinic acid. It has been shown that rate liver slices possess the ability to convert tryptophan into nicotinic acid and quinolinic acid. However, quinolinic acid does not seem to be an intermediate in the conversion of tryptophon to nicotinic acid since a deficiency of Vitamin B_6 reduces the nicotinic acid formed from tryptophan in rat liver slices but increases the conversion to quinolinic acid; addition of Vitamin B_6 reduces the amount of quinolinic acid.

Deficiency of Vitamin B_6 also affects the induced formation of tryptophan peroxidase in the rat. Phenylalanine, tyrosine, indole, cystine and cysteine exert an inhibitory action on tryptophan peroxidase. A rapid spectrophotometric method for the quantitative determination of 3-hydroxykynurenine in the presence of kynurenine has been described; a deficiency of Vitamin B_2 reduces the hydroxylation of kynurenine to 3-hydroxykynurenine in rat liver mitochondria.

Investigations on amino acid metabolism in the silk glands include the transamination between various amino acids and α-keto acids, the pathway of formation of alanine from aspartic acid, two isoenzymes of aspartate-glutamate transaminase and the combined action of transaminase and glutamate dehydrogenase. The occurrence of asparaginase and branched chain amino acid transaminase only in *Attacus ricini* has also been reported. Further studies have revealed some differences in the enzyme systems responsible for the synthesis of the chief amino acids.

The effect of analogues of juvenile hormones on the regulation and control of the enzyme systems connected with the synthesis of glycine and alanine is being pursued.

By the depletion and repletion of proteins in the diet of the rat, comparisons have been made of the activity levels of certain enzymes with a view to investigate the relation of protein metabolism and enzyme synthesis in the liver. The effect of riboflavin deficiency on protein metabolism has been carried out in different laboratories.

Since ammonia is a product of protein metabolism, the disposal of ammonia will be affected when the liver function becomes abnormal. Using experimentally induced liver cirrhosis hepatomas of the rat as a model, studies have been made on the detoxification and metabolic derangement of ammonia.

Research in the metabolism of nucleic acids, vitamins, lipids, and hormones has been carried out. Relatively systematic studies on the various enzymes and enzyme systems concerned with the metabolism of *Schistosoma japonicum* have also been made.

Active research has been conducted on metabolic control and regulation, especially on the metabolism of cancer tissues. Observations on the changes in enzymatic activity of various metabolic processes in hepatomas, cancer of the skin, and thymo-lymphosarcomas have helped to gain a deeper insight into the mechanism of carcinogenesis.

In the area of plant biochemistry, a good deal of work has been done on photosynthesis, respiratory mechanism, nitrogen metabolism and the formation and conversion of starch; good progress has been made especially in photosynthesis. The advances in these aspects have recently been reviewed and will not be dealt with here.

5. Structure and function of biomembranes

Research on biomembranes is intimately related to several fundamental problems of the life phenomena, such as the origin of the cell, energy transduction, active transport, neurotransmission, metabolic regulation, cell recognition and immunity, hormone and drug action. Important work has been done on respiratory chain, oxidation phosphorylation of the mitochondria and photophosphorylation of the chloroplasts. Some of the components of the respiratory enzyme system have been isolated and purified; their properties, structure, reconstitution and roles in electron transport have been studied.

Studies have also been carried out on the energy-linked reduction of NAD^+ in a submitochondrial particle preparation of rat liver; the addition of a soluble factor from a mitochondrial extract to the washed particle preparation markedly increased the activity. The properties of ATPase and the mechanism of oxidative phosphorylation coupled with succinate oxidation in rat-liver mitochondria and the oxidative phosphorylation system in *Brevibacterium ketoglutaricum nov. sp.* 1990-6, have been investigated. Work on photophosphorylation has been concerned with the quantum yield and the "light intensity effect", and also the nature of the intermediate accumulated in the chloroplasts during short illumination by strong light.

In recent years, research on the structure and function of mitochondrial and chloroplast membranes has been pursued at the molecular level. Reconstitution of the submitochondrial membrane vesicles from the soluble mitochondrial ATPase (F1) and the depleted inner mitochondrial membrane has been attained. Also successful was the reconstitution of hybrid submitochondrial membrane vesicles from the soluble ATPase of rat liver mitochondria with the inner mitochondrial membrane of human primary liver carcinoma or with the plasma membrane of *E. coli*. Furthermore, work on the dissociation and reconstitution of mitochondrial ATPase (F1) has been carried out. These results do not only increase our understanding of the structure and function of ATPase but may also provide some clues for the understanding of the evolution of membrane macromolecules and the interrelation of the two genetic systems (cytoplasma nucleus system and mitochondria or chloroplasts). Analogous to the hybridization of mitochondrial ATPase, the action of the coupling factor of photophosphorylation from different species of higher plants has been tested and a "complementary action" with certain species was observed.

Experiments with insect mitochondria have led to the conclusion that in the process of differentiation of insect thoracic cells, the mitochondria is sequentially formed, but the orders of assembly for different enzyme systems in the inner membrane of the mitochondria are not entirely similar.

Apart from mitochondria and chloroplast membranes, recent objects of investigation include various membranes from the simplest mycoplasma of prokaryotes to those possessing characteristics of nuclear membrane structural components of eukaryotes (such as the cell membranes of heart muscle, tumour, cytoplasm, peptide hormone receptors and liposomes). Besides the resolution and reconstitution of membrane enzymes, studies on the fine structure and fluidity of membranes have also been undertaken.

6. Molecular genetics and genetic engineering

This is one of the most active areas of molecular biology in recent years. It is an area of utmost importance from both the theoretical and applied aspects; it opens up vast possibilities to the solution of certain major problems in industry, agriculture and national defence. Aside from some research on induced mutation in microorganisms, our work in this area is just beginning. However, several laboratories are already actively engaged in it, and within a relatively short period of time, progress will be made in the preparation of the necessary enzymes, like the restriction endonucleases (DNA-ligase, reverse transcriptase, etc), and also the vectors commonly employed in recombinant DNA research. The *in vitro* construction of a recombinant plasmid containing pBR 322 and λ-phage DNA segment has been reported. Similar *in vitro* construction of a plasmid DNA by joining Eco R1 generated fragment of separate plasmids pSC101 and pCRl was earlier published. The whole ribosomal RNA gene of *Attacus ricini*, as well as some of its fragments, has been successfully cloned in *E. coli*.

Work on the control of gene expression is being actively pursued in several laboratories. In investigations on the mechanism of the control of gene expression in the carcinogenesis of hepatomas, liver mRNA and α-fetoprotein (AFP) synthesizing-polyribosome have been prepared and their translation in a cell-free wheat germ system studied. By observing the changes in the polyribosomal RNA, nuclear RNA and polysomal RNA complementary to the non-repetitive DNA sequences in the process of carcinogenesis, studies have been made of the transcriptional properties of the non-repetitive sequences in rat liver carcinoma cells. Observations have been made on the changes in the various components of liver chromatin in carcinogenesis. Transformation with DNA has been employed for producing new strains from *Bacillus subtilis*. New strains produced by cross-breeding studies of different species of grain crops or fresh water fish have also been analyzed from the angle of molecular genetics.

II. Biochemical research in connection with industry, agriculture and medicine.

Biochemistry is a discipline which is closely linked with practical applications; it is intimately related to industrial and agricultural productions, medical, pharmaceutical and public health practices. This has been an important driving force in the development of biochemistry and because of its wide application, it has already had great influence in the above mentioned areas.

1. Biochemical research related to medicine

Since liberation, definite progress has been obtained in the adoption and improvement of various biochemical methods for clinical diagnosis, the determination of various indices for normal and pathological conditions, the development and application of plasma substitutes and preparation of vaccines. Since the 1960's, active research has been conducted on the etiology, pathogenesis, therapy and prevention of diseases which constitute the most serious threats, such as cancer and cardiovascular disorders. Outstanding success has been obtained in the early diagnosis of hepatoma. A large scale survey of several hundred thousand people in districts with high incidence of hepatoma, revealed and confirmed the reliability of using AFP levels in blood for the early diagnosis of hepatoma. A rapid and sensitive method for AFP detection has been worked out and applied in the screening of millions of people. AFP has been purified and its physico-chemical properties and physiological function investigated. The effect of the mRNA from normal liver on the differentiation reversal of hepatoma cells *in vitro* has also been studied.

More attention has been given to the study on abnormal haemoglobins. Immunobiochemical research has had a good beginning; immunological techniques are being established. The preparation of transfer factor from leucocytes, improved methods for the isolation of pure IgG from human serum, and secretory IgA from human colostrum have been worked out. The anti-cancer effect of immune-RNA has also been studied. Observations have been noted on the changes in the concentrations of certain amino acids and enzymatic activities in the blood, the neurotransmitters of the brain tissue during acupuncture anaesthesia and also the relation between endorphins and acupuncture anaesthesia. The biologically active proteins and polysaccharides of Chinese herbs are being investigated with regard to their structure and properties. Research of this nature will definitely help to promote the union of Chinese and Western medicine.

2. Biochemical research in relation to industry and agriculture

Early in the 1950's, our scientific workers have been involved in certain areas of microbial and organic biochemistry, in particular, antibiotic production. At the same time they have expedited the development of our fermentation industry, including the production by fermentation of various antibiotics, glutamic acid and other amino acids, inosinic acid and other 5'-nucleotides and 3'-nucleotides, the continuous fermentation of acetone-butanol and the fermentative production of fumaric acid.

The production of biochemical reagents is also expanding very rapidly. The Shanghai Dong Feng biochemical reagent factory has produced over 500 different kinds of biochemicals. Biochemical reagents are being produced by biochemical drug factories in various cities and also by chemical reagent factories and other related factories. Production advanced not only in terms of output and variety but also in the quality of the products. The large scale production of various peptide hormones, enzyme preparations, nucleotides and their derivatives has paved the way for their wide clinical applications.

The industrial use of enzyme preparations started with proteolytic enzymes and has thus replaced the older processes of tanning in food industries. Research on immobilized enzymes has made good progress in recent years, and over ten different immobilized enzymes have been prepared with new supporting and cross-linking reagents introduced. Immobilized enzymes of microbial cells, such as nuclease P_1, penicillin amidase, amino acylase and glucose isomerase have proved to possess special merits in industrial production.

Plant virus research has received its due attention and a great amount of work has been done on the identification and early diagnosis of viruses of grain crops, fruit trees and melons.

Xinjiang maize dwarf mosaic disease, Hami melon mosaic disease, wheat mosaic disease and barley streak diseases and their pathogens have been identified; a new bullet-shaped virus has been found in maize streak dwarf plants from Dunhuang, Gansu. Two types of mycoplasma have been isolated from the citrus yellow shoot disease and the witches' broom disease of sweet potato. Based on the assumption of DNA-hybridization, use has been made of isozyme analysis in an attempt to explain on a molecular basis, the rich experience of crossing between distantly related species of grain crops widely practised in our country.

Before 1966, the late Professor Chu Si had conducted experiments to solve the problem of self-spawning of the four main species of fresh water fish in still waters. This is of great importance to our freshwater pisciculture. He used a combination of human chorionic gonadotropin (HCG) and fish pituitary to accomplish this purpose. The limitations of the sources of HCG and fish pituitary and the refractory effect in the parent fish after repeated use of these hormones, have led to a replacement of the synthesized analogue of the hypothalamus luteinizing-hormone-releasing hormone, thus bringing great advances in the development of fish spawning in recent years.

III. Biochemical techniques and other aspects

1. Biochemical techniques and biochemical apparatus

The manufacturing of instruments used for biochemical research has received much attention. Many types of instruments varying from small apparatus routinely used in the laboratory, such as fraction collectors, peristaltic pumps, rotatory evaporators, ultra filtrators, refrigerated high speed centrifuge, to sophisticated instruments like UV monitors, UV-visible spectrophotometers, infrared spectrophotometers, spectrofluorometers, double-beam spectrophotometers, Mössbauer spectrometers, liquid scintillation counters, electron microscopes, etc. are already being fabricated or in the trial-production stage.

Techniques commonly used in biochemical research abroad ten years ago are being adopted in our laboratories; eg., affinity chromatography, gel electrophoresis, iso-electric focussing, plastic-membrane chromatography, immuno-enzyme labelling, radioautography, radio-immunological methods, countercurrent immunoelectrophoresis, radio-rocket electrophoresis, enzyme electrode, etc. Isotopes and elctron microscopes have been widely used in biochemical work. Nuclear magnetic resonance is being employed for the conformation of histones and nucleic acids and their interactions. Flourescence polarization and spin labelled electron paramagnetic resonance are beginning to be used in biomembrane work. UV difference spectrophotometry and flourescence spectrophotometry have been employed in conformational investigations on proteins in solution. It is a good sign that electronic computers are beginning to take a place in our biochemical research.

2. Other aspects.

As a result of the continuous penetration of mathematics and physics into biochemistry, theoretical molecular biology has gradually developed, and has attracted due attention, such as the analysis by graph theory of molecular orbitals on the steady-state enzyme kinetics, biochemical reactions and dissipative structures.

Finally, mention must be made of the interesting contribution by biochemical workers to the biochemico-archaeological investigations of the ancient Mawangtui cadaver of the West Han Dynasty; the work was carried out under the instructions of the late Premier Zhou Enlai.

It is impossible to cover all the results of biochemical and molecular biological research of the past 30 years within the space of this short article. Important omissions are unavoidable. Work done before 1958 had been reviewed (see "Ten Years of Chinese

Science" Biology, Section III) and is therefore only treated briefly here. In our mind, a review of the past serves to give a clearer view of the future. We ought to affirm our achievements in some aspects of certain areas, which compares well with those of the advanced ranks, yet when biochemistry and molecular biology are taken as a whole, our foundation is still comparatively weak. Gaps still exist in many important areas. The serious interference and damage by Lin Biao and the "Gang of Four" has set us back for more than ten years and has widened the disparity that exists between us and the advanced nations. We are facing the heavy task of making up for the time lost and, at the same time, to keep up with the rapid strides with which this branch of science is advancing. Our biochemical workers as a whole are determined to implement the "eight-word line" of the Party Central Committee (readjustment, improvement, overhaul, raising standards) and push on our scientific research solidly and sturdily in order to raise our biochemistry and molecular biology to a new level, and contribute to the realization of the Four Modernizations.

CHAPTER 7

RECENT ADVANCES OF CHINESE PALAEOANTHROPOLOGY

by Wu Ruking (Woo Jukang) and Wu Xinzhi

Little anthropological research was carried out in China before 1949; papers were mainly published in foreign journals thus reflecting the condition of science in a semicolonial country like China.

Since the liberation of China, anthropology, like the other sciences, was revived because of the profound concern of the Communist Party and the Government. In 1953, a teaching team in anthropology was established in the Department of Biology at Fudan University in Shanghai. The team later started to specialize in selected areas of anthropology. A number of anthropologists were trained and anthropological studies on living people have been carried out for the past 30 years. In 1953, the Laboratory of Vertebrate Paleontology (including human paleontology) was established by Academia Sinica based on the former Cenozoic Laboratory. Some years later, it expanded into the Institute of Vertebrate Paleontology and Paleoanthropology. In recent years, some geological institutions and provincial museums also began to conduct research on the chronology and related sciences of human fossils.

Anthropology was listed as an independent item in the National Long-term Programme of Sciences of China formed in 1956, 1962 and 1977. This greatly promoted the development of anthropology, in particular, paleoanthropology in China. This paper deals only with paleoanthropology.

I. Discovery and Study of Important Human Fossils

Before liberation, only three localities yielding human fossils were discovered. These were the localities for Peking Man, Ordos Man and Upper Cave Man. More human fossils have been unearthed in the last 30 years. The more important discoveries are discussed below.

The two incisors of *Homo erectus* from Yuanmou (1956) were shovel-shaped with basal tubercle and finger-like protuberance, similar to those of Peking Man.

The skull-cap of *Homo erectus* from Gongwangling, Lantian County, Shaanxi Province (1964) and the mandible of *Homo erectus* from Chenjiawo of the same county (1963) were obviously more primitive than those of Peking Man. It had a smaller cranial capacity (780 cm^3) with less height of the calotte, a thicker bone wall, more robust brow ridges, flatter frontal squama, a more acute inclination angle of mandibular symphysis. It had some features similar to those of Peking Man, such as the fronto-nasal suture which is continuous with the fronto-maxillary suture forming a horizontal line, a flat and broad region around nasion, short and broad basal bone and large teeth etc. (Fig. 1)

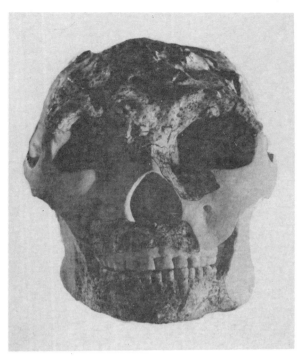

Fig. 1. The reinstated cranium of *Homo erectus* from Lantian County.

The new materials of Peking Man include 6 teeth, fragments of humerus, tibia, mandible, frontal and occipital bones. Based on the research of limb bones of Peking Man, the phenomenon of nonequilibrium between the development of skull and limb bones in the evolution of man was noticed. In Peking Man we can see that the limb bones approached the level of modern man earlier than the skull and brain. The study of the skull fragments found in 1966 showed that during the period which Peking Man lived in the cave, the skull had undergone some changes. Examples include a thinner bone wall and brow ridges with the distance between the external and internal occipital protuberances becoming shorter etc.

The studies of the teeth of *Homo erectus* from Yunxian and Yunxi Counties, Hubei Province and Nanzhao County, Henan Province are still going on.

A cranium from Dali County, Shaanxi Province, is the most complete specimen found among the early *Homo sapiens* in China. The detailed study of its morphology is still going on. (Fig. 2)

Other human fossils belonging to the paleoanthropic stage also include a skull-cap from Maba, Guangdong Province; three teeth and a fragment of parietal

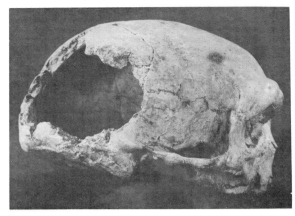

Fig. 2. Cranium from Dali County.

bone from Dingcun, Shanxi Province; several fragments of skull from Xujiayao, Shanxi Province; a fragment of maxilla and a premolar from Changyang, Hubei; two teeth from Tongzi, Guizhou; a premolar from Loc. 4 at Zhoukoudian (Choukoutien).

The human fossils from the above mentioned 7 sites are intermediate between *Homo erectus* and late *Homo sapiens*, both morphologically and chronologically. Some features such as the sagittal keelings in the Dali cranium and Maba calotte, shovel-shaped incisors in Dingcun and Tongzi specimens and short nasal spine in Chanyang maxilla, indicated consanguinity between them and Peking Man and the modern Mongoloid.

There are also many human fossils of late *Homo sapiens* of which the most important specimens are from Liujing, Ziyang and Wushenqi (Ordos).

A complete cranium and a postcranial skeleton were unearthed in Liujing County, Guangxi Zhuang Autonomous Region. The zygomatic bones are large and protruded forward; the nasal bones are low and broad; the prenasal fossae are shallow and the nasal spine is small. It has a moderate alveolar prognathism and shovel-shaped incisors.

From Ziyang, we have a cranium and a fragment of maxilla, and from Wushenqi, Inner Mongolia, fragments of parietal bones and femur etc, have been unearthed.

In addition, human or other fragments have been found in Shiyu, Shanxi; Leibin and Lipu of Guangxi; Xintai of Shantung; Shuicheng of Guichou; Jiande of Zhejiang; Lijiang, Xichon and Chenggon of Yunnan etc.

The discovery of these fossils provided direct evidence for a long history of man of 1.7 million years in China. These discoveries demonstrated the continuity of the history in China and provided a detailed knowledge of the history of the remote inhabitants in this country.

These fossils have a lot of common features such as the sagittal keeling on the vault, the protruding molar surface of the zygomatic bone, shovel-shaped incisors; the occurrence of torus is usually very low. It is known that these features occur in Mongoloids at a high frequency. This is a strong piece of evidence indicating the continuity of fossil men in China and their close relationship with modern Mongoloids.

There is evidence indicating that China might be within the area of human origin. At first, ten *Dryopithecus* teeth were found from the lignite bed of Xiaolungtan, Keiyuan, Yunnan in 1956 and 1957. In 1976, a *Ramapithecus* mandible and several teeth were unearthed from Shihuiba, Lufeng, Yunnan. It was named *R. lufengensis*.

The specimens from Keiyuan had been dated as Early Pliocene, but later it was changed to Late Miocene. The specimens from Lufeng were dated as Early Pliocene.

For the past ten years, *Ramapithecines* was considered as a probable ancestor of man by most anthropologists. Specimens of this kind have been discovered in the neighbouring areas of India and Pakistan, Western Asia, Southeastern Europe and East Africa. Therefore this vast area (including Yunnan of China) has been considered as the region of human origin. Some anthropologists in recent years have suggested the re-evaluation of these beings. It was felt that the *Ramapithecines* may not necessarily be the direct ancestor of man. There is, however, a great potential for the discovery of more ape fossils as the neogene lignite beds are widely distributed in Yunnan. If postcranial skeletons of *Ramapithecines* and other ape fossils could be found, the transition from ape to man would be more easily explained.

A cranium, a mandible and teeth of the *Sivapithecus* have been found in association with the *Ramapithecus* in Lufeng.

In addition, the discovery of Australopithecoid teeth associated with the *Gigantopithecus* was reported in 1975. However, the exact classification of these teeth must wait until more fossils are uncovered. The *Gigantopithecus* is the largest primate known so far. The first reported specimen was a lower molar recovered from a drugstore in Hong Kong. It has been suggested as being unearthed from the caves of Southern China and was dated as Middle Pleistocene. Since 1956, the Laboratory of Vertebrate Paleontology (later named IVPP) of Academia Sinica explored and excavated many caves in the Guangxi region. This effort resulted in the discovery of four *Gigantopithecus* sites; Lenzhaishan of Liucheng is the most important site as three *Gigantopithecus* mandibles and more than a thousand isolated teeth were uncovered. The other three (Daxin in 1956, Wuming, 1965 and Bama, 1973) yielded only a few teeth. The last two sites are probably of the Middle Pleistocene age. In 1968, a number of *Gigantopithecus* teeth were found in western Hubei. In 1970, a *Gigantopithecus* site of the late Early Pleistocene was found in Jianshi County Hupei Province.

In 1967, a *Gigantopithecus* mandible smaller than the Liucheng specimen was found in India. According to the study of the above mentioned specimens, it is generally agreed that the size of the *Gigantopithecus* experienced a development process of enlargement and it became extinct in Middle Pleistocene. Whether this extinct *Gigantopithecus* belongs to the hominid or pongid lineage is still in debate.

Besides the research mentioned above, some new studies were done on the human fossils found before the Liberation of China.

In 1959, in conjunction with the production of the documentary film, "The Peking Man", a new reconstruction of Peking Man was made, correcting many of the shortcomings of F. Weidenreich's reconstruction.

In the same year, Upper Cave Man was studied in detail. The Upper Cave Man was found in 1933 and a preliminary report was published by F. Weidenreich in 1939. He was of the opinion that the three skulls represented primitive Mongoloids, Melanesoids and Eskimoids respectively. He mentioned that they failed to shed any light on the origin of the Chinese and they could have been attacked and exterminated by the real representatives of Chinese race, who were the native population of this region. A new study revealed that all of the skulls of the Upper Cave Man have a series of common features such as a long head, a low face, moderate prognathism, low orbit, broad nose and the existence of prenasal fossae. There was a close relationship between the Chinese, Eskimos and American Indians. It is possible that Upper Cave Man was close to the common ancestor of the latter modern races.

II. Chronological Studies of Human Fossils

Before liberation, the chronological study of Chinese human fossils was solely based on traditional methods (correlation of fauna and stratigraphical data) to obtain the relative dating. After liberation the study of the Quaternary fauna and stratigraphy was more involved. In recent years we began to use radioactive isotopes, paleomagnetism and other new techniques for dating. In addition, some scholars have also combined the stratigraphical study with the analysis of pollens to obtain a correlation with the ice ages of the world.

According to the data available, *Homo erectus* from Yuanmou is the most ancient representative of *Homo* in China. It is Early Pleistocene in age; this being determined by correlating with the associated fauna. In 1978, two papers on paleomagnetic study were published which dated Yuanmou Man as 1.63–1.64 million years (m.y.) and 1.7 m.y. BP. respectively.

Many scholars are of the opinion that the dates of the two sites yielding human fossils in the Lantian County are more or less in the same period, i.e. early Middle Pleistocene. This was based on studies on the correlation of associated fauna and strata. Some anthropologists considered Gongwangling as belonging to early Middle Pleistocene and Chenjiawo to late Middle Pleistocene. Two papers were published in 1978. Cheng Kuoliang *et al* suggested that the ages of Gongwangling and Chenjiawo were 0.95 m.y and 0.53 m.y. BP. respectively. The amino acid racemization dating published in 1979 is 0.51 m.y. BP. for Gongwangling. The dating is based on the supposition that the paleotemperature at the period of the fossilization was comparable to the average temperature of that area in the recent period. Therefore there is some discrepancy between the date (0.51 m.y. BP.) and the actual dates.

Peking Man has been dated as being of Middle Pleistocene age (about half a million years ago) by the analysis of related fauna before liberation. Recently the paleomagnetic study indicated that the strata higher than the 13th layer are less than 0.69 m.y. BP.

According to the amino acid racemization dating (with the supposition that the paleotemperature at the period of fossilization was comparable to that of recent years), the date of the third layer is 0.37 m.y. BP, that of 8–9 layers about 0.39 m.y. with the 11th layer being 0.46 m.y. BP. Based on the analysis of the paleoclimate, it is generally agreed that the period of Peking Man corresponds to the Mindel-Riss Interglacial and probably even till the end of the Mindel Glacial period.

Judging from the comparison of the available information of the upper and lower layers of human fossils, animal fossils and absolute datings, Peking Man lived in that cave for a rather long time.

Homo erectus fossils uncovered at other sites were estimated by current dating methods to be Middle Pleistocene; while Dali Man is probably of the late Middle Pleistocene age.

There are two different views on the age of the Dingcun specimens. One is of late Middle Pleistocene as deduced from the lithological studies of the strata. The other is deduced not only from the strata studies but also from fauna studies which indicated that Dingcuan is of early Late Pleistocene in age. It is more reasonable to assign this site to early Late Pleistocene judging from the advanced morphology of the human fossils. Again, a difference of opinion exists when assigning this site in relation to the Glaciation Period of the World. The Riss Glaciation was one suggestion; while another opinion corresponded to the Riss-Würm Interglacial period.

Human fossils from Maba and Changyang were dated by the study of fauna when they were discovered. In the late 1950's, some fossils apparently belonging to Late Pleistocene (such as Liujiang Man) were discovered which associated with this fauna. It was then seen that the latter could have survived till Late Pleistocene. The dates of Maba and Changyang specimens were thus revised to Late Pleistocene. These two sites were assigned to the Middle Pleistocene period as the fossils related to fauna (Stegodon-Ailuropoda) of that period.

Human teeth from Tongzi were reported as being Middle or Late Pleistocene in age.

Human fossils from Xujiayao were originally dated as between Middle and Late Paleolithic period; they were recently revised to Middle Paleolithic. A human premolar from Loc. 4 of Zhoukoudian (Xindong) was dated as Late Pleistocene.

Some fossils of late *Homo sapiens* were dated by radioactive carbon (^{14}C). The results are: 28135±230 BP. for Shiyu, 1830±410 BP. for Upper Cave, and less than 11460±230 for Djalainor.

A skull of Ziyang was dated as Late Pleistocene by fauna analysis in the 1950's. Recently, ^{14}C dating of the plant fossils in the same strata showed that it is about 7000 years BP. But the condition of the deposition of fossils is very complicated; hence the exact dating of the Ziyang Man is not yet known.

The Ordos Man was said to be of the Middle Paleolithic period. However, a new survey and excavation in the early 1960's revised it to be of the Late Paleolithic period.

Other fossils of the late *Homo sapiens* were dated as Late Pleistocene by fauna analysis.

There are controversies on the period assignment of some fossils sites with the time scale of World Glaciation. Some believed that the Ordos Man lived during the Riss-Würm Interglacial, while others correlated it with the Würm Glacial period. The Quaternary glaciologists of China generally believe that the

Upper Cave Man and the Shiyu occipital belong to the Würm Glaciation. But the *Paguma larvata, Cynailurus* cf. *jubatus, Hyaena ultima, Struthio* and other animals also lived in the warm region. Thus, this is not in accordance with the climate of the Ice age and more research is required for further clarification.

The use of new techniques to date fossils was only recently introduced in China. Therefore it is not unusual that discrepancies exist in the correlation of various human fossils with the glacio-interglacial cycle of Europe. All these problems can only be clarified by intensive and more painstaking studies.

III. Discussion on some theoretical problems

The anthropologists of China began to learn the opinions on the origin of man expressed in the works of Marxist-Leninism at the same time as the deepening of the learning of this theory in this country after liberation. We have discussed some theoretical problems on the origin of man and have achieved a relatively unified view according to which there was a rather long period (probably more than 10 million years) of transition. Various characteristics of the human being were gradually formed in this period.

Some anthropologists are of the opinion that the creatures of the transitional period should belong to the lineage of man (hominid), while others considered them belonging to the lineage of ape (pongid).

Some anthropologists believed that various essential characteristics of man were not formed at the same time. They generally believed that during the transition from ape to man, both erect posture and bipedal walking were the initial human characteristics formed. These characteristics were formed as a result of the use of natural objects for labour and thus marked the beginning of a qualitative change from ape to man. There is a qualitative difference in the use of unprocessed natural objects as instruments by early man and by the modern chimpanzee. The former used the natural objects as tools regularly for various kinds of work. Survival depended on the use of these instruments. On the contrary, the chimpanzee utilizes the natural tools incidentally and simply, and they can survive without the use of the instruments. The regular use of instrumentation for long periods of time (instinctive or elementary labour) produced "unpure" consciousness, elementary primary language and sprouted man's conscious dynamic role which developed into the ability for manufacturing tools (the beginning of true labour) and the appearance of the first human society. This then marked the end of the transitional period from ape to man.

Other people did not agree on the working of natural tools. They argued that only artefacts can be called tools and the natural objects used for production could not be called tools. So that labour only began with the manufacturing of tools in their opinion.

In the past, the social organization of Peking Man and other *Homo erectus* have been assigned to the stage of "primitive group" in many books and periodicals. Recently, some anthropologists suggested that "primitive group" should refer to the primitive band of creatures during the transition from ape to man. It was not until the end of this transition that there was the first human social organization — the appearance of consanguineous families.

Due to the lack of direct evidence for this transition of ape to man, many views proposed were mainly based on the understanding of man from the classical works by K. Marx, F. Engels etc. Therefore, most of the problems have not yet been solved even after a rather long period of contention.

CHAPTER 8

BRILLIANT ACHIEVEMENTS OF PALAEONTOLOGICAL RESEARCH IN CHINA

by Lu Yanhao, Zhou Mingzhen, Hao Yichun and Li Xingxue

Palaeontology is a branch of science that spans biology and geology, as it not only deals with the origin, development and extinction of organisms in geologic ages, but also includes the correlation of fossil-bearing strata, palaeogeography, palaeoclimate with the formation and distribution of mineral deposits such as coal, petroleum, etc. In this context it is generally regarded as a basic science; but, as an applied science, it is indispensable for the prospecting and exploitation of mineral resources.

It is well known that fossils are the objects of palaeontological study. In as early as the Tang and Sung Dynasties, the Chinese people took notice of them from the angle of naive materialism. In an essay "Note on the Altars to the Immert on Ma-Ku Mountain", Yan Zhenqing, a famous scholar of the Tang Dynasty, described the formation of fossils as follows: "Even in the stones and rocks on the lofty mountain, there are shells of oysters and clams to be seen. Some people think that they were transformed from mulberry-groves and fields once under the water". Shen Kuo, a great man of learning in the Sung Dynasty, wrote in his "Meng Chi Pi Tan" about the bamboo-shoots in Yenchow and the snails and bivalves on the Taihang Mountain. He thought then that Yenchow was low and damp and the Taihang Mountain was by the sea, and that both had undergone transitions from the sea to the mulberry fields or vice versa.

Palaeontology as a branch of science has a history of more than 200 years. Owing to its close relation with production, this branch of science is still full of vitality. The frequent contacts among international scientists and the introduction of new techniques have created favourable conditions for its development; and the involvement of the different disciplines of science makes it possible to venture into new research areas such as molecular palaeontology (palaeobiochemistry), biomineralogy and nanopalaeontology.

The study of palaeontology in China has been highly acclaimed worldwide. The discovery of Peking Man *(Sinanthropus pekinensis)* in Zhoukoudian is recorded as an important development in the history of palaeoanthropology. The late famous geologist, Li Siquang, had a high reputation in the thirties for his work on the classification of fusulinids. The late palaeontologist, Yang Zhongjian, one of the founders of Palaeontological Society of China, who is renowned for his study on dinosaurs, made great contributions to palaeontology in China. However, in the pre-liberation days, the progress of palaeontology, like other branches of sciences was halted on account of the negligent development of science and technology by the Kuomintang government. At that time there were in China no more than 30 persons who were engaged in the study of palaeontology, with 5 – 6 working on vertebrate palaeontology.

After the birth of New China, under the leadership of the Party and along with the development of socialist construction and education, a new attitude prevails in the palaeontological circle. At present, there are more than 1,000 people working in palaeontology, in the various research institutions of the Chinese Academy of Sciences and other institutions such as the Nanjing Institute of Geology and Palaeontology, Institute of Vertebrate Palaeontology and Palaeoanthropology, Institute of Geology, Institute of Botany, Institute of Oceanology, Beijing University, Nanjing University, Wuhan Geological College, Changchun Geological College and Chengdu Geological College. Prospecting teams under government departments of geology, petroleum and the coal-industry are also involved in palaeontological studies. In addition, there is a large palaeontological library in Nanjing, housing about 70,000 volumes of publications.

Fifty years have elapsed since the founding of the Palaeontological Society of China in 1929. With the aim of reviewing recent achievements in palaeontological research and to promote the exchange of academic views between scientists, the 3rd National Conference and the 12th Symposium of PSC were held simultaneously in Suzhou from 16th–23rd April, 1979. This meeting was to mobilize Chinese palaeontologists in response to the Party's call, and to strive for the modernization of the country. In addition to senior scientists, young palaeontologists who grew up after the liberation also attended the meetings. The symposium witnessed the presentation of 352 papers by both the middle-aged and young scientists, indicating that there is no lack of successors to carry on palaeontological studies in China.

At present, about 1,500,000 species of animals, 400,000 species of plants and 100,000 species of micro-organisms are found on the earth. In the process of evolution, many more organisms that had lived on earth have ceased to. Palaeontology is generally divided into palaeobotany, invertebrate palaeontology (including micropalaeontology) and vertebrate palaeontology. The main achievements obtained in palaeontological research in the past thirty years are summarized below.

Palaeobotany

Vascular plants are the chief objects of palaeobotanical study. Since liberation much progress has been made in this field. The Devonian plants are so rich and well-preserved in China that they can give us a clue to the origin of terrestrial plants. For instance, *Zesterophyllum* (fig. 1), recently found in the

Fig. 1. *Zosterophyllum*

Lower Devonian in Southwest China, is considered to be a genus of the earliest vascular plants on earth. In the study of Carboniferous and Permian floras, we have found the Tournaician and Namurian plants to be widely distributed in this country, and have established the plant assemblages and succession in the Late Carboniferous and Early Permian coal-bearing strata of North China. We have also discovered some Gondwana plant remains in Qubu of South Xizang (Tibet) and Angara floras from the late Early Carboniferous to Late Permian in Northwest and Northeast China. In addition, we have done much work in the study of the composition, character, geological and geographical distribution of Cathysian flora, particularly its representative genus *Gigantopteris*. Some important results obtained from the research of Mesozoic plants of China are the discovery of some Early and Middle Triassic plants (e.g. *Pleuromies*), the systematic description of the Late Triassic plants from South China, North China and North-east China, the establishment of the Late Triassic and Early Cretaceous plant assemblages and the finding of the Late Cretaceous angiosperms in Northeast China and South China. Only a few studies on cenophytes have been recorded in Old China. We have, to a certain extent, laid a fairly good foundation in this field, for we have made investigations of the Tertiary flora at places in East China, South China, Northeast China, Northwest China and in the

Chinghai-Tibet Plateau as well. By and large, on the basis of plant megafossils, we have established the floral succession of various geological ages in this country, with correlations of the plant-bearing strata to provide an outline of the development and evolution of land plants since the Devonian period.

In the days before liberation, very little palaeobotanical studies on lower plants were made. However with researches of stromatolites, charophytes, calcareous and non-calcareous algae, diatoms, dinoflagellates (including hystrichospherids), fungal remains and acritarchs, this deficiency has been corrected. Furthermore, we began to study the eukaryotic organisms, probably the oldest known organic remains on earth. Here it is necessary to point out that rapid progress has been made in the study of charophytes. At present, there are Chinese specialists who devote themselves to this subject and have attained considerably advanced results in their researches. Figure 2 shows two genera of charophytes, *Peckichara* and *Sicidium* found in China.

Fig. 2-a. *Peckichara* (Tertiary)

Fig. 2-b. *Sicidium* (Devonian)

We ushered in the study of palynology shortly after the founding of New China. In the last 30 years, rapid advances have been made in this field. There are presently about 400 palynologists (including technical assistants) in China, and the Palynological Association of China has been founded. Besides, a great deal of research work has been done on the spore-pollen and microfloras of different ages, particularly of the pre-Cambrian, the Devonian and the Tertiary. Moreover, in recent years, rapid progress has also been made in the study of the Quaternary palynomorphs, which is closely related to other subjects, such as the origin of loess, vegetation, climatic changes, palaeoanthropology, geomorphology, engineering geology and Quaternary glaciation. It needs to be pointed out that we have begun to establish the Late Palaeozoic and the Tertiary palynological assemblages and succession and have made a relatively penetrating study of the late Triassic – Early Cretaceous palynological assemblages and the Late Cretaceous – tertiary flora and climatic zones.

In recent years we also conducted palaeobotanical studies in the remote border regions, namely, Xinjiang, Qinghai, Xizang (Tibet), Nei Monggol, Northeast China and Yunnan. We published many papers concerning the morphology, anatomy and classification of fossil plants and other related problems, such as palaeogeography, palaeoclimatology, palaeoecology, palaeolatitude, palaeoaltitude, etc. Among the significant contributions are the monographs "Older Mesozoic plants from the Yenchang formation, Northern Shensi" (Sze, 1956) and "Fossil plants of the Yuehmenkou Series, North China" (Lee, 1963). The articles which dealt with the fossil plants from the Chinghai-Tibet Plateau revealed the changes of vegetation, climate and altitude during the different geological ages and provided evidence for the formation of the plateau. Furthermore, we compiled three volumes of "Fossil plants of China" (Palaeozoic, Mesozoic and Cenozoic), which are of good use for popularizing the knowledge of palaeobotany and for understanding the general aspects of the Chinese fossil floras.

Invertebrate palaeontology

The study of invertebrates in palaeontology is a most complicated subject. Since the founding of New China, we have greatly extended our research fields, especially in the study of animal macrofossils. Nevertheless, there is still much to be done.

Since 1962, we have successively compiled a series of treatises, "All groups of fossils of China". There are currently 17 published volumes, including 14 volumes dealing with invertebrate palaeontology (the other 3 volumes concerning palaeobotany). In these treatises, about 10,000 species of fossils are recorded, which is certainly useful in the interpretation of the development of ancient life and in the division and correlation of stratigraphic successions of various geological ages.

In the early days of post-liberation, we compiled the publication, "Handbook of index fossils" to meet the urgent needs of geologists in their geological prospecting. In the seventies, we published a "Handbook of stratigraphy and palaeontology of Southwest China", in which about 1,000 species of fossils (belonging to about 20 groups), with over 700 new species and 100 new genera were described. In recent years we have been compiling "Atlas of fossils" of the various provinces or large administrative regions. Some of these have been published.

Besides, in the last thirty years, we have also published many monographs such as "The new materials of the dendroid graptolites of China", "Fusulinids from Penchi series of Taitzeho Valley, Liaoning", "Contribution to the geology of Mt Qilian" (4 fascicules dealing with palaeontology, 1960–1962), "Report of Scientific expedition to the Mt Qomolongma region" (3 fascicules concerning palaeontology, 1975–1976), "Carboniferous brachiopods from Mt Baoluohuoluo of Xinjiang", "Middle and Upper Triassic brachiopods from Central Guizhou", "Mesozoic fossils of Yunnan", "Contribution to the Palaeozoic strata of the western part of Mt Dabo", "Contribution to the palaeontology of the coastal area of Bo Hai (Pohai Sea)", "Cretaceous-Tertiary estracods from the Liaohe-Songhuajiang plains", "Devonian Symposia of South China" and "Mesozoic and Cenozoic red bed of South China". Also, due to the accumulation of a large number of fossils, we have been involved in the studies of theoretical problems of palaeontology in recent years.

Relatively speaking, in spite of the fact that much research had been done in the pre-Liberation days on trilobites, graptolites, corals and brachiopods, more progress has been made in these subjects after Liberation. Take brachiopods for example — about 1,000 genera and 4,000 species have been described. Furthermore, two significant contributions in the researches of trilobites are the two monographs "Ordovician trilobites faunas of Central and Southern China" and "Cambrian trilobite faunas of Southwest China". Figure 3 shows a specimen of *Coronocephalus,* the characteristic form of Silurian trilobites of China, which was found in Changning of Sichuan. In the study of graptolites, we have described a large number of species and, on this basis, we have set up 52 graptolite sequences, of which there are 4 in the Cambrian, 24 in the Ordovician, 19 in the Silurian and 5 in the Early Devonian.

Fig. 3. *Coronocephalus* (Silurian)

The Cephalepoda (nautiloid and ammonoid) play an important role in stratigraphic divisions. In recent years we carried out investigations of the Cambrian nautiloids in North China and of the Late Permian ammonoids in South China. Figure 4 shows *Pleuronodoceras multinodosum,* a species of ammonoids which is known to occur in the uppermost part of the Permian in the world.

With regard to Archaeocyatha and Conchestraca, only one paper was reported on each subject in pre-Liberation days. Recently, a great number of Archaeocyatha specimens have been secured in this country, which makes it possible to divide the Early Cambrian Archaeocyatha faunas into 4 assemblages. The lowest assemblage occurring in the Chiungchussu Formation may be considered to be one of the earliest Archaeocyatha assemblages in the world. The important results we have obtained in the Conchestracans studies were included in the publication "Fossil Conchestraca of China" (1976), in which 63 genera and 401 species were described systematically, a work which embraced such abundant material never encountered before. Figure 5 shows a species of *Euestherites bifurcatus* found in the Late Cretaceous oil-bearing bed in the Daqing oil-field.

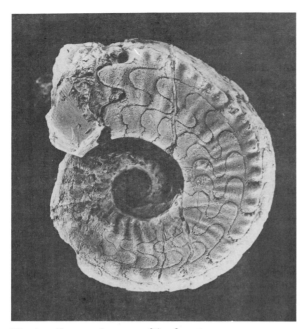

Fig. 4. *Pleuronodoceras multinodosum*

Fig. 5. *Euestherites bifurcatus*

We began to study Stromatoporoidea and Bryozoa after Liberation and, in the seventies, we included Hyelitha and Tentaculite in our fields of study. It should be pointed out that through the researches of Early Lower Cambrian Hyelitha, Monoplacophora, Gastropoda and other small shelly fossils, we have obtained ample evidence for a differentiation between Cambrian and Sinian.

Results obtained in the study of Lamellibranchia, are recorded in the publication "Fossil Lamellibranchia of China" (1976), which included 337 genera and 1,100 species of bivalves ranging from Ordovician to Quaternary. The study of Bradorida was started in the mid-1950's and the specimens gathered have been richer both in number and taxa compared to those found in other parts of the world. After studying these fossils in detail we not only could trace their origin, but could also determine their evolution and classification.

We proceeded to study micropalaeontology (exclusive of fusulinid fossils) only after the birth of New China. The Chinese geologists, the pathbreakers of socialist construction, have collected large quantities of microfossils from the drilling cores during their geological prospecting, thus making it possible to describe them in a systematical way. The rapid development of micropalaeontology in China is due to the fact that the microfossils help to solve stratigraphic problems like the subsurface formations. There are currently in China about 500 micropalaeontologists (including technical assistants), a 100 times the number before Liberation. For the purposes of encouraging the exchange of academic views and the promotion of friendship among scientists, the Micropalaeontological Society of China was formed in 1979.

The microfaunas we are now studying include fusulinids, feraminifers, ostracods, radiolarians, conodonts, etc. Among them, the fusulinids were a group of fossils that had been studied in the days of pre-Liberation, and we thus have a good knowledge of them. The study of conodonts in China began in the early 1960s and is now in full swing. Moreover, in this country, the Mesozoic and Cenozoic continental deposits are widely developed and the fresh-water ostracods are very rich. Several papers on these topics have been published in the last thirty years. Therefore, the study of ostrocods as a subject of palaeontology is quite promising.

Since the seventies, the problems on the biogeographical provinces of various geological ages have been looked into through the study of invertebrate palaeontology. It should be noted that we have put forward new ideas in some subjects, for example, in the study of Cambrian biostratigraphy, "the bioenvironment control hypothesis" is applied and in the study of Ordovician bio-stratigraphy, "Graptolite biogeographical provinces and ecological differentiation" is raised. Articles like these have been highly credited in the international palaeontological circle. This indicates that in the last thirty years the study of palaeontology in China has developed to a new level, i.e. subsequent to the description of a large number of fossils, we have delved further into the investigation of theoretical problems of palaeontology. In fact, the study of palaeobiogeographic provinces paves the way for a penetrating study of biostratigraphy. We deem it possible that in the near future the study of the biostratigraphy of various ages will lead to more successes, and which will naturally give impetus to the development of other branches of geology.

Vertebrate Palaeontology

In the last thirty years, we have achieved great success in the study of vertebrate palaeontology. Now, in China, there are more than 150 scientific workers who are engaged in this study. Fossil-bearing sites have disclosed fossils from the Silurian to the period Holocene and these localities are scattered around every corner of this country. The fossils we have collected include almost all the main groups of vertebrates. The most important of all are the animal groups (exclusive of fossil man) of three stages, viz. the fishes in the Devonian, the reptiles in Permian-Triassic, and the mammals in the Lower Tertiary. In addition, we have also made advances in the study of Dinosaurs and Osteiehthyans, and of Cenozoic mammals.

The Devonian period is considered to be the first important stage for the radiation and development of early fishes. In South China, the Devonian continental (or partly marine) strata contain rich remains of various piscine groups, many of which are still nicely preserved and also show striking endemic features. The Agnathans and the Antiarchs (represented by Yunnanlepis) form the key basis in the determination and correlation of the Lower Devonian of South China. In addition Arthrodires are found mostly in the Devonian and in the Silurian as well, but they occur in lower beds when compared with those discovered outside this country. The Crossopterygians, which were practically unknown for a long time in China, possess the features of an ancestral group that gave rise to the earliest land-dwelling tetrapods. Recently, they have been found in abundance in various Devonian sites in this country and, unexpectedly, some of them with well-preserved endocrania.

The *Lystrosaurs* found in the Lower Triassis of Xinjing bear such close similarities to those of South Africa and the Antarctic that they give substantial arguments supporting the theory of continental drift.

Here special emphasis is laid on the increasing discoveries of a large number of Palaecone mammals, which enable us to establish a complete mammal-bearing stratigraphic sequence, fill the gap in the chronological scale of continental strata of China and, greatly enrich our knowledge of the Earliest Tertiary mammals of the world. Furthermore, advances have also been made in the study of Early and Middle

Fig. 6. *Bemalambda*

Eocene mammals. Figure 6 shows the fossil of *Bemalambda* found in the Palaeocene in Nanxiong of Gungdong.

It should be mentioned that the findings of Yuanmou Man (*Homo erectus yuanmouensis*), Lantian Man (*Homo erectus lantianensis*) and the Quaternary giant ape (*Gigantopithecus*) are of great significance. It is also interesting to point out that the systematic study of the mammals and faunas of different Pleistocene stages enables us to revise our views on what was formerly known as the great panda — stegodont fauna, which leads us to believe that they may be separated into three units: Early Pleistocene Liucheng fauna, Middle Pleistocene Yanjinggou fauna and Late Pleistocene Mapa (or Liukiang) fauna.

Finally, special mention should be made of the palaeontological investigations in the Chinghai-Tibet Plateau, which is also the world's youngest plateau, still being steadily uplifted at the present time. Here towers the lofty Qomolangma Peak renowned as "the third pole of the earth", which has, for a long time, been attracting the attention of many palaeontologists both at home and abroad. Since the fifties, Chinese scientific workers under the sponsorship of the Chinese Academy of Sciences, conducted comprehensive investigations of this area, (especially in 1966–1968 and 1973–1976), and, as a consequence, numerous stratigraphical and palaeontological data were collected and a series of valuable scientific papers was published. The sediments outcropping in the Mt Qomolangma region range in age from the Cambrian to the Tertiary with a total thickness of 11,000 m; except in the Cambrian period, fossils are known to occur in the sediments of various geological ages and many of them have been found for the first time in this country. As a result of these investigations, it is believed that the Mt Qomolangma region has a sedimentary history of about 500–600 million years, a figure at least 200 million years earlier than those supposed by previous investigators.

In the last thirty years palaeontology as an inseparable branch of geology has made great strides in China; in some research fields we have caught up with the international level. By and large, what we have been doing is concentrated on the description of new discoveries, the division of biotic sequences and stratigraphic correlation. In this connection, we still fall behind in many respects, particularly in the research of systematic biology, evolution morphological functions and organism community and ecosystem, and in the application of new techniques and new methods.

China is a great country with a vast territory. She is endowed with rich fossils, which are certainly advantageous to the palaeontological studies. We firmly believe that under the leadership of the Party, Chinese palaeontologists will do their uttermost and make greater contributions to modernization process of the country.

CHAPTER 9

A SURVEY ON THE DEVELOPMENTS OF MATHEMATICS IN NEW CHINA

by Su Buchin and Sun Laixiang

I. Developments of Mathematics in New China

A distinguishing feature of the mathematical work in New China during the past thirty years was marked in the course of its development in the following aspects: an improvement was made in its foundations to a sounder one; an enlargement was made in its domain of research from a narrow one; an increase was made in the number of its professionals from a small one; and a change was made in its means of approach from the separation of theory and practice to an equal stress laid on both pure and applied mathematics.

The legacy in mathematics handed down from China before liberation was utterly poor. We wish to quote a passage from the report made by the American Pure and Applied Mathematics Delegation to China in 1976 after returning home, on the mathematical work of China before liberation, which reads: "In the thirties, a small number of Chinese mathematicians, fresh from earning their doctorates abroad, began to assume teaching posts. The general level of mathematical education improved. For example, the geometer Su Buchin and the analyst Chen Chienkung, both trained in Japan and both teaching at Chekiang University, were forming schools at that time. Although the Sino-Japanese War of 1937-1945 disrupted this development of mathematics, a few Chinese mathematicians continued their work and left their mark on world mathematics — among them S. S. Chern, W. L. Chow, P. L. Hsu, Hua Luogeng, and C. C. Lin." At the time of liberation, only P. L. Hsu among the above-mentioned mathematicians stayed in China; such was an indication of weakness in the foundations of the Chinese mathematical circles in those days. However, there were still some other mathematicians who persisted in staying and working in China, and soon afterwards Hua Luogeng and others returned from abroad. Hence not only the progress made by the Chinese mathematical circles was never to stagnate, but under the leadership of the Chinese Communist Party and at the endeavours made by a few aged veteran mathematicians and a number of mathematics workers of the younger generation, those mathematical branches such as differential geometry, topology, theory of numbers, theory of foundations of mathematics etc., which had laid comparatively good foundations in the past, and some of which had formed schools in their respective specialities, had been flourishing with full vitality. Some mathematical monographs and books were published in this period, thus laying a good foundation for fostering a large batch of mathematical workers and teachers at the later stage. In 1950, the Institute of Mathematics of the Chinese Academy of Sciences was established, and in 1956, the Mathematical Research Section at Fudan University was set up as its branch, which later developed into the Institute of Mathematics at the same university. The Chinese Mathematical Society first published its periodical "Acta Mathematica Sinica" in 1951 and then "Progress in Mathematics" in 1955.

The years 1956-1966 constituted an exceedingly important decade in the mathematical development of New China. In 1956, the State Council convened the National Conference on the 12-year Programme for Sciences, at which it was proposed, for the development of mathematics, that new paths of development for mathematical research be set up under the guidance of the principle of integration of theory with practice. In view of the fact that, due to the narrowness in its range, the mathematical research work at the early stage of liberation was unable to meet the requirements of socialist construction, a development programme was formulated at the conference for the mathematical sciences, aimed at enabling mathematics, on the one hand to combine itself with practice in production, and on the other hand, to strengthen its researches on basic theories in a down-to-earth manner. At the endeavours jointly rendered by scientific workers of the whole nation, such a magnificent programme was finally realized ahead of schedule in 1962. As far as mathematics was concerned, a number of mathematical branches made comparatively systematic progress in their respective research work up to the year 1966, as reflected in the publication of quite a few mathematical monographs and a great number of treatises in the periodicals *Acta Mathematica Sinica* and *Progress in Mathematics* in this period.

It is worth mentioning here that the Great Leap Forward of 1958 had speeded up the development in mathematics, which was mainly reflected in the following three aspects:

1. More specialized organizations with respect to mathematics such as computing technique research institutes and computing centers were established and some well-conditioned universities began to set up new specialities of mathematical logic, computational mathematics, mechanics, applied mathematics, etc.

2. Mathematical workers were beginning to make an extensive integration of theory with practice in production. Most of the mathematical knowledge that was more frequently and more directly used in production were in the fields of differential equations, mathematical statistics, computational mathematics, operational research, etc. Reciprocally, the integration of theory with practice had given an impetus to the systematic developments of these branches.

3. A newly emerging force, including especially a still greater number of young mathematics workers who had already taken part in new domains, grew rapidly to maturity. Mathematics courses were taught ahead of time by young teachers, some of whom even gave lectures on those mathematical subjects which had never been given in the past. One after another, young teachers in mathematics joined the various professional organizations, set up discussion classes, and participated in mathematical seminars, thus raising the academic levels in their respective specialities.

How should the Chinese mathematical level be evaluated up to the year 1966 after all? We tentatively hold that, with some of its branches having made fairly outstanding achievements, our mathematical level, after making great efforts in the past 17 years, was approaching the international level.

In the above-mentioned report made by the American Pure and Applied Mathematics Delegation to China, we read the following comment: "The period between liberation (1949) and the beginning of the Cultural Revolution (1966) was the formative period in China's progress towards mathematical independence and maturity. China was more active mathematically in these 17 years than in any previous such period of its history ... Nevertheless, seen from the available evidence, the achievements of the Chinese mathematicians from 1949 to 1966 were altogether creditable. The level of technical competence in the papers of *Acta Mathematica Sinica* at that time reveal a healthy increase in mathematical activity, both in quantity and in quality."

The years 1966-1976 constituted the decade of the "Cultural Revolution". As it is well known, the mathematics in China, just as in other sciences, was seriously disrupted. As a result, the mathematical developments were at a low tide, with theoretical research work at a standstill, and mathematics departments and institutes in the various universities were almost abolished. For example, the Institute of Mathematics at Fudan University, which had already 34 and 9 members in its full-time and part-time research personnel respectively and which had already developed 6 or 7 research orientations since its inauguration in 1965, was disbanded during the "Cultural Revolution". Its staff became scattered and only seven members remained at the time of its resumption in 1976. The already lessened disparity between the Chinese and the international mathematical levels was once again widened.

In spite of such setbacks, many mathematicians persisted in their theoretical work and made academic achievements at first-rate levels. Chen's Theorems concerning the Goldbach conjecture, for example are the best results obtained up to the present; on the classical Nevanlinna theory, Yang Le and Zhang Guanghou obtained new and fruitful results concerning the Borel directions and deficient values of a meromorphic function; Hou Zhenting derived a uniqueness criterion for non-conservative Q-process in his study of the theory of structures of the homogeneous denumerable Markovian process.

It should also be mentioned that during this decade many valuable contributions were made in applied mathematics. Some mathematicians, headed by Hua Luogeng, had done much in the nation-wide work of popularizing the Optimum Seeking Method (OSM) and the Critical Path Method (CPM), which had been extensively used by production departments. Since 1974, Su Buchin participated in the mathematical theory of the lofting of ship hulls at the Jiangnan Shipyard, Shanghai, and at the same time devoted himself to the theoretical research on singular points and inflexions of a parametric spline curve, and amalgamated this work with the research of computational geometry. There was a great deal of work in applied mathematics, which will not be enumerated here however.

1976 was the year in which a significant change took place in the political life of the Chinese people. From then on, order was being brought out of chaos, and the work of reinstatement and rectification was being set about. In 1977, a new programme for the development of mathematics was worked out in Beijing; the Chinese Mathematical Society and its branches in other parts of the country were restored, and the research on basic theories was strengthened. In 1978 the Third National Mathematical Annual Symposium was held in Chengdu, at which more than 500 representatives were present, more than 400 papers were read and academic achievements were as dynamic as bamboo shoots springing up after a spring rain. Nevertheless, work of reorganization was just being started and successes were quite preliminary. Although some nation-wide periodicals such as *Scientia Sinica, Acta Mathematica Sinica, Acta Mathematicae Applicatae,* etc. and the academic journals of various universities have come out during the past two years, yet a large number of papers are still unable to get published at present. This is a disadvantage for the development of sciences.

When we look back on the mathematical development of China, it seems necessary to raise the problem of training research workers. It is clear that scientific research must be closely linked with the work of manpower training and in particular, the mathematics departments of our universities are shouldering the responsibility of sending out talented personnel to other departments. Hence the ways and means by which talented personnel and highly qualified students are trained and brought up have a bearing on the development of mathematics research.

We are not going to describe in detail the developments in mathematical education during the past thirty years with the limited space in this paper. It must be pointed out, however, there has been continual effort in this direction. After the rectification of colleges and universities in 1952, the educational system inherited in China before liberation was reformed, and teaching programmes, teaching materials, educational objectives, etc. were preliminarily set up. Due to the lack of experience, the curricular, teaching materials and reference books adopted were still very much Soviet oriented. In 1958, all the mathematics departments in colleges and universities throughout the country carried out educational reforms to various degrees, and certain success was achieved. Nevertheless, the educational system which was established at its preliminary stage and because of the very limited educational practices, was thoroughly disrupted by the "Gang of Four" during the "Cultural Revolution". It is only after the reinstatement over the recent two years and in years ahead that it can return basically to the road which had been shaped in the past.

The scale of development in mathematics education, compared with that at the beginning of liberation, can be said to have been increased to a large extent. The present condition of the mathematical specialities which are fostering mathematical talents throughout the country is roughly as follows: mathematics departments and mathematical institutes were set up in 30 universities, with more than 1,300 teachers in the faculties and some 4,000 students enrolled annually; mathematical institutes were set up in 45 normal colleges, with more than 2,700 teachers in the faculties and more than 15,700 students enrolled annually; mathematical institutes were also set up in 39 teachers colleges, with more than 600 teachers in the faculties and more than 6,400 students enrolled annually.

II. Academic Achievements during the Past Thirty Years

There are so many branches in mathematics that it

is impossible to give a detailed and comprehensive account of the academic achievements in every branch. What we try to do here is mainly to give a fuller description of those branches which are comparatively familiar to us, with some comments on the other branches. Moreover, due to the limitations of our knowledge, descriptions therein would be far from complete, and omissions inevitable.

1. Mathematical Logic

During the first decade after liberation, Chinese mathematicians Hu Shihua, Mo Shaokui, Wang Shiqiang and others investigated some problems in the field of deductive logic such as bi-valued and multi-valued logic, the axiomatization of model logic, etc. and obtained some results on such problems as normal forms in the theory of recursive functions. Later on, Hu Shihua, Wang Shiqiang, Zhang Jinwen and Yang Dongping investigated the algorithmic language, the model theory, the set theory and the α-recurrence, respectively. Wu Wenjun made contributions to the mechanical proof of elementary geometry (without the axiom of order) and differential geometry. For years, Mo Shaokui of Nanjing University probed into logical operations (the essence of advanced functions, the improvement on the system of natural inference), the axiomatics of recursive arithmetic, the axioms of set theory, especially the theorem of compactness, and obtained certain new and extensive results.

2. Theory of Numbers

In the first decade after liberation, Hua Luogeng treated by means of the exponential sum, the topic of "major arc" in the Waring's problem satisfactorily. In a certain sense, he got a result which cannot be improved further. Min Sihe and others dealt with the problems concerning lattice points, divisors, the distribution of prime numbers and the estimation of average order of arithmetic functions in the classical analytic theory of numbers. Applying the sieve method, Wang Yuan obtained a fruitful result, denoted by (2, 3): Any sufficiently large even number is the sum of two almost prime numbers, one of which is a product of at most two prime factors, and the other a product of at most three prime factors. Ke Zhao made some progress in Diophantine equations and in the determination of the class number C_n for positive definite quadratic forms of n variables with integral coefficients and unit determinant. He proved $C_{12} = C_{13} = 3$, $C_{14} = 4$, $C_{15} = 5$, and determined the representative form in each class as well as the number of their automorphic transformations. Later on, progress was made continuously in the Goldbach conjecture. In 1962, the result (1, 5) was obtained by Pan Chengdong and N.B. Barban independently.

In 1963, Pan Chengdong, N. B. Barban and Wang Yuan further proved the result (1, 4). After making some new and important improvements on the sieve method in 1966, Chen Jingrun obtained the best result (1, 2) known as Chen's Theorem. The original proof of Chen has been simplified recently by Ding Xiaqi, Pan Chengdong and Wang Yuan. In their joint work on the algebraic theory of numbers, Hua Luogeng and Wang Yuan made use of the properties of cyclotomic fields as well as some profound results of the analytic theory of numbers in the problem of numerical integration. Besides, many departments including the Institute of Mathematics of the Chinese Academy of Sciences, Shandong University, etc. attained very important achievements on the analytic theory of numbers with their applications. Ke Zhao and others of Sichuan University obtained some important results not only on quadratic forms, indefinite equations, analytic theory of numbers, etc., but also on the application of the combinatorial theory, the theory of numbers (for example, a study on the transformations of the theory of numbers), etc.

3. Algebra

During the first decade after liberation, Hua Luogeng proved the fundamental theorem of one-dimensional projective geometry in his research on the problem of semi-automorphism of a field. He also did some work on automorphism of typical groups. Making use of the technique in matrix calculus and starting with the rather difficult low dimensions, he obtained some results in the case of high dimensions by further application of induction. More complete results are obtained by this method than by any other known method. Wan Zhexian and Yan Shijian made a study of the problem of generalizing the results of automorphism in typical groups over a field to a more general ring.

With regard to Lie algebra and Lie groups, Duan Xuefu and Chevalley proved the following fundamental fact concerning the algebraic Lie algebra: the Lie groups in the algebra of L. Maurer can be precisely described by the algebraic Lie algebra which is, in turn, defined through the infinitesimal tensor invariants and replica of matrices. Yan Zhida and others arrived at an application of the graphical method for root systems to determine the classification of real semi-simple Lie algebras and their automorphism. They also studied the problem of the existence and conjugacy of some subalgebras in a Lie algebra.

The theory of group representation finds its application in the theory of functions of several complex variables, by means of which Hua Luogeng proved the

Abelian summable convergence theorem for the Fourier expansion of a continuous function over a compact group.

In the theory of algebras and the valuation theory, Wang Xianghao obtained some important results including an accurate statement of Grunwald's Theorem in both the case of prime numbers and the general case, its proof and its generalization. Moreover, a preliminary study had been made in the structure of multiplicative groups of simple algebra over an algebraic number field, and research had also been carried out on the structure of a complete field of discrete valuation and its multiplicative groups.

Besides, a certain amount of research work on the field of the theory of rings and the theory of algebras was done by Xie Bangjie; on the field of the theory of general groups by Cao Xihua and Zhang Yuanda and on the field of lattice theory by Wang Shiqiang.

Later on, Wan Zhexian carried out the research work on typical groups. Yan Zhida made a study of semisimple Lie groups, Lie algebras and symmetric spaces including, for example, the classification of Cartan subalgebras, the classification of non-compact symmetric spaces, etc.

Over many years, Wan Zhexian also investigated the coding problem of binary recurring series.

In recent years, Xu Yonghua made a study of the problem regarding a relation between the maximal and minimal conditions of rings, the relative results of which include the familiar theorems of Hopking and of Akitzuki. He further made a unified study of rings (in algebra) by means of a general method, and put forward the concept of "non-associative and non-distributive rings" so as to derive a theorem concerning the structure of non-associative and non-distributive rings under the minimal condition. A further study on the structure of primitive rings was also made.

4. Differential Geometry

(1) The metric and projective differential geometry is one of the mathematical branches which had laid a comparatively good foundation in China before liberation, and the work in this field continued since liberation. Bai Zhengguo made an extension of Fenchel's Theorem regarding curves in global space, and clarified the relation between the modern and the classical Jacobian theorems. In this study of the projective minimal surface, Su Buchin acquired new characterisations and a series of Laplace sequences in the five-dimensional space.

(2) In their study of differential geometry of general space, Su Buchin, Gu Chaohao and others had put many topics under systematic discussions, especially the geometry of K-spread spaces, which include those problems relative to the theory of projective motions, integrability conditions, plane axioms, a new theory established by the method of implicit functions, etc. Gu Chaohao and others carried out a series of research work on the problems of imbedding and deformation, including those, for example, of imbedding in the large in the Finslerian manifold and spaces with affine connection, from which they procured more explicit solutions.

(3) Yan Zhida, Wu Guanglei and others carried out research on the topological properties of Riemannian manifolds and the geometry of Lie groups, from which they obtained some results regarding the Betti number and certain integrals of the Gauss-Bonnet type. They also made a study of homogeneous spaces including those determined by pseudo-analytic groups, especially the problem of classification and some geometric problems arising out of the development of the theory of functions of several complex variables in our country, which include, for example, those with respect to the research of unitary geometry over a non-continuable domain, the properties of unitary curvature and analytic equivalence of domains, etc.

Later on, in the field of differential geometry, a comparatively strong contingent of research workers was gradually formed at Fudan University, and their work has been extended in the following three aspects:

(i) Research on the conjugate nets of high dimensional projective spaces. In 1959-1966, Su Buchin, using the method of exterior differential forms, made a systematic study of the theory of conjugate nets of high dimensional projective spaces. He brought in such a fundamental concept as the kth class of conjugate sequences of a Laplace sequence, from which he succeeded in deriving a series of new configurations, including such important results as the theorem of existence, the imbedding theorem and the fundamental properties of periodic and pseudo-periodic Laplace sequences.

(ii) Regarding the research on the infinitely continuous quasi-groups of transformations, Gu Chaohao and Hu Hesheng arrived at many results, among which we note particularly the following one: there are only ten classes of real irreducible linear groups among all the possible isotropic groups of the infinitely continuous groups.

(iii) In the area of the groups of motion in Riemannian spaces, Gu Chaohao carried forward Cartan's method in 1959-1960; besides, Hu Hesheng derived several theorems in 1963-1966 concerning the orders of both the complete groups of motion and the groups of motion in the homogeneous Riemannian space V_n, from which a method is proposed to determine all the gaps in the complete groups of motion and groups of motion in the homogeneous Riemannian space.

Since 1974, Yang Chengning, a Chinese-born American physicist, in collaboration with Gu Chaohao and others of Fudan University, made a study of gauge fields, the mathematical structure of which is fairly complicated. Some of its rather fundamental problems have close relations with the theory of fibre bundles in modern differential geometry and the theory of nonlinear partial differential equations. They made a thorough investigation into the structure of gauge fields in differential geometry, which includes a geometric interpretation of the gravitational instanton and its Riemannian space with local duality, the symmetric defect of the gauge fields and the determination of the value of quantization of dual charges, the decisive effect on gauge fields exerted by their strength, the gauge fields with spherical symmetry and their determination, etc. Much importance has been attached to their work in international circles.

During the recent years, Su Buchin and others made considerable progress in utilizing projective geometry and algebraic geometry to promote the study of the computational geometry, and applied it satisfactorily in the shipbuilding industry. In the application of the method of moving frames to the study of gears, Yan Zhida and Wu Daren simplified and developed the classical principle of meshing gears, showing that differential geometry has found its way into applications in the mechanical industry.

5. Topology

On the basis of having proved the topological invariants of Stiefel-Whitney characteristic classes in compact differentiable manifolds, Wu Wenjun, Liao Shantao and others proved that the reduced Pontrjagin characteristic classes with mod 3 and mod 4 are topological invariants. They introduced some non-homotopic topological invariants, which have brought about the necessary condition for an n-dimensional Euclidean space R^n to be realizable in a topological space and the necessary and sufficient condition for an n-dimensional ($n \neq 2$) complex to be realizable in R^{2n}. They determined completely the dimension of a normal space with a countable base, and proved that any differentiable realization of the n-dimensional ($n > 1$) compact differentiable manifold in R^{2n+1} are differentially isotopic. With respect to a space X of periodic transformation with prime P, Liao Shantao brought to light the relation among the homology of X with special modulus, the multiplicative structures of X and of the invariant point set of X under modulus P, and proved that there is always a fixed point under any periodic transformation in the four-dimensional Euclidean space. Liao Shantao further introduced the symmetrization method of fibre bundles in dealing with the problem of obstruction class of fibre bundles, and as a result, derived a formula for the secondary obstruction class in sphere bundles and a formula for the relative extension and homology classification of the secondary obstruction class.

Zhang Sucheng discovered seven kinds of fundamental A_2^n ($n > 2$) polyhedrons, so as to form the normal forms of A_2^n polyhedrons. By first establishing the algebraic theories of regular isomorphism and subsequently introducing the homology invariants, he obtained the multiple torsions as well as the characteristic polynomials. These invariants are applicable both to the theory of normal forms and to the computation of cohomotopy groups. He further expressed a certain necessary condition for the existence of sections in the fibre space in terms of an inequality between the base space and the multiple torsions of fibre space. Zhou Xueguang and others expressed the homology group $\pi_{n+2}(x)$ of the $(n-1)$-dimensional space x ($n > 4$), in terms of homotopy groups and cohomology operations.

Later on, Wu Wenjun and Wang Quiming made a study of the work by Dennis Sullivan on homotopy, and further carried out the work in this field, obtaining a spectral sequence by filtering a CW complex with its skeletons. Jiang Zehan, Jiang Boju and Shi Genhua did systematic work on the theory of fixed-point classes. Zhang Sucheng and Shen Xinyao carried out the computation of homotopy groups. In 1965, Wu Wenjun published his work entitled *The Problem of Realizing the Triangulable Manifolds in Euclidean Space*, an important achievement in algebraic topology.

6. Theory of Functions

This is a branch of modern mathematics in which the researches were begun earlier than any other branch in our country. For this reason, there were more mathematical workers in this field than in others, and a majority of the mathematical papers published during the first decade was in this field. In a summary, the contributions consist of the following:

(1) Theoretical research projects were launched and many achievements were made in the following areas: fundamental theories concerning the integral and the meromorphic functions in connection with their derivatives; the theory of normal families of analytic functions (the above by Xiong Qinglai, Li Guoping and Zhuang Qitai); geometrical theories of

functions of complex variables (Chen Jiangong, Xia Daoxing, Gong Sheng and others); the pseudo-convergence of orthogonal polynomial series; uniqueness of multiple trigonometric series and integrals (Chen Jiangong, Cheng Minde and others). Yu Jiarong arrived at some results in generalizing the Dirichlet series.

(2) In the field of approximation theory of functions, Chen Jiangong, Li Guoping, Cheng Minde and others engaged in the research on the constructive theory of functions, the approximation of holomorphic functions by polynomials, the approximation of continuous and holomorphic functions by linear combinations of integral functions, the best approximation of periodic functions of several variables by trigonometric polynomials, the best approximation of special functions, and so on.

(3) In the field of the theory of functions of several variables, Chinese mathematicians have proposed some elegant methods. Their principal achievements may be summarized as follows: Hua Luogeng solved the problem of expanding the regular functions of a function of several complex variables concretely and non-locally into a series in a canonical domain, and expressed explicitly, by means of the representation theory of groups, these regular orthogonal functions with the elements of irreducible representation of the canonical groups. Hua Luogeng and Lu Qikeng also contributed to the theory of harmonic functions in the canonical domains. It is worth noticing that the Dirichlet problem in this domain was solved by considering its connection with the elliptic equations degenerate on the boundary of a domain and with certain special differentiable manifolds. These results are applicable to the theory of group representations, the theory of partial differential equations and functional analysis.

The important monograph of Hua Luogeng, *Harmonic Analysis of Functions of Several Complex Variables in Canonical Domains* (1965), was translated into English by the American Mathematical Society. Furthermore, Chen Jiangong summarized his own research on the approximation theory of functions by trigonometric series in the monograph: *Theory of Trigonometric Series* (1965).

The Teaching and Research Groups of the Theory of Functions at Fudan University and Hangzhou University, headed by Chen Jiangong, made comparatively systematic studies on the approximation theory of functions, the summation process of trigonometric series, the parametric methods, the method of extremal length, the variational method and the method of area principle in the theory of Schlicht functions as well as the coefficients and covering problems of Schlicht functions. In 1958–1965, the members of the Teaching and Research Group of the Theory of Functions at Fudan University also made a study on quasi-conformal mappings in plane, and in recent years they resumed their research on the evaluation of the coefficients of a Schlicht function and the problem of coefficients in univalent functions with quasi-conformal extension.

In the valuation distribution theory of functions of complex variables, Yang Le and Zhang Guanghou arrived at the following theorems: (1) Given ρ, $0 < \rho < \infty$, and a nonempty closed set E, there exists a meromorphic function f of order ρ, with E as its Borel set of directions; (2) Let f be a meromorphic function of order ρ $(0 < \rho < \infty)$, and let f have at least one deficient value. If f has q Borel directions, then either $q = 1$ and $\rho < 1/2$, or $q > 1$ and there exist two Borel directions separated by an angle $\leq \pi/\rho$.

7. Functional Analysis and Integral Equations

During the first ten years, Chen Chuanzhang, Zhang Shixun and Hu Kunsheng obtained some results on the distribution of characteristic values and singular values, the characteristic values of normal kernels, the normalizable kernels and the existence and uniqueness of solutions of the systems of integro-differential equations for a certain type.

In the field of functional analysis, Zeng Yuanrong introduced the virtual solutions of inconsistent equations and the concept of generalized traces of linear operators and made a classification of closed linear operators. He also obtained some concrete results on biorthogonal systems. Li Wenqing and others made a study on the theory of a Banach space and its linear operators. Feng Kang put forward a new way of dealing with the generalized functions, and made a study on the generalized Melin transformations. In the field of topological linear space, Jiang Zejian and others introduced the concept of (BS) space in combination with the classical resonance theorem. Chen Wenyuan, Guo Dajun and others had a discussion on the continuity and total continuity of some types of nonlinear operators. Guan Zhaozhi, Lin Qun and others investigated the methods of solving functional equations by approximation, while Yang Zongpan made a study on the semi-sequential linear space.

The work on functional analysis since 1959 was reflected in the following aspects:

(1) Xia Daoxing and others made a systematic study on the theory of measures and integrals over the infinite-dimensional spaces, and laid emphasis

upon the quasi-invariant measure and the denumerable additiveness with respect to harmonic analysis and column probability. The monograph of Xia Daoxing, *The Measures and Integrals over Infinite-dimensional Spaces* was published in 1964, and was translated into English and published in U.S.A. eight years later. Now he is carrying on his research work in combination with the applications in the quantum field theory.

(2) Since 1961, the study made by Xia Daoxing with respect to the non-normal operators had some influence upon the research work of Pincus, Putram and others of the United States. Xia Daoxing and Yan Shaozong further investigated the spectral theory of operators in the indefinite metric space, which has been applied in the quantum field theory. Jiang Zejian of Jilin University, Wang Shengwang of Nanjing University and others studied the spectral and the generalized spectral operators, while Tian Fangzeng, Yang Mingzhu and others of the Chinese Academy of Sciences applied the spectral theory of operators to the problem of transfer equations of neutrons, from which some results are deduced. Guan Zhaozhi and others applied the spectral theory to the problem of controlling the distribution of parametric systems.

(3) Emphasis was placed on the study of nonlinear functional analysis by a great many mathematical workers, among whom Zhang Gongqing of Beijing University made some achievements in the application of bifurcation theory of points to the nonlinear partial differential equations and in the research on Urison's operators and the topological degree of nonlinear operators.

(4) Zhang Gongqing dealt with the application of the method of functional analysis to the study of linear partial differential operators, and put forward his research on the partial differential operators.

8. Ordinary Differential Equations

During the first ten years, there appeared in the theory of stability the work on stability with saddle point conditions, stability with time lag, stability with finite time, etc.; there also appeared works on the existence and uniqueness of periodic solutions and the work on the distribution of limit cycles in the quadric differential systems.

The principal achievements obtained from then on are as follows: the work of Liao Shantao with respect to the proof of the closure lemma, the typical systems of equations, the theory of obstruction sets and its application to the necessary and sufficient condition of structural stability; the work of Qin Yuanxun with respect to the theory of approximate analytic solutions; the further study of Ye Yanqian in the distribution of limit cycles in the quadric differential systems and their numbers, furthermore, the work of Lin Zhensheng with respect to the Floquet theory of almost periodic equations and the theory of integral manifolds; the results obtained by Zhao Suxia on the stability of automatic adjustment systems; the research work of Ding Tongren, Ye Yanqian and Wang Xian on the nonlinear differential equations of the lock phase technique in the theory of electronic focal beams, etc. A certain amount of research work on the stability of nonlinear adjustment systems and relay systems was made by the mathematicians of Fudan University.

9. Partial Differential Equations

Since 1954, on the initiative of Wu Xinmou and others, research work was launched in the field of partial differential equations in China, and a research contingent was gradually formed in our country.

In the fifties, the research work on partial differential equations centered mainly on the equations of the mixed type. Wu Xinmou, Ding Xieqi and Wang Guangyin investigated the uniqueness of the Tricomi problems. Dong Guangchang, Qi Minyou and others studied the stability of Cauchy problem of equations of the mixed type and of equations containing singular lines. In the sixties, Gu Chaohao obtained some results in his research with respect to the differentiable solutions of higher order in the systems of positive symmetrical equations, asymptotic properties of solutions of the systems of quasilinear positive symmetrical equations, etc. He made comparatively outstanding achievements in developing the theory of positive symmetrical equations, and started a new approach to the mixed-type equations with several variables, (including both the linear and the nonlinear cases) which led to some recent results in the way of proposed boundary conditions. At the same time, Wang Rouhuai obtained comparatively good results on the analyticity of solutions in the linear and nonlinear elliptic equations.

On the basis of some problems in mechanics of one-dimensional gas flow, the plane flow around a body, etc., Gu Chaohao investigated at the end of the fifties the problem of boundary values of quasilinear hyperbolic equations, and obtained some profound results in the case of two equations. On the basis of Gu's work, Li Daqian and Yu Wenci developed the method of *a priori* estimate and the method of dealing with undetermined boundaries in this class of boundary value problems, and established comparatively systematic results, which have been applied to the local construction of piecewise smooth solutions.

Ding Xiaqi, Zhang Tong and others made a study on the global solutions of quasilinear hyperbolic equations, and procured results on the existence of solutions under certain conditions in the case of systems of equations for two unknown functions. Wang Guangyin, Jiang Lishang and others also obtained some results in their study on the problem of indefinite boundary with respect to nonlinear parabolic equations, etc.

Recently, Wang Guangyin initiated some new research on the theory of differential operators.

A large amount of work has been done on the application of partial differential equations. For example, during recent years, Li Daqian and others, in summing up their research on the practical problems of electromagnetic fields and thermal fields, put forward the problems of boundary values of isoplethic planes in the quadric self-conjugate elliptic equations and the problems of complementary boundary values, etc., and made a study on stability and the method of solution.

10. Theory of Probability

The study on the theory of probability consists mainly of the theory of stochastic process. Xu Baolu (P.L. Hsu) did a lot of work on the Markovian process, the limit theorem and mathematical statistics, and it is to be regretted that his works were not published during his lifetime.

With respect to the study of the Markovian process, Wang Zikun, in his investigation of the structure of birth and death processes and integral functionals, put forward the transitional method by limits, solved the problem of structure in the birth and death process, and later on did a lot of work on the analytic properties, the probability properties and the classification of states of the intermittent Markovian process and the limit theorem of integral functionals. In the 1970s, Hou Zhenting further developed Wang Zikun's method in his study on the theory of structure of the homogeneous Markovian process, set up the method of minimal nonnegative solutions and obtained the uniqueness criterion of the non-conservative Q-process. A summary on the above work has been included in his monograph published recently.

Jiang Zepei, in his research on stationary processes and the analysis of time series, mainly on the asymptotic properties of regression coefficient estimate and the prediction theory of stationary process, generalized the work of U. Grenander and others to the cases in which disturbances become the stochastic fields, and the multivariate stochastic process. He obtained the regular singular conditions and Wold's decomposition of the multivariate stationary process. Hu Guoding of Nankai University made a study on Shannon's information theory. The mathematicians of Fudan University did some work on the generalized stochastic process, the prediction theory of stationary process and the statistics of processes. With respect to the additive process, Wang Jiagang supplemented the results obtained by Skorohob, and derived the absolutely continuous or singular conditions for measures induced by the additive process and the Lebesgue's mutual decomposition.

The earliest result with respect to the limit theorem was the work by Wang Shouren and Zhang Liqian, which was related to the nonlinear test made by Kolmogorov-Smirnov. The mathematicians of Fudan University obtained some independent and parallel results through the consideration of the limit theorem of random sums. Zheng Zengtong and others obtained certain results on the limit distribution of the sum of stochastic variables.

During the "Cultural Revolution" some research on applications was made, e.g., Wu Lide and others applied the method of probability and statistics to the prediction of earthquakes and groundwater levels, and the method of time series to geological prospecting.

11. Operational Research

In 1956, operational research was included in the long-range programme for scientific developments formulated at that time, and in 1958, in order to further explore new ways of solving practical problems with mathematical theories, many departments all over the country initiated the learning of and research on operational research.

As from the year 1964, at the active promotion and motivation by Hua Luogeng, twenty-two provinces and municipalities of China engaged in the activities of popularizing the Optimum Seeking Method and the Critical Path Method, from which certain economic benefits were derived by some factories and enterprises. Hua Luogeng and others also made an outstanding study on the relevant theories, making contributions towards the development of applied mathematics in our country.

With respect to network programming, Guan Meigu put forward the "problem of the Chinese postmen" and suggested the "method of operation on charts with odd and even points". In 1962, Xie Litong and others gave a new proof by employing the method of combinatorial topology, and generalized the problem to that of finding extreme values in an n-dimensional complex. Zhu Yongjin and Liu Zhenhong found an algorithm for minimal tree derivations. In addition, Wu Wenjun made a study on the network problem by means of combinatorial topology, and also investigated the problem of designing plates of printed circuits with the aid of some theoretical work which he had done himself in his early years.

With respect to linear programming, the Institute of Mathematics of the Chinese Academy of Sciences put forward the method of operation on charts, for which they derived a fundamental theorem. The Institute of Mathematics of the Chinese Academy of Sciences, Shandong University and other departments solved the problem of allocating and transporting goods and materials by the application of linear programming, and obtained comparatively good results.

With respect to nonlinear programming, Yue Minyi and Han Jiye made improvements on the reduced gradient method. Yu Wenci made a study on the theory of the direct search method. Also, the application of nonlinear programming to optimum designs was investigated by the Operational Research Section of the Institute of Mathematics and many other departments.

In queueing theory, Yue Minyi and Xu Guanghui independently studied the instantaneous states of the classical systems. Xu Guanghui investigated the steady states for bulk service systems and the distribution of waiting times under different queueing rules. Hou Zhenting completed a proof of Palm's assertion.

12. Control Theory

In 1958, Qian Xuesen published the Chinese edition of his monograph *Engineering Cybernetics*. Prompted by his lectures, mathematical workers started to investigate the mathematical problems in the control theory in engineering. Qin Yuanxun and others made a study on the effects caused by the time lag on the stability of control systems. In 1962, the Control Theory Research Section was set up in the Institute of Mathematics of the Chinese Academy of Sciences, and launched extensive research work on the general theory of control systems and on the theory of optimal control, with Guan Zhaozhi as the leader. The paper read by Song Jian and others at the second congress of the International Federation of Automatic Control (IFAC) regarding their research on the time optimal control of linear systems by means of isochronal regions created great interest. Recently they made a further study on the distributed parameter system of ordinary differential controllers. At the seventh congress of IFAC, Chen Hanfu presented his paper on the controllability and observability of stochastic control systems. Guan Zhaozhi, Zhang Xueming, Zhang Siying and others investigated the problem of tranquillity of elastic vibrations, the problem of optimal control of the distributed parameter systems and the centralized parameter systems respectively. Still many more mathematical workers engaged in the research work of mathematical models and designs of control equations of production process controlled by computers, thus integrating the control theory with practice in production.

Mathematicians at Fudan University also made a study on the absolute stability of adjustment system and the time optimal control. Recently, Li Xunjing and others further generalized the results on the time optimal control of linear systems to the infinite-dimensional distributed parameter systems.

13. Computational Mathematics

The vigorous development of this branch of mathematics flourished only as from 1958. In the fifties, with respect to the method of numerical solutions of algebraic equations, Zhao Fangxiong suggested an iterative method called the "gradient method", for solving systems of any number of simultaneous equations of any type. He further put forward the direct method for solving systems of linear equations, and used the gradient method in solving for the eigenvectors of symmetric square matrices.

In 1965, Feng Kang developed the finite element method independently in his paper entitled *The Difference Scheme on the Basis of Variational Principle*. Recently he further put forward a theory of elliptic partial differential equations in the region connected by manifolds of different dimensions, and extended the finite element method to this type of equations. In order to expedite the utilization of the finite element method, he designed a computer algorithm to be used in the triangulable plane regions. In order to deal with the problem of singularity, he also made a modification in the finite element method by the use of an infinite number of similar elements.

As from the year 1959, the Mathematics Department of Fudan University began to engage in the research work in the field of computational mathematics. Jiang Erxiong and others investigated the approximation properties of the inherent functions in the Fredholm's integral equations, the stability of difference equations, the improvement on the p-condition of positive definite symmetric matrices, and the algorithm of the transformation matrices in the Jordan canonical forms.

During the recent years, there appeared in the field of computational mathematics a lot of outstanding results, which include, for example, the method of infinite similar elements for computing the strength of factors as proposed by Ying Longan.

The academic achievements made in the thirteen branches of mathematics have been briefly described in the foregoing paragraphs. While they give a rough picture of the developments of mathematics research in China during the past 30 years, it should be mentioned that there are other achievements and breakthroughs that are not covered in this survey, due to the limitations of space and time.

Due to the disruption caused by Lin Biao and the "Gang of Four", various scientific undertakings stagnated for nearly ten years, and mathematics was no exception. Our present mathematical level as a whole, with the exception of some fields, thus lags behind the advanced countries in the world by ten to twenty years. However, looking back at the past, it is clear that the research work on mathematics in China has already made some encouraging progress. It is our belief that, through the efforts of our large numbers of mathematics workers, the field of mathematics will show greater vitality in its growth and development in the near future.

CHAPTER 10

A THEORY OF POLYMERIZATION OF SILICIC ACID IN AQUEOUS SOLUTION

by Tai Anpang and Chen Yungsan

An important characteristic property of silicic acid in a solution of sufficient concentration is its polymerization to form a gel. Since the time of T. Graham, this behaviour of silicic acid has been much studied and many hypotheses have been advanced to explain the mechanism of the reaction. It is generally thought that polymerization occurs due to a condensation of silicic acid molecules with the elimination of water:

$$(HO)_3Si\text{-}OH + HO\text{-}Si(OH)_3$$
$$= (HO)_3Si\text{-}O\text{-}Si(OH)_3 + H_2O.$$

Iler maintains that polymerization involves an ionic mechanism. Above pH 2, the rate of polymerization is proportional to the concentration of the OH^- ion, and below pH 2, to the concentration of the H^+ ion and F^- ion. More recently, N_{BaHOB} proposed that the polymerization reaction proceeds as follows:

$$H_3SiO_4^- + H_3SiO_4^- = H_4Si_2O_7^{2-} + H_2O$$
$$H_4Si_2O_7^{2-} + H_3SiO_4^- = H_5Si_3O_{10}^{3-} + H_2O$$

$$H_{n+2}Si_nO_{3n+1}^{n-} + H_3SiO_4^- = H_{n+3}Si_{n+1}O_{3n+4}^{(n+1)-} + H_2O.$$

In spite of the various attempts made to understand the polymerization reaction of silicic acid, the comment given by Vail is still appropriate: "Although silica is the most abundant of terrestrial substances and its behaviour in the presence of water is of vital importance in many fields of science and technology, we are still without a completely clear and convincing picture of the exact mechanism by which it reacts as a weak acid." Here we present the rudiments of a theory of polymerization of silicic acid in solution. We believe it is more coherent and quantitative and explains more experimental facts than any of those so far proposed.

1. Protonation of Silicate Ion.

When sodium monosilicate is dissolved in water, the anion is mostly $H_2SiO_4^{2-}$. Upon acidification, protonation takes place according to the following steps:

$$\begin{bmatrix} \text{O} \\ | \\ \text{HO-Si-OH} \\ | \\ \text{O} \end{bmatrix}^{2-} \underset{-H^+}{\overset{H^+}{\rightleftarrows}} \begin{bmatrix} \text{OH} \\ | \\ \text{HO-Si-OH} \\ | \\ \text{O} \end{bmatrix}^{-} \underset{-H^+}{\overset{H^+}{\rightleftarrows}} \begin{bmatrix} \text{OH} \\ | \\ \text{HO-Si-OH} \\ | \\ \text{OH} \end{bmatrix} \rightleftarrows \begin{bmatrix} \text{OH} \\ H_2O\diagdown |\diagup \text{OH} \\ \text{Si} \\ HO\diagup | \diagdown OH_2 \\ \text{OH} \end{bmatrix} \underset{-H^+}{\overset{H^+}{\rightleftarrows}} \begin{bmatrix} \text{OH}_2 \\ H_2O\diagdown |\diagup \text{OH} \\ \text{Si} \\ HO\diagup | \diagdown OH_2 \\ \text{OH} \end{bmatrix}^{+}$$

$$\begin{matrix} \text{(I)} & K_2 & \text{(II)} & K_1 & \text{(III)} & & \text{(IV)} & K_0 & \text{(V)} \\ A^{2-} & & HA^- & & H_2A & & H_2A & & H_3A^+ \end{matrix} \quad (1)$$

As a hydrogen ion combines with an oxygen of the silicate ion, the length of the Si-O bond increases. When all the coordinate bonds around silicon increase in length, its coordination number increases from four to six. Species (III) and (IV) are in equilibrium. Upon further protonation, (IV) becomes positively charged, and this has been proved by transference experiments. K_0, K_1 and K_2 are acid dissociation constants respectively as indicated in equation (1).

2. Two Mechanisms of Polymerization.

The pH at which (III) and (IV) are predominant and in equilibrium may be called neutral points (N) of silicic acid. On the alkaline side of N, polymerization proceeds by oxolation of an anion (II) and a neutral molecule (III) of silicic acid; and on the acid side of N, it proceeds by olation of a neutral molecule (IV) and a cation (V) both of coordination number 6:

$$\begin{bmatrix} \text{OH} \\ | \\ \text{HO–Si–OH} \\ | \\ \text{OH} \end{bmatrix}^{-} \underset{-\text{H}^{+}}{\overset{\text{H}^{+}}{\rightleftharpoons}} \begin{bmatrix} \text{OH} \\ | \\ \text{HO–Si–OH} \\ | \\ \text{OH} \end{bmatrix} \rightleftharpoons \begin{bmatrix} \text{H}_2\text{O} & \text{OH} \\ \diagdown | \diagup \\ \text{Si} \\ \diagup | \diagdown \\ \text{HO} & \text{OH}_2 \\ & \text{OH}_2 \end{bmatrix} \underset{-\text{H}^{+}}{\overset{\text{H}^{+}}{\rightleftharpoons}} \begin{bmatrix} \text{H}_2\text{O} & \text{OH}_2 \\ \diagdown | \diagup \\ \text{Si} \\ \diagup | \diagdown \\ \text{HO} & \text{OH} \\ & \text{OH}_2 \end{bmatrix}^{+}$$

$$\text{(II)} \qquad\qquad \text{(III)} \qquad\qquad \text{(IV)} \qquad\qquad \text{(V)}$$

$$\begin{bmatrix} \text{OH} & \text{OH} \\ | & | \\ \text{HO–Si–O–Si–OH} \\ | & | \\ \text{OH} & \text{OH} \end{bmatrix} + \text{OH}^{-} \qquad\qquad \begin{bmatrix} \text{HO} & \text{OH}_2 \\ \text{HO} \diagdown | \text{HO} \diagdown | \diagup \text{OH} \\ \text{Si} \qquad \text{Si} \\ \text{HO} \diagup | \diagdown \text{OH} \diagup | \diagdown \text{OH} \\ \text{H}_2\text{O} & \text{OH} \end{bmatrix}^{+} + 2\text{H}_2\text{O}$$

$$(2) \qquad\qquad\qquad (3)$$

Experiments showed that there was a slight increase of pH in reaction (2) due to the release of OH⁻, and there was no appreciable change of pH in reaction (3) as indicated in the equations. However, polymerization does not stop at the formation of dimers. On the alkaline side of N, neutral dimers may combine with anionic monomers to form trimers. And a neutral polymer on account of its higher acidity becomes an anion on dissociation. It may then combine with a neutral polymer to further form a larger polymer and so on until colloidal size is reached and gelation takes place. On the acid side of N, in like manner the cationic dimer may olate with neutral molecules of silicic acid forming still larger polymers. When colloidal size is reached, gelation sets in. Gelation is due to the aggregation of the colloidal particles of the polymerized silicic acid. Since the same factors similarly affect polymerization of monomers and low molecular weight polysilicic acid, it is probable that even the conversion of colloidal silicic acid to a gel is by the same two mechanisms, oxolation of neutral and anionic species and olation of neutral and cationic species.

3. Rate of Polymerization of Silicic Acid.

The rate of polymerization of silicic acid is inversely proportional to its gelation time, assuming that in different solutions when gelation sets in, the acid is of about the same degree of polymerization. When HA⁻, H₂A and H₃A⁺ are used to represent anionic, neutral and cationic silicic acids respectively, rate of polymerization v, and time of gelation t, are expressed by the following equations: for oxolation,

$$v_1 = \frac{k_1}{t_1} = k'_1 [\text{H}_2\text{A}][\text{HA}^-] \qquad (4)$$

and for olation,

$$v_2 = \frac{k_2}{t_2} = k'_2 [\text{H}_2\text{A}][\text{H}_3\text{A}^+] \qquad (5)$$

wherein K_1, K_2, K'_1 and K'_2 are rate constants. For the sake of simplicity, K'_1 and K'_2 are assumed to be nearly equal; then the total rate of the reaction will be:

$$v = v_1 + v_2 = k[\text{H}_2\text{A}]([\text{HA}^-] + [\text{H}_3\text{A}^+]) \qquad (6)$$

and the log of the total time of gelation:

$$\log t = \log C - \log\{[\text{H}_2\text{A}]([\text{HA}^-] + [\text{H}_3\text{A}^+])\} \qquad (7)$$

where C is the gelation rate constant. At a given concentration of silicic acid, the amount of each species in solution varies according to the hydrogen ion concentration and depends on its ionization constants according to the following relations:

$$K_1 = [H^+][HA^-]/[H_2A] \quad (8)$$

$$K_2 = [H^+][A^{2-}]/[HA^-] \quad (9)$$

$$K_0 = [H^+][H_2A]/[H_3A^+] \quad (10)$$

Substituting (8) − (10) in (7), we obtain

$$\log t = \log C - 2\log[H_2A] - \log\left(\frac{K_1}{[H^+]} + \frac{[H^+]}{K_0}\right) \quad (11)$$

Let α_{H_2A} represent the partial concentration of H_2A, that is, $\alpha_{H_2A} = [H_2A]/T_A$, where T_A is the total concentration of all the species of silicic acid in solution and equals to $[A^{2-}] + [HA^-] + [H_2A] + [H_3A^+]$. Let $\beta_1 = 1/K_2$, $\beta_2 = 1/K_1K_2$, and $\beta_3 = 1/K_2K_1K_0$. Then

$$\alpha_{H_2A} = \frac{\beta_2[H^+]^2}{1+\beta_1[H^+]+\beta_2[H^+]^2+\beta_3[H^+]^3} \quad (12)$$

Substituting (12) in equation (11), we obtain:

$$\log t = \log C - 2\log T_A$$
$$+ \log\frac{(1+\beta_1[H^+]+\beta_2[H^+]^2+\beta_3[H^+]^3)^2}{\beta_2\beta_3[H^+]^3}$$
$$- \left(\log\frac{\beta_1}{\beta_3} + [H^+]^2\right) \quad (13)$$

In strongly acidic solutions, β_1/β_3 in the parenthesis of the last term of the above equation, which is equal to K_1K_0, is negligible when compred with the second term in the parenthesis, and so (13) becomes

$$\log t_a = \log C - 2\log T_A$$
$$+ \log\frac{(1+\beta_1[H^+]+\beta_2[H^+]^2+\beta_3[H^+]^3)^2}{\beta_2\beta_3}$$
$$+ 5\,pH \quad (14)$$

In less acidic solutions, $[H^+]^2$ is negligible as compared with β_1/β_3 and (13) may be written as

$$\log t_b = \log C - 2\log T_A$$
$$+ \log\frac{(1+\beta_1[H^+]+\beta_2[H^+]^2+\beta_3[H^+]^3)^2}{\beta_2\beta_3}$$
$$+ 3\,pH \quad (15)$$

Fig. 1 gives the relationship between log of gelation time of water glass and initial pH of the solution using acetic acid as acidifying agent. The points are

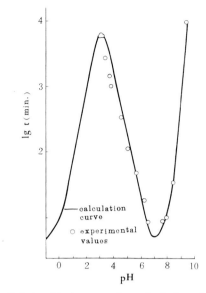

Fig. 1. Relationship between time of gelation of water glass and initial pH of solution using acetic acid as acidifying agent

experimental values and the curve is calculated according to equation (13) with $K_2 = 10^{-9.0}$, $K_1 = 10^{-7.2}$, $K_0 = 10^{0.8}$, $T_A = 0.0765$ M and $C = 7.36 \times 10^{-3}$. The agreement of the theoretical curve with the experimental one may be considered as fairly satisfactory.

When sodium monosilicate is acidified with acetic acid, monosilicic acid is obtained, whose successive dissociation constants are $K_1 = 10^{-8.5}$, $K_2 = 10^{-13}$ and $K_0 = 10$. Taking both T_A and C as 1, a more complete $\log t - pH$ curve can be constructed (Fig. 2). At point a, $[H_3A^+] = [H_2A]$; at b, $[H_2A] = [HA^-]$ and c, the neutral point, $[H_3A^+] = [HA^-] \ll [H_2A]$.

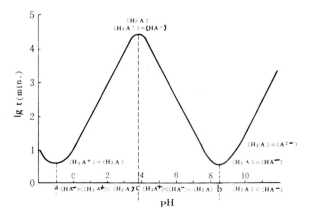

Fig. 2. Gelation Time of Monosilicic Acid at Different pH

Equation (13) predicts that in the log t – pH curve, there ought to be one maximum and two minimum points. Using sulphuric acid as acidifying agent, we studied the gelation time of a solution of sodium silicate which contained potassium fluoride whose concentration was one tenth that of the silicate. A more complete log t – pH curve was obtained with a minimum on the acid side of N in addition to the one on the basic side. Experimental realization of this new minimum point as predicted by our theory of polymerization of silicic acid in solution provides an important proof of its truth.

Since $K_1 = [H^+][HA^-]/[H_2A]$, $pK_1 = pH + p([HA^-]/[H_2A])$, therefore, at b, where $[HA^-] = [H_2A]$,

$$pH_b = pK_1 \qquad (16)$$

Since $K_0 = [H^+][H_2A]/[H_3A^+]$, $pK_0 = pH + p([H_2A]/[H_3A^+])$, hence at a, where $[H_2A] = [H_3A^+]$,

$$pH_a = pK_0 \qquad (17)$$

Also $K_0 = ([H^+]^2[HA^-]/K_1[H_3A^+])$, $pK_0 = 2 pH - pK_1 + p(\frac{[HA^-]}{[H_3A^+]})$ that is, at c,

$$pH_c = (pK_1 + pK_0)/2 \qquad (18)$$

Also since $K_2 = [H^+][A^{2-}]/[HA^-]$, $K_1 K_2 = [H^+]^2[HA^-][A^{2-}]/[HA^-][H_2A]$, whence $pK_1 + pK_2 = 2 pH + p[A^{2-}]/[H_2A]$, and hence on the more basic side of b, where the slope, log t/pH = 1 and $[H_2A] = [A^{2-}]$,

$$pH_1 = (pK_1 + pK_2)/2 \qquad (19)$$

From the above relations, successive dissociation constants of silicic aid, pK_1, pK_2 and pK_0 are readily obtained (Table 1 and 2).

Table 1. pK Values of Silicic Acid formed by Acidification of Sodium Silicates (35°C)

Acidifying agent	Approximate molecular complexity	pK_0	pK_1	pK_2
HCL	monomeric	−5.2	8.0	13.2
	dimeric	−	7.7	−
	trimeric	−3.9	7.3	9.7
HAc	monomeric	−0.7	8.5	12.9
	dimeric	−0.6	8.0	10.6
	trimeric	−0.8	7.9	9.6

Table 2. pK values and Concentrations of Silicic Acid (30°C)

Acidifying agent	Concentration of monosilicate (M)	pK_0	pK_1
HCL	0.240	−5.32	7.82
	0.339	−5.64	8.14
	0.479	−5.89	8.43
	0.677	−6.29	8.79
HAc	0.243	−0.49	7.99
	0.339	−0.47	8.27
	0.480	−0.27	8.55
	0.678	−0.17	8.83

These constants are apparent constants, because their values depend on experimental conditions, such as concentration and complexity of the acid and nature of acidifying agent. Minimum points a and b are characteristic of the gelation time curves. At a, $[H_3A^+] = [H_2A]$ and at b, $[H_2A] = [HA^-]$. Partial concentration of each is equal to ½ and the sum of the last two terms of equation (13) is 0.602. Therefore, when (log t − 0.602) is plotted against log T_A, a straight line should be obtained. The intercept is log C and was found for monosilicic acid to be about −2, regardless of the acid used as acidifying agent. The slope of the straight line ought to be −2 according to equation (13), but it was found mostly to be around −3. This discrepancy is to be explained presently.

4. Salt Effect on the Rate of Polymerization of Silicic Acid and Modification of Gelation Time Equation.

The finding that the slope of the log t − log T_A curve was not as expected by the theoretical equation is found to be due to the effect of the salt formed or/ and added in the process of acidification of sodium monosilicate. When the concentration of sodium monosilicate used was kept constant and different amounts of a salt corresponding to the acid used as acidifying agent were added, the gelation time curves obtained showed that the added salt delivered a little promotion effect on gelation at the maximum point. The pH value did not change except when acetic acid was used as an acidifying agent, in which case the pH value of the maximum point was shifted to the less acidic side.

However, with the minimum points of all the curves, they all shifted toward a higher pH value and exhibited a marked promotion effect on the rate of polymerization. When log t was plotted against log T_s, the total amount of salt added, a straight line was also obtained with a slope of about -1, indicating that the gelation time of monosilicic acid is inversely proportional to the first power of the concentration of the added salt. When the concentration of the added salt was kept constant and that of silicic acid varied, no matter which acid was used as an acidifying agent, the pH of the maximum points of all gelation curves remained constant, while that of the minimum points varied slightly. And in all cases, the slope of the log t – log T_A curve was -2, indicating that the experimental results were in complete accord with the theoretical equation when the effect of added salt was excluded. When the effect of the salt present is not excluded, the expression should be modified slightly by including the part of the salt effect in the coefficient of log T_A and it takes the following form:

$$\log t = \log C - (2 + n') \log T_A$$
$$- \log \frac{(1+\beta_1 [H^+] +\beta_2 [H^+]^2 +\beta_3 [H^+]^3)^2}{\beta_2 \beta_3 [H^+]^3}$$
$$- \log (\frac{\beta_1}{\beta_3} + [H^+]^2) \qquad (20)$$

wherein n' expresses the order of inverse proportionality of log t with log T_s. In the case of monosilicic acid, n' = 1, hence coefficient of log T_A should be -3. When the salt effect is eliminated, n' = 0, coefficient of log T_A becomes -2, then the expression (20) becomes (13).

5. Effect of Added Ions on Rate of Polymerization of Silicic Acid.

The effect of hydrogen ion concentration on rate of polymerization is reflected by the N-curve of gelation. On the more alkaline side of the minimum point where $[H_2A] = [HA^-]$, the hydrogen ion added combines with HA^-, whose amount is more than H_2A ion, to form the latter and so increases the rate of gelation. After this minimum point, further addition of hydrogen ion expands the inequality of H_2A and HA^- and so decreases the rate until the maximum point, where H_2A predominates and $[H_2A^-] = [H_3A^+]$ in minute amounts is reached. Then, addition of H^+ increases the rate until the minimum point on the acid side of N is reached where $[H_3A^+] = [H_2A]$. After this point, still further addition of H^+ again slows down the rate of polymerization of silicic acid. Therefore, it is the mechanism that determines the rate and the dissociation constants of the different species of silicic acid that are involved in the reaction (the internal causes) and hence are the controlling factors. The hydrogen ion concentration of the solution is the external cause which becomes operative only through internal causes.

At a pH higher than the minimum point on the alkaline side, cations like the hydrogen ions exert a similar effect but to a lesser extent on the anionic species of silicic acid by combining with them to form neutral species and thus enhance the attainment of the minimum point and increase the reaction rate. On the more acidic side of the maximum point, cationic silicic acid comes into existence. Cations have no effect on it but anions have a retarding effect by pairing with the positively charged silicic acid. The fluoride ion is a notable exception. At the maximum point and lower pH, the fluoride ion exists almost entirely as HF. Then its reaction with neutral silicic acid which is the predominant species in solution may be represented as follows:

$$\begin{bmatrix} & OH_2 & \\ HO & | & OH \\ & Si & \\ H_2O & | & OH \\ & OH & \end{bmatrix} + HF + H^+ = \begin{bmatrix} & OH_2 & \\ HO & | & FH \\ & Si & \\ H_2O & | & OH \\ & OH & \end{bmatrix}^+ + H_2O \qquad (21)$$

whereby a neutral silicic acid molecule is transformed into a cation, and, in consequence, polymerization according to reaction (3) is enhanced:

$$\begin{bmatrix} & OH_2 & \\ HO & | & OH \\ & Si & \\ HO & | & OH_2 \\ & OH & \end{bmatrix} + \begin{bmatrix} & OH_2 & \\ HO & | & FH \\ & Si & \\ HO & | & OH \\ & OH & \end{bmatrix}^+ = \begin{bmatrix} & OH_2 & H & OH_2 & \\ HO & | & O & | & FH \\ & Si & & Si & \\ HO & | & O & | & OH \\ & OH & H & OH & \end{bmatrix}^+ + 2H_2O \qquad (22)$$

The fluoro-containing disilicic cation may further olate with another neutral molecule to form a higher polymer. Thus, in highly acidic solutions, fluoride ions not only promoted the rate of polymerization of silicic acid, but also shifted the maximum point to a higher pH. However, this is not a catalytic action as

Iler has contended, since the fluoride ion itself not only participated in the reaction, but also formed a part of the reaction product. Syneresis liquid of the silicic acid gels formed did not contain all the fluoride ions that had been present in the reacting solution and in some cases did not contain any at all, indicating that fluoride ions had become a constituent of the polymerized silicic acid.

6. Effect of Temperature on the Energy of Activation of Polymerization of Silicic Acid

In the temperature range 0–70°C, all log t — pH curves assumed the N-shape, and at a constant pH within the range tested, the rate of gelation always increased with temperature. The activation energy calculated thereof by the Arrhenius equation was found to vary with the acidifying agent and concentration of silicic acid. However, when the total salt concentration of the solution was maintained constant, the activation energy would not change with the concentration of the acid, and it did only with the change of pH (Table 3). There was a maximum value at a certain pH which corresponded to that of the maximum point of the gelation curve. There was no minimum value for the activation energy at a pH where the time of gelation was at a minimum. It is therefore probable that the cause of variation of activation energy of the polymerization of silicic acid with pH is due to the fact that there are two mechanisms of polymerization, olation and oxolation. Both processes are participated by neutral molecules of two different forms, 4 coordinated and 6 coordinated, which are the predominant species at N and are in equilibrium with each other. The reaction of one form seems to be affected somehow by the presence of the other. Therefore, the value of activation energy increases with the amount of neutral silicic acid which attains maximum at N with the activation energy reaching maximum correspondingly. On either side of N, only one kind of mechanism operates throughout, hence there exists no minimum.

7. Order of Polymerization Reaction of Silicic Acid.

It has been pointed out above that the order of the polymerization reaction of silicic acid as calculated from the gelation time data is usually higher than 2 and this is due to the effect of salt formed and/or added in the process of acidification of silicate solution. When this effect is deducted, the order will be 2 as indicated in equation (13).

A modified silicomolybden yellow colorimetric method has been devised in this laboratory to measure monosilicic acid more specifically. With this method, the initial rate of polymerization was determined and it was found that the reaction was of second order in agreement with that found by gelation experiments as described above.

8. Summary

From the above discussion, it is seen that the theory of polymerization of silicic acid in solution which we have proposed correlated more with experimental results. The following points of the theory are worth noting:

1) In solutions of high acidity there exist cationic species of silicic acid which have been demonstrated by transference experiments.

2) Upon protonation of the silicate ion, silicon changes its coordination number from 4 to 6 at the neutral point where the two forms of silicic acid molecules are in equilibrium; and in the polymerization reaction the presence of one form interferes somehow with the other.

3) The pH at which neutral silicic acid species predominates and the anionic and cationic species are equal but in minute amounts, is the neutral point (N) of silicic acid and it varies with the acidifying agent used.

4) There are two different mechanisms of polymerization of silicic acid, oxolation of neutral molecule and anionic species on the alkaline side

Table 3. Activation Energy of Gelation of Monosilicic Acid (Constant Salt Concentration)

Acidifying Agent: H_3PO_4

Concentration (M)		pH	Activation Energy (K_{cal}/mol)
Monosilicate	NaH_2PO_4		
0.12 – 0.48	0.96	1.5	12.8
		2.0	15.8
		2.5	17.0
		3.0	16.2
		4.0	15.3
0.12 – 0.24	0.48	6.5	15.1
		7.0	13.5
		7.5	12.0
		8.0	10.8
		8.5	9.2

of N and olation of neutral molecules and cationic species on the acid side.

5) An equation is derived to express quantitatively the relationship between gelation time and pH of the solution.

6) The prediction of a minimum point in the gelation time-pH curve on the acidic side of N has been substantiated by experiments.

7) As to the effect of salts on the rate of polymerization of silicic acid, on the alkaline side of N where anionic species are present, the cation of the salt is effective and on the acidic side where cationic species exist, the anion is more effective.

8) Fluoride ions in right amounts have a marked acceleration effect on the rate of polymerization of silicic acid by combining with the neutral molecules to form cationic species.

9) The reaction of polymerization of silicic acid is of second order when the presence of salt is eliminated.

10) With the log t − pH curve, successive apparent dissociation constants of silicic acid may be estimated.

CHAPTER 11

RECENT DEVELOPMENT IN THE STUDY OF THEORETICAL ORGANIC CHEMISTRY IN CHINA

by Sheng Huaiyu

The purpose of this article is to review recent developments in theoretical organic chemistry research of the past three decades in our country. During this period significant progress in theoretical organic chemistry has been achieved abroad. The traditional classification of the chemical sciences (i.e. organic, inorganic, physical and analytical chemistries) can no longer be applied because of the rapid developments in chemistry. At present in many important fields of chemical research, such as coordination chemistry, organo-metallic chemistry and elemento-organic chemistry, the boundary between organic chemistry and inorganic chemistry cannot be clearly distinguished. A classification based on the nature of chemical problems has even been proposed. Personally, I fully agree with Professor Jiang Mingqian on his views expressed in his paper "Trends in the Development of Comtemporary Chemistry". According to him, the three characteristics of contemporary chemistry are: a transition from descriptive science to deductive science; a transition from qualitative science to quantitative science; and a transition from macroscopic theory to microscopic structural theory. In these transitions, theoretical organic chemistry is still one of the most important branches of contemporary chemistry.

Research in organic chemistry has a long history in our country. However, before liberation, research developments in universities and institutes had been restricted by financial difficulties, influenced by wars of resistance against foreign aggression, as well as the war of liberation from 1937 to 1949. Only a few of the research workers could continue their studies during the war. Much of the research was in the fields of nature products and the component analyses of Chinese medical herbs. Little work was done in the field of structural science.

Colleges, universities and the institutes of scientific research were reorganized soon after liberation in 1953–1954. A series of lectures on material science was then given by Professors Tang Aoqing, Lu Jiaxi, Wu Zhengkai, Xu Guangxian and Tang Youqi. Since then, all departments of chemistry of the various universities started courses on structural and theoretical chemistries. At the same time a number of laboratories in theoretical chemistry was established. A steadily increasing number of research workers were engaged and the scope of the research was widened. Thus rapid progress in this research field was achieved in the 1950's. However, progress was once again slowed down because of the Cultural Revolution in 1966–1976.

The contents of these contributions dealt with a wide range in theoretical chemistry, which included molecular orbital theory, the relationship between structure-chemical and the physico-chemical properties, factors influencing reaction rates and equilibria, various reaction mechanisms including free radical reactions, photochemical reactions, radiation chemistry and laser chemistry. In the latter, much work was done in the realm of reaction kinetics. Some of the main results are summarized below.

I. Achievements in Quantum Chemistry and the Theory of Chemical Bonds

(1) Hybrid orbital theory is one of the most important contributions of quantum mechanics to the theory of chemical bonds. Many chemists worked in this field in the 1950's. Hybrid orbital theory was put forward by Pauling and Slater and then developed by Hultgren and Duffey, but their methods were complicated mathematically. Professors Tang Aoqing, Lu Xikun, Liu Ruzhuang and other scientists developed a simpler and more systematic method for the construction of bond functions. This new method could be used to treat all hybridization problems. Professor Sun Juchang, Tang Aoqing and Dai Shushan extended the group theoretical analysis of hybrid orbital theory to systems containing f-orbitals and proposed a method for the construction of s-p-d-f-hybrid orbitals. Twenty-five figures were used to illustrate the various types of chemical bonds.

(2) In the application of the molecular orbital theory to the discussion on the structure of organic compounds, Professor Xu Guangxian et al. proposed a self-consistent field multi-centered molecular orbital method for the treatment of the hydrogen molecule and the calculation of the binding energy of hydrogen. In order to solve the secular determinant, the method of successive approximation has been adopted, and in this procedure they used not only the LCAO but also the LCNAO (Linear combination of non-atomic orbitals) SCF method. The nature of chemical bonds in π-conjugate triatomic molecules of AB_2 type (such as ClO_2, O_3, SO_2, NO_2, CO_2, N_3^-, NO_2^+, N_2O, $HgCl_2$) had been discussed from the viewpoint of molecular orbital theory. The absorption spectra of the amylose-iodine complex had been explained quantum mechanically by using a three-dimensional free electron model. The valence electrons of the conjugate iodine chain were assumed to move freely in a cylinder, with a width equal to that of the effective internal diameter of the amylose helix; the results obtained were in good agreement with the experimental observations.

(3) Graph theory of molecular orbitals: in the last thirty years, many attempts were made in the application of quantum theory to solve chemical problems. The Schrödinger equation based on quantum mechanics could clearly describe the correlation between molecular structure and the properties of electrons. One of the basic problems in molecular orbital theory was to solve the secular equation. Graph theory of molecular orbitals (MO) developed by Tang Aoqing and Jiang Yuansheng is characterized by its clear intuitive picture, its simplicity in calculations, and a systematic solution of two basic problems in simple organic molecular orbital theory, namely the treatment of eigenpolynomials and molecular orbitals. In addition, it is possible to discuss the stability of polyenes using graph theory, and, from the E_π expanding formula of conjugated molecules, to establish a 4-parameter criterion of stability for straight-chain conjugated molecules in the form.

$$E_\pi = 0.523 S_0 + 0.495 S_2 - 0.049 S_4 + 0.0025 S_6$$

PERE (π electronic resonance energy)

$$= \frac{E_\pi (\text{HMO}) - E_\pi}{n}$$

It successfully elucidated the ($4n+2$) and $4n$ rule for monocyclic conjugated molecules, and also gave a better explanation for the stability of polycondensed-rings and poly-linked-rings. Strict mathematical verification of such a theory had also been pursued by other workers, and applications made to even more complicated organic conjugated molecules. The EP orbital energy levels and the coefficients of MO have been obtained easily this way.

(4) In 1965, Woodward and Hoffmann proposed their famous theory on the conservation rule of molecular orbital symmetry. Since then, there has been great interest in the development of symmetry rules for predicting the course of a chemical reaction and the stable structure of a molecule. The symmetry rules can presumably be used as a guiding principle for organic synthesis as well. It is thus not surprising that a number of alternative theoretical formulations had been set forth. Professor Tang Auchin and his colleagues in the Department of Chemistry of Kirlin University discussed and evaluated these theories (correlation of molecular orbitals, frontier orbital theory and Mobius structure theory). They considered that the symmetry rules of Woodward and Hoffmann more correctly reflected the objective

reality. This principle of symmetry was based upon a correlation of the electron energy levels of the initial states of the reactants with those of the final states of the products. The correlation diagrams correctly represented the symmetric relationship between these energy levels. According to this principle, the energy levels of the initial and final states were connected with straight lines and lines with the same symmetry were not allowed to cross each other. This theory, however, cannot describe the whole course of chemical reactions and quantitative results on activation energies cannot be obtained. Moreover, in cases of olefines containing any substituents, all the symmetric elements of the molecules concerned will be lost, and there will be no symmetric planes or axes. In some reactions such as the o-bond migration and reactions of concerted cycloadditions, all the symmetries will be lost and correlation diagrams of energy levels could not be drawn. Professor Tang et al. in their studies on electrocyclic and ring-open reactions have calculated the energy levels of the electrons by MO theory. They improved the calculation method and derived equations which aimed to conserve the existing molecular orbital symmetry, and raised it from a qualitative description to a semi-quantitative one. This improved rule of correlation of energy was applied to σ-bond migration and organic catalytic reactions with satisfactory results. Professor Deng Conghao derived an expression for the rate constant for various reactions from the Schrödinger equation, and verified the empirical (and approximate) rule of the conservation of orbital symmetry.

(5) Studies on the stability and reactivity of organic compounds with quantum chemical methods: the mechanistic and kinetic studies of acid-catalyzed isomerization of xylene had been published in the literature, but conclusions and predictions differed considerably among publications. Researchers at Nanking University studied this isomerization reaction by the use of the CNDO/2 method. The stabilities of xylene isomers and their carbonium ions (formed from protonation) were calculated. Using this approximation they deduced the mechanism of this reaction. The calculated values fitted well with the experimental results. It was found that the stabilities of the xylene isomers and their carbonium ions decrease in the following order: localized σ-complex $>$ delocalized π-complex $>$ localized π-complex $>$ isomers of xylene $>$ CH_3 delocalized complex.

The reaction was suggested to proceed as follows:

II. Achievements in the Research on the Relationship between Molecular Structure and the Physical Properties

The structural theory can elucidate the origin of various material properties from a viewpoint of the micro-phenomena and pinpoint the various factors which affect them. It can also relate the microscopic models with macro-phenomena by using the statistical theory. A good deal of research work in this field has been done by our chemists. Achievements in the 1950's has been reviewed elsewhere. We report here only the more recent achievements published in the last two decades. In a series of papers published by Professor Jiang Mingqian et al., a general quantitative relationship expressing the regularity of homologous gradation in the orbital energies and related physicochemical properties was formulated and designated as the rule of homologous linearity.

1. The rule of homologous linearity for the energies of molecular orbitals and related properties.

The quantitative relationship between the molecular structure and the homologous gradation of physico-chemical properties of various types of organic structures had been anticipated and widely studied. However, a quantitative theory on a sound structural basis and applicable to all closely related properties of the various types of homologues has yet to be constructed.

Usually a given physico-chemical property P of a homologue series is a function of the homologous ordinal number n, as expressed in the following formula:

$$P = a + b F(n) + \ldots \quad (1)$$

Here the question is how to find a function $F(n)$ of homologous ordinal number n, such that a linear relationship is obtained between P and $F(n)$. For this, the energy level difference ΔX of the HMO theory, where $\Delta X = 4\pi \sin [\pi/(4n + 2)]$, as well as the energy level difference ΔE of FEMO theory, where $\Delta E = k(2n + 1)/(n + 1)^2$ have been used. Although all had a linear relationship with P, the accuracy was low. Quite recently, Professor Jiang, who studied the electronic spectra of conjugated polyenes, polyphenyls and condensed aromatic compounds, proposed a generally accurate and specific rule of homologous linearity. By using the homologue factor $(1/\alpha)^{2/n} = (1/2)^{2/n}$ of the frontier orbitals (HOMO and LUMO), the homologous structure was correlated with the physical properties. He pointed out that all the overtone bands as well as the main peaks in the electronic spectra of almost all types of polyene compounds conform very well to this rule. Furthermore, the linearity of the homologous lines obtained was shown to be superior in precision to that based on the transition energies $(M_{m+1} - M_m)$ calculated by the Hückel molecular orbital method. He emphasized that a real homologous series must possess not only the correct general structural formulae, but the homologue compounds must also have the same number of aromatic sextet. (Aromatic sextet means the complete conjugation of the structure type of benzene, denoted by ⬡). According to the Clar theory, the number of aromatic sextet in condensed benzene compounds greatly influences the colours and chemical reactivities of their benzologs. If the number of the condensed benzene rings is increased, while the arrangement of aromatic sextets in each benzene ring remains unchanged, the absorption peaks will exhibit a regular bathochromic shift. On the contrary, if the number of condensed benzene rings remain the same, while the arrangement of the aromatic sextet is different, the characteristics of their spectra will be different, indicating that they do not belong to the same homologous series. For instance, the general formula for phenanthrene series is B and not A, and that for benzopyrene series is D and not C. This is because in forms B and D the connection parts have no aromatic sextet and the number of aromatic sextets remains unchanged in each homologous series.

Jiang considered that the intrinsic nature of double bonds in aromatic sextets should be quite different from those in other structures. Thus the equivalent number for each terminal benzene ring was assigned as 2 (denoted by $t = 2$) and the equivalent number for each double bond in a non-aromatic sextet was assigned as 1. Hence, for the benzologous series the serial number for homologues should be calculated according to the formula

$$N = n + 2d + e = n + t. \quad (2)$$

Here d is the number of aromatic sextets of the terminal group, e is the number of double bonds of the terminal group besides the aromatic sextet, and t is the terminal equivalent. According to this treatment, he proposed what he deemed the correct general formulae of polyphenyls and condensed benzene compounds. The values (N) of the serial number for the respective homologues are as follows:

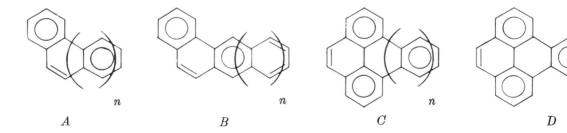

The spectra of the four series of condensed aromatic hydrocarbons consist of α-, P- and β-absorption bands. Some also possess β' and β'' absorption bands. The various absorption maxima belonging to any one of the α-, P-, β-bands constitute by themselves a distinct group of neighbouring homologous peaks. Each group possesses certain characteristics with regard to the position and the slope of their homologous lines as well as the spacing of the neighbouring groups.

($N = 2n+4$)
$t = 4$

($N = 2n+4$)
$t = 4$

($N = n+2$)
$t = 2$

($N = n+5$)
$t = 5$

($N = n+5$)
$t = 5$

($N = n+6$)
$t = 6$

The absorption peaks which existed within the same band conform well to the approximate superposition rule given in equation (4) below, and the linearity of the homologous lines of each group satisfies equation (3):

$$\tilde{\gamma} = a + b(1/\alpha)^{2/N} \quad (3)$$

$$\tilde{\gamma} = a_1 + b_1(1/\alpha)^{2/\beta N} \quad (4)$$

On the other hand, the peaks belonging to different bands do not obey this rule. This reveals that the various absorption maxima of the same band may owe their origin to essentially the same structural factors, whereas those of different bands may be caused by quite different structure factors.

After a study on the quantum chemical bases of the rule of homologous linearity mentioned above, Professor Xu Guangxian and his colleagues found another simple and precise function of homologous linearity $F(n)$ with an improved Hückel molecular orbital method (MHMO):

$$F(n) = X_1 = \sin(\pi/m).$$

Here m is an effective atomic number of conjugated chain of the homologue molecule, and

$$F(n) = X_1 = \sin\frac{\pi}{2n+1}, \quad t = 0$$

or

$$F(n) = X_1 = \sin\frac{\pi}{2N+1}, \quad t \neq 0, \quad N = n + t.$$

Here t is the terminal effect; N is the effective serial number of homologues. It is further proved that the function of homologous linearity, X_1, is the energy level factor of the highest orbital and is called the homologous energy level factor of frontier molecular orbital theory. When this factor is plotted against the properties of homologues, its degree of precision and the wide applicability of these linear relationships are just as good as those from homologous factor of frontier orbital $(1/\alpha_1)^{2/n}$. For any energy level of molecular orbital of homologues, they suggested the following equation:

$$E(n \cdot k) = a + bX_k,$$

$$X_k = \sin\frac{k\pi}{(2N+1)}, \quad N = n + t.$$

For conjugated even-polyene, according to MHMO Method and the conjugated integral ratio of the long and short alternated bonds, they finally obtained a formula for the energy levels which can be put in the following condensed form:

$$\eta = \beta_2/\beta_1 = \cos\frac{\pi}{2N+1}.$$

When they treated the problem of eigenvalues of the long and short alternated polyenes, they applied the graph theory of molecular orbitals, and proposed a complementary theory, from which the function X_k for any energy level k could be obtained. Thus the energy level factor of the MHMO method was

$$X_k = X(n \cdot j = n - k + 1) = \pm\sqrt{1 + \eta^2 - 2\eta\cos\frac{(2n-k)\pi}{2n+1}},$$

or
$$X_k = \frac{2k-1}{k} \sin \frac{k\pi}{2n+1}.$$

The coefficients of the eigenfunctions could be calculated using this equation. Thus the calculations of the electron population, bond order and vibration intensities of electron absorption of polyenes were improved. In order to check whether $X_k(m)$ can represent the regular change of electronic absorption peaks of various homologous series, they compared the calculated results of polyphenyls, condensed benzene compounds, polyenes, substituted polyenes, polyynes and substituted polyynes with the calculated HMO energy levels as well as with the experimental data.

Professors Gao Zhenheng and Zhao Xue Zhuang of Nankai University have calculated the energies of frontier orbitals of para- and meta-polyphenyls with graph theory of molecular orbitals. The molecular graph constituted a linearly connected polycyclic polyene of N rings:

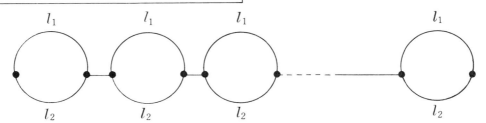

In each ring there are $l_1 + l_2 + 2$ carbon atoms, i.e. for para-polyphenyl $l_1 = l_2 = 2$ and for meta-polyphenyl $l_1 = 3$, $l_2 = 1$. The eigenpolynomial is

$$G_N = \eta^N g_N(\mu/\eta) + gl_1 gl_2 \eta^{N-1} g_{N-1}(\mu/\eta) \quad (5)$$

where g is the Geganbauer polynomial, $\mu = G_1 - gl_1 gl_2$, $\eta = gl_1 + gl_2$.

From (5), the energy of each orbital energy-level can be calculated: $\in = \alpha + \chi\beta$. α and β represent the Coulomb and exchange integrals respectively. The subscript H of $\in_H$ and χ_H denotes the HOMO molecular orbital. Professors Kao and Chao plotted the calculated results ($\chi_H = (\in_H - \alpha)/\beta$) and the wavenumber of maximum absorption bands ($\bar{\gamma}$) against $(1/2)^{2/n}$. The results illustrated that both of these two homologous series complied with the rule of homologous linearity. Since the slope of the parahomolog was steeper, they considered that the maximum absorption band depended upon the energy of the HOMO or the transition energy ($2\chi_H \beta$) between the HOMO and LUMO orbitals. Since the slopes of these two homologous series were different, they believed that the para- and meta-polyphenyls belonged to two separate homologous series.

2 Inductive effects and chemical reactivity

There are many achievements in the study of the relationship between inductive effects and chemical activities. Professors Jiang Mingqian and Dai Cuichen studied the relationships between their own inductive effect index and a series of organic compounds (hydrocarbons, alkyl halides, alcohols, mercaptans, amines, carboxylic acids and carbonyl compounds). In addition, Professor Zeng Guangzhi reported that the validity of the Hammett equation was further substantiated with dissociation constants of substituted aniline as well as substituted phenols. When it was found that the pKa values of eighty of the substituted anilines, seventy of the substituted phenols and twenty one of the substituted N, N-dimethylaminobenzenes had a good linear correlation with the substitution constant (σ), Professor Zeng proposed a complete set of $\sigma^-_{ArNH_3^+}$ values as another parameter for this reaction series in which the funtional group was capable of donating electrons conjugatively. When it was used together with the $\sigma^-_{I.R.}$, a linear free energy relationship was found for nucleophilic substitution reactions. He also found that a good linear relationship existed between the dipole moments of aminobenzenes and their substitution constants. Professor Chen Qingyun, Jiang Xikui and their coworkers studied the inductive effect of perfluoroalkyl groups. The new compounds, perfluro-t-butyliodide (I) and perfluro-t-butyl bromide (II) were prepared, and their structures were proved by ultraviolet and F^{19} nuclear magnetic resonance. When (I) or (II) reacted with dimethylamine (or alcoholic potassium hydroxide), the first step in the reaction was an $S_N 2$-like reaction, in which the nucleophile directly attacked the halogen (iodine or bromine atom):

$$NX - C(CF_3)_3 \rightarrow N - X + C(CF_3)_3$$
$$\text{(I) or (II)} \qquad\qquad\qquad \text{(III)}$$

The carbonion (III) eliminated easily a fluorine atom to form the isobutane (IV) which reacted further with the nucleophile (dimethylamine) to give $(CF_3)_2C=CF-N(C_2H_5)_2$ (V). The product (V) was identical with that obtained directly from the reaction between perfluroisobutane and diethylamine. The ultra-violet solvent shifts $\delta(H\text{-}B)$, were determined for $(CF_3)_3Cl$, $(CF_3)_2CFI$ and $CF_3CF_2CF_2I$ (here the H represents the hydrocarbon solvent; B the Lewis base solvent). The $\delta(H-B)$ values provided evidence for an inductive order ($-I$) for perfluroalkyl groups such as

$$(CF_3)_3C- > (CF_3)_2CF- > CF_3CF_2- > CF_3-$$
(order 1)

Arguments were presented for the above order and against the generally accepted view

$$CF_3- > CF_3CF_2- > (CF_3)_2CF- > (CF_3)_3C-$$
(order 2)

The authors pointed out that in saturated compounds of the poly- and perfluoro-type there exists an "α-fluorine substitution effect". A direct consequence of this effect is that the sum of all the $-I$ of the three fluorines in the CF_3 group cannot be equal to three times the $-I$ of a single fluorine in a group such as FCH_2^-. On the other hand, the sum of $-I$ of the three CF_3's in the $(CF_3)_3C-$ group may approach three times the $-I$ of the single CF_3 in a group such as CF_3-CH_2-. The generally accepted views had not taken the above consideration into account.

Acceptance of order 1 in conjunction with the phenomenon of hyper-conjugation will naturally lead to the following order for the stability of the carbonions:

$$(CF_3)_3C^\ominus > (CF_3)_2CF^\ominus > (CF_3)CF_2^\ominus > CF_3^\ominus$$

Professor Jin Songshou et al. of Hangzhou Univ. investigated the activation energy of reactions and the structure of reactants (he introduced the term "structural adaptability of reactants"). He compared the empirical formula of activation energy widely applied in recent years (including the Jiang Mingqian formula for the calculation in free radical reaction), to the empirical formula proposed by them:

Exothermic reaction:
$$\Delta E = 0.14(\Sigma D_R + \Sigma D_P) - 0.25q$$

Endothermic reaction:
$$\Delta E = 0.14(\Sigma D_R + \Sigma D_P) + 0.75q$$

ΣD_R is the energy required for bond breaking, ΣD_P is the heat of the reaction.

This empirical formula when applied to many reactions had calculated values essentially equal to those of experimental values. From the recently published empirical formulae (including the formula proposed by Prof. Jin Songshou) they found that there existed two kinds of reactions where the calculated results were respectively higher and lower than the real values. After studying in detail these two kinds of reactions, they realized that the activation energy of a reaction was also dependent on the structural adaptability of reactants, i.e. whether the electronic properties of the steric configurations of the reaction centers could be adapted to form the transition state. Several examples were used to elucidate their supposition.

3. The polar and solvent effects in free radical addition reactions.

Professor Jing Xikui et al. (Shanghai Institute of Organic Chemistry, Academica Sinica) reported their studies on the solvent dependency of the regioselectivity (R) of the cyclopropyl-radical addition to fluoroolefins. A solvent effect was found for the free radical additions which would discriminate between the two termini of a double bond. This special solvent effect appears unusual and intriguing. The "bidirectional addition" phenomena of the electron donor radical, cyclopropyl, to the fluoroolefins (vinylidene fluoride, trifluoroethylene and vinyl fluoride) was discussed. A great many experiments were conducted to elucidate the geometric factors of the solvents in controlling the orientation of the free radical addition. Fifteen solvents with different polarities and different geometries were used for the measurement of the R values in the addition reaction of cyclopropyl radical to vinylidene fluoride at a given temperature (80°C). It was found that there was no correlation between R-values and any of the known solvent property parameters (including dielectric constants $\in$, dipole moments μ, parameters which are considered to be useful cybotactic probes of the polarity of the solvents e.g. A_N, Z, B_T, π, X_R values, and the parameters which are related to the electron pair donicity and acceptability of the solvent e.g. D_N, A_N). They found that the regioselectivities, R, varied in various isomers of hydrocarbon solvents. In the various polar and non-polar solvents, the corresponding R-values might be divided into several ranges. In a series of hydrocarbon solvents with equal or approximately equal number of carbon atoms but different structures the R-values were smaller in straight chain and cyclic hydrocarbons ($R = 0.25-$

0.37) than those of highly-branched-chain hydrocarbons (R = 0.55-0.82). This result indicated that there might be a subtle "geometry factor" in operation. All the reaction products of the process described below were determined by quantitative GLC analysis.

The ratio k_{a-1}/k_{a-2} is taken to be a measurement of the regioselectivity R. Here $R \cong (6)/(7)$ further demonstrated that the geometry factor of the solvent has an important solvent effect in this addition reaction. They used the binary mixture of different compositions as solvents for measuring R-values (n-octane and iso-octane, methyl cyclohexane and iso-octane), and derived the simple expressions. Their calculations based on these equations could be compared with experimental results. They assumed that the molecules of the neighbouring solvent and solute in a solution continously interact and change their modes of patterns of mutual packing. The probability of existence of the different packing patterns depends not only on the equilibrium conformation states of the solvent molecule, but also on the interaction between various parts of the solvent and the solute molecules. Not only is there a small difference in size between the two ends of the solute $CF_2 = CM_2$ molecule, but there is also a difference in attractive and repulsive forces between the two terminals in their interaction with the solvent molecules. Therefore, during the dynamic process of inelastic collision and mutual packing, the interactions with the solvent molecules are different for the two dissimilar parts of the solute molecules.

III. Mechanisms and Kinetics of Organic Reactions

The mechanisms and kinetics of organic reactions were both widely studied in our country. In the area of reaction mechanisms, attention was focused on the oxidation-reduction, addition and elimination, rearrangement reaction and double bond shift, free radical reaction, the reactions in elemento-organic syntheses, coordination chemistry and organo-metallic synthetic reactions. In the field of chemical kinetics, the published papers concentrated on the topics of homogeneous phase reaction, kinetics of high-molecular reactions and the multi-phase catalyses. All achieved considerable success.

1. Progress in the chemical reaction researches

The applications of the Wolff-Kishner reduction modified by Professor Huang Minglong were continously studied and extended in recent years. In the case of the carbonyl compounds, α-diones, α-ketols, α-keto acids and ɤ-kato acids, the normal reduction products were obtained. But in the case of the β-diones, (such as dibenzoylmethane) only the pyrazole derivative (I) was obtained; β-katoacid (such as benzoylacetic acid) gave only pyrazolone (II), δ-dione (such as dibenzoylethane) gave only the dihydropyridazine derivatives (III), α, β-unsaturated acid (IV) gave the pyrazol derivative (V). The latter underwent alkaline decomposition and gave the cyclopropane derivatives (VI). Thus these structures are suitable for ring closure and no further reduction proceeded.

Professor Xu Xiangong et al. (Zhongshan Univ.) reported their research on the electro-withdrawing effect of the nitrogen atom of pyridine. The nucleophilic substitution rate constants between δ-chlorophyridine and p-chloronitrobenzene which were reacted with sodium methoxide-methanol or sodium ethoxide-ethanol at 75°C respectively were determined and the substitution effect was compared. They concluded that the relative magnitude of electron withdrawing ability was larger with the NO_2 group than that of the nitrogen atom of pyridine (–N=)–. The electrowithdrawing ability of the latter was only 1/9 – 1/14 that of $-NO_2$ group. The yields and the rates of reaction were quite different in these cases. It was shown that the nucleophilic power of sodiumethoxide was stronger than that of sodium methoxide.

Professor Yuan Kaiji et al. (Shanghai First Medical College) reinvestigated the Knoevenagel reaction of methylmalonic acid and its derivatives with aromatic aldehydes. Reactions of this kind were rarely studied in the past. They compared the different conditions of condensation between the Genzler process and the Fujiwara process. They recognized that the former process had a wider range of application than that of the latter, and its yields were also higher. Three aromatic aldehydes (benzaldehyde, p-methoxybenzaldehyde and p-nitrobenzaldehyde) were condensed with methylmalonic acid under different reaction conditions (those of the Genzler and Fujiwara processes). The product (I) was obtained by the Genzler process, while the product (II) was obtained by the Fujiwara process.

(I).

$O_2N-\langle\ \rangle-\overset{OH}{\underset{H}{C}}-\overset{CH_3}{\underset{}{CH}}-COOH$
(I)

$O_2N-\langle\ \rangle-CH=\overset{CH_3}{\underset{}{C}}-COOH$
(I)

The cause was attributed to a difference in reaction mechanism. The reaction which proceeded under the Genzler condition underwent the Knoevenagel or Rodionov mechanism, whereas the reaction which proceeded under the Fujiwara process passed through the Lapworth mechanism.

2. Achievements in free radical chemistry

Professor Liu You Cheng et al. (Lanzhou Univ.) published several papers on the structural chemistry of free radicals. These included the correlation between structure and the reaction activities in the free radical addition of thiols to olefines, isomerization of free radicals in solution, the reaction of Grignard reagents with alkylhalides in the presence of cobaltous chloride and investigations on the mechanism of free radical reactions. Several stable amine-oxygen free radicals of piperidones were synthesized. These free radicals were very stable and could be preserved for a long period, and their reactivity was retained after they were used in some reactions. The influence of the polarity of solvents on the ESR of the above mentioned free radicals was observed. The g values of free radical III decreased, while the superfine splitting constant a_N increased as the polarity of the solvent was increased. The a_N value had linear relationships with microscopic polarity parameters E_T and Z of model resetions. It was found that an increase in the polarity of solvent increased the number of superfine splittings. They explained that it was due to the interaction of the nuclear spin of the 12 protons of four methyl groups with the unpaired electrons of the free radical = N-O.

product (unknown structure)

The reaction mechanism of the dissociation of acyl peroxide radicals was also studied. By using DPPH, the change of optical density at 434 mm which is the maximum, absorption peaks were measured, and the self-dissociation rate constants were calculated for those peroxide radicals.

Professor Chen Yaozu (Lanzhou Univ.) studied the reaction of free radical addition, such as, the free radical addition of silane catalyzed by acylperoxide or irradiated by ultraviolet light (radical $SiCl_3$ added on to octene-1). It was discovered that the aliphatic azo compound could also induce this free radical addition, for example:

$$(CH_3)_3-\underset{\underset{CN}{|}}{C}-N=N-\underset{\underset{CN}{|}}{C}(CH_3)_3 \longrightarrow 2(CH_3)_2\underset{\underset{CN}{|}}{C}\cdot + N_2$$

$$\underset{CH_3}{\overset{CH_3}{>}}\underset{\underset{CN}{|}}{C}\cdot + HSiCl_3 \longrightarrow \underset{CH_3}{\overset{CH_3}{>}}\underset{\underset{CN}{|}}{C}-H + \cdot SiCl_3$$

$\cdot SiCl_3 + \bigcirc \longrightarrow \bigcirc\!\!-SiCl_3$

$\bigcirc\!\!-SiCl_3 + HSiCl_3 \longrightarrow \bigcirc\!\!-SiCl_3 + \cdot SiCl_3$

Professor Liang Xiaotian (Institute of Materia Medica, Chinese Academy of Medical Science) investigated the sterochemistry of free radicals, methanolic Kolbe electrolysis of optically active methylethylacetic acid in methanol solution and the Kolbe mixed electrolysis of methylethylacetic acid and succinic acid monomethylester, giving various products, such as in the former case: methyl s-butyl ether; 3, 4-dimethylhexane; s-butyl methylethylacetate and an ester of unknown structure. The free radical (s-butyl radical) formed during the electrolysis was found to be unable to retain its steric configuration. One mechanism for Kolbe electrolysis is oxidization of the organic acid at the anode to the diacyl peroxide; it then undergoes decomposition to form the observed products. The authors found that the diacyl peroxide from (+) -methylethylacetic acid decomposed to give (+) -s-butyl (+) -methylethylacetate instead of the ester with a racemic alcohol moiety, which would be the case with Kolbe electrolysis. Kolbe electrolysis also strongly disfavors the peroxide mechanism. The nucleophilic substitution reactions such as the hydrolysis of o-nitroaniline and the formation of phenols from the α- methylpyridine were also studied. The course of the latter reaction is shown as follows:

(I) → (II) → (III) → (IV)

Professor Liu Yunpu et al. (Tienjin Univ.) investigated the hydration rate of trimethylethylene and its reaction mechanism. The reaction rates were determined at different temperatures and in a wide concentration range of hydrochloric acid (0.1–2.6M), percholoric acid (0.1–2.5 M) and sulphuric acid (0.25–2.5M). From the reaction rates, the activation energy and the entropy of activation were calculated. Based on the idea of free water entertained by Bascombo and Bell, a kinetic function was derived and a reaction mechanism was thus proposed:

Hydrochloric acid $\log k/CH_3O^+ = 1.930 + 0.41\, C_{HCl}$

Nitric acid $\log k/CH_3O^+ = 1.930 + 0.222\, C_{HNO_3}$

Sulphuric acid $\log k/CH_3O^+ = 1.930 + 0.390\, C_{H_2SO_4}$

Perchloric acid $\log k/CH_3O^+ = 1.930 + 0.402\, C_{HClO_4}$

$\Delta H^\# $ HCl 22.1; $HClO_4$ 21.5; H_2SO_4 21.4; HNO_3 22.5 K_{cal}

$\Delta S^\# $ HCl + 0.9; $HClO_4$ − 1.4; H_2SO_4 − 1.3; HNO_3 1.3 e.u

They assumed that in the hydrochloric, perchloric and sulphuric acids, the proton existed as a tetrahedral hydrate $H_9O_4^+$, while in nitric acid it existed as H_3^+O.

$$\begin{cases} -\overset{|}{C}=\overset{|}{C}- + H(H_2O)_m^+ \longrightarrow [Tr^{\#}] \xrightarrow{\text{slow}} -\overset{|}{\underset{\oplus}{C}}-\overset{|}{\underset{H}{C}}- + mH_2O \\ \\ -\overset{|}{\underset{\oplus}{C}}-\overset{|}{\underset{H}{C}} + H_2O \underset{\longleftarrow}{\overset{\text{fast}}{\longrightarrow}} -\overset{|}{\underset{H_2O^+}{C}}-\overset{|}{\underset{H}{C}} \\ \\ -\overset{|}{\underset{H_2O^+}{C}}-\overset{|}{\underset{H}{C}}- + mH_2O \xrightarrow{\text{fast}} -\overset{|}{\underset{H_2O^+}{C}}-\overset{|}{\underset{H}{C}}- + H(H_2O)_m^+ \end{cases}$$

Professor Huang Baotong et al. (Institute of Applied Chemistry, Academia Sinica) studied the autooxidation of hydrocarbons.

Due to the importance of research on the mechanism of thermal decomposition of cumene hyperoxide for understanding the initiation process of polymerization, they carried out the thermal degradation of cumene hydroperoxide in isopropylbenzene solution in different concentrations and at different temperatures. It was found that the rate of reaction was essentially a first order reaction, but the rate constant increased with increasing initial concentration. The difference between rate constants was more pronounced when the temperature was raised. In order to elucidate this phenomenon they assumed that two kinds of reactions took place in the course of thermal dissociation of cumene hydroperoxide in isopropylbenzene. One was the first order reaction of $-O-O-$ bond breaking; the other was the inductive dissociation induced by the free radical. The latter became more prominent at higher temperatures. The kinetics of this dissociation was considered as follows:

$$\frac{-d(CHP)}{dt} = K_1(CHP) + K_i(CHP)^{3/2}$$

Here, CHP is the cumene hydroperoxide, K_1 is the dissociation constant of the inductive dissociation, and K_3 is the rate constant of the termination reaction, and $K_i = K_2\sqrt{K_2/K_3}$. The activation energy E_1 is smaller in comparison with E_i. From the experimental results, it was found that an increase in the temperature was indeed favourable for the inductive dissociation. They also studied the rate constant of thermal dissociation in ethylbenzene solution, and discovered that the rate of dissociation was smaller than that of previously mentioned reactions in isopropylbenzene. It means that the formation of $C_6H_5-C(CH_3)_2$ radicals in isopropylbenzene was much easier than that of $C_6H_5-CHCH_3$ in ethylbenzene, so that the radicals formed from the solvent molecules played an important role in this chain-reaction of thermal dissociation.

3. Research on the kinetics and mechanisms of chemical reactions

Professor Sun Chenge (Beijing Univ.) and his coworkers studied the Menschutkin reaction, the reaction of the quinolinemethyl iodide in salicylic aldehyde by conductometric method, and the relationship between solvent effect and the rate of quarternary ammonium salt formation. The second order velocity constants of the reactions were found as follows: $4.37 \times 10^{-3} dm^3 mol^{-1} s^{-1}$ at 25°C; $9.57 \times 10^{-3} dm^3 mol^{-1} s^{-1}$ at 35°C; and $1.98 \times 10^{-3} dm^3 mol^{-1} s^{-1}$ at 45°C respectively. The activation energy of the reaction was 1.42 Kcal/mole, and that of the entropy of activation -32.0 cal per mole per degree at 45°C. They also studied the kinetics of Menschutkin reaction of quinoline with methyl, ethyl and n-propyl iodide in nitrobenzene and n-propyl alcohol.

Professor Wang Xu et al. (Department of Pharmacy, Beijing Medical College) studied the chemistry of 1,2,4-triazines. 1,2,4-triazine (I) (3-methylmercapto-5-hydroxy-6-1,2,4-triazine) was formed by the alkylation of 3-mercapto-5-hydroxy-1,2,4-triazine derivative which gave compound (II) after tosylation. The latter compound underwent a series of substitution reactions as follows:

Professor Long Kanghou et al. (Zhongshan Univ.) published their work on the reaction of some monoterpenes with dichlorocarbene. It was found that dichlorocarbene reacted readily with α-pinene, dipentene and camphene, and the yields of such adducts were about 62-70%. The reactions of some monoterpenes with nitrogen trioxide were studied. For preparative purposes the condition of this addition reaction was investigated. The addition product, α-pinene nitrosite, underwent autophotolysis and was converted into carvone.

Professor Gao Jiyu et al. (Nanjing Univ.) improved the conditions of Nesmejanov reaction for the preparation of arylmercury compounds. The old method used Cu powder as the halogen acceptor;

$$ArN_2Cl \cdot MgCl_2 + 2Cu \longrightarrow ArHgCl + N_2 + 2CuCl$$

It is a heterogeneous, endothermic and violent reaction. The reaction must be well controlled, otherwise the yields of arylmercury compounds would decrease considerably. The authors discovered that by using some organic compounds (such as diethyl or triethyl phosphite) as the halogen acceptor, catalytic amounts of cupric chloride and acetone as the solvent, the reaction proceeded smoothly in a homogeneous medium. Thirty one arylmercuric chlorides were synthesized by these two methods. The yields were higher and purification of the reaction products would be simpler as they usually crystallize out from the reaction medium during the course of reaction. Two mechanisms were proposed to elucidate the Nosmejanov reaction, i.e. a free radical and an ionic reaction mechanism. Prof. Gao et al. prepared $(C_2H_5O)_2P(O)HCl$ from the phosphite and mercuric chloride, dissolved it in acetone, and catalytic amounts of cupric chloride and a solution of $C_6H_5N_2Cl$ were then added. The reaction took place immediately, and the yield of benzenemercuric chloride was 81%. It was found that when the benzene solution of benzoyl peroxide was heated with $(C_2H_5O)_2-P(O)HgCl$, it also gave the same product, but the yield of benzenemercuric chloride was only 18%. Preliminary investigation indicated that the reaction probably proceeded by a free radical mechanism.

Professor Huang Yaozhen et al. (Shanghai Institute of Organic Chemistry, Academia Sinica) reported their studies on the application of organic compounds with elements of the fifth and sixth groups in the periodic table for organic syntheses. They were particularly interested in the reaction of organo-arsenic compounds with substituted olefins:

$$Ph_3\overset{\oplus}{As}-\overset{\ominus}{C}HCOR + R'CH=CR^2COOCH_3 \longrightarrow Ph_3As + \begin{array}{c} COR \\ (R')H \diagup H \diagdown R_2 \\ (H)R' \quad COOCH_3 \end{array}$$

They postulated that the reaction began with a nucleophilic attack on the double bond and the intermediate formed with a preferred lower energy conformation, which then underwent the subsequent cyclization to give the stable final product. They also found that carbomethoxymethylene triphenylarsorane (I) was able to react with fluoroolefins in very mild conditions with moderate yields. The product (III) was not the corresponding cyclopentane derivative. Compound (III) reacted easily with water to form the unstable product (IV), easily losing HF to give the final product (V).

$$Ph_3\overset{\oplus}{As}\overset{\ominus}{C}HCOOCH_3 + CF_2=CFR_F \xrightarrow{-HF}$$
$$(I)$$

$$\underset{(II)}{Ph_3\overset{\oplus}{As}-\overset{\ominus}{C}-COOCH_3 \atop |\atop CF=CFR_F} \xrightarrow{H_2O} \left[\underset{(IV)}{Ph_3\overset{\oplus}{As}\overset{\ominus}{C}-COOCH_3 \atop |\atop CF-CFHR_F \atop |\atop OH}\right] \xrightarrow{-HF} \underset{(V)}{Ph_3\overset{\oplus}{As}-\overset{\ominus}{C}-COOCH_3 \atop |\atop CO-CFR_F}$$

The mechanism of ring formation described above was discussed and a mechanism of molecular rearrangement of the latter reaction was also proposed. All the above-mentioned products were ascertained by various analytical methods. It was confirmed that the arsoranes were more reactive than the corresponding phosphoranes, and the reaction could proceed in very mild conditions, hence having great practical importance in organic synthesis.

The general survey of the research in theoretical organic chemistry reflects some of the achievements in our country in the last three decades. As mentioned at the beginning of this article, virtually no research was carried out in this area before Liberation. Therefore we cherish every scientific achievement, any increase in the research personnel, any progress in experimental techniques and the exploration of any new field in chemistry, such as the success in the synthesis of insulin and the new achievements of research in nucleic acid chemistry. Great advances had been made in the fields of organic synthesis and theoretical organic chemistry. These advances reflected that the level of research has been raised continually and rapidly. Many pieces of modern physico-chemical apparatus have also been constructed. It thus allows for quantitative treatments of the relationships between structures and chemical activity, and also the proofs for structure determinations and structural effects. Much has been achieved using the chemical structural theories (such as inductive effects, conjugation effects, linear free energy relation, etc.), which were derived from macroscopic observation and were established since the 1960's. It should be emphasized, however, that a microscopic theory (quantum chemistry) should deserve special attention in the future as well.

CHAPTER 12

A SURVEY ON ASTRONOMY RESEARCH IN NEW CHINA

by Yi Zhaohua and Qu Qinyue

Before liberation, there were only about 20 astronomers scattered over Nanjing, Shanghai, Guangzhou, Kunming, Chengtu and Jinan in China. The largest telescope at the Zijinshan Observatory, with a diameter of only 60 cm, was destroyed during the anti-Japanese war and was not restored until after liberation. Very little work in astronomy was carried out except in the area of the history of astronomy, mostly by Chinese astronomers overseas or by foreign scientists in China. Almost nothing was pursued in a systematic manner. There were only about 40 graduates in astronomy from the higher institutions in China then, and less than one-third of them were engaged in research.

Under the leadership of the Central Committee of the Chinese Communist Party and the people's government since liberation, astronomy research in China has recovered and developed rapidly. There are now five observatories (the Zijinshan, Shanghai, Beijing, Shaanxi and Yunnan Observatories), one factory (the Nanjing Astronomical Instruments Factory), and three observation stations (the Urumqi and Changchun stations and the satellite-tracking station in Guangzhou). There are also four universities (Nanjing University, Beijing University, Beijing Normal University and the Chinese University of Science and Technology) engaged in astronomy research as well as the training of research workers in the various specializations.

At present, Chinese astronomy is progressing in almost all aspects, and there are relatively few gaps. Systematic research is being carried out in the various specializations, and results of varying degrees have been achieved. We shall describe the many developments below.

I. The Manufacture of Astronomical Instruments

Astronomy is mainly a science based on observations and relies a great deal on sophisticated instruments. Before liberation, most of the astronomical instruments were imported, and the installation and maintenance greatly depended on foreign expertise. Much attention was paid to the manufacture of astronomical instruments after liberation, and cooperation between research institutions and manufacturing factories was encouraged. It is now possible to produce our own small and medium-size instruments, such as radio telescopes for wavelengths of 3.2 cm and 10 cm, and the 16-element compound radio interferometer at the Beijing Observatory. In fact, since the establishment of the Nanjing Astronomical Instruments Factory, most of the optical instruments needed were produced locally, including the 60 cm reflector, the 18 cm photospheric-chromospheric double telescope, the solar magnetograph, 43/60 cm and 60/90 cm Schmidt satellite astrographs, etc. A 2.16 m reflector is presently being attempted. Particularly noteworthy developments are:

(i) the photoelectric astrolabe employing photoelectric recording and reflection optics was successfully designed and produced by the Astronomical Instruments Factory. The effects of anomalous refraction within the instrument as well as that of atmospheric refraction were successfully eliminated by creating a vacuum within the interior of the astrolabe, thereby improving the instrument's precision. Based on the observational results obtained using the type II photoelectric astrolabe, the achieved precision was found to be superior to that attained by astrolabes produced abroad. At present, the type III photoelectric astrolabe is under design and manufacture. The factory has also been successful in producing the photographic zenith tube which employs a programmed automatic control observation technique and the photoelectric magnification technique. A vacuum is again used to eliminate effects of anomalous refraction. The precision level attained, matches that of similar instruments produced abroad.

(ii) The Astronomical Instrument Factory has also come up with new ideas in optical instrument design. A method for automatic optimization using computers was established, which allow the image quality of optical systems to be improved significantly.

Furthermore, a reflector with a 1.56 m focal length is now under construction at the Shanghai Observatory, in collaboration with other factories. A 60 m high precision porcelain mirror, manufactured at the Zijinshan Observatory, has been used for observations, and a 46 cm solar tower is being installed and tested at Nanjing Observatory. The peripheral equipment is also being improved; a photoelectric integrated photometer was produced by the Zijinshan Observatory. It employs punched tapes for output and can be combined with a computer to handle the observational data. The 3-channel photoelectric photometer made by the Beijing Observatory is also equipped with a special computer for data handling.

II. Astronometry

As work on time and latitude is closely related to the various constructions as well as communications in the country, developments in astrometric research have received much emphasis since the founding of the new China. Due to the united efforts of the Shanghai observatory and others, the accuracy of the Chinese time signal was within ± 0.0023 sec in 1963, comparable to the international standards. The accuracy was further improved to ± 0.0012 sec in 1975, using better observational data obtained with Chinese-made photoelectric astrolabes and atomic clocks. It should be mentioned that the difference between our results (obtained from 8 time-measuring instruments) and BIH, which employs 50 instruments around the world (except China), is much less than that between the USSR results (based on 30 instruments) and BIH. In addition to the short wave transmission of standard time-signals, a synchronized system for long wave transmission was also established at the Shaanxi Observatory.

In the areas of latitude variation and polar motion, polar coordinates have been calculated at the Tianjin Latitude Station since 1959, using our own observational data. Later, a study group involving the various observatories was founded to carry out systematic investigation of polar motion. A system of polar coordinates (JYD) based on the year 1968.0 was established in 1977, with an internal consistency of ± 0.01".

Based on the large amount of observational data from time and latitude measurements we obtained, improvements on the right ascension and declination of the common stellar positions were made at the Shaanxi, Shanghai, Yunnan and Beijing Observatories, and at the Astronomy Department of Nanjing University. This work laid the foundation for our further research on star catalogues, and also provided data for the international catalogues of common stars.

Work on photographic astronometry also developed actively. Photographic measurements using artificial satellites were actively pursued at Zijinshan Observatory, and measurements of stellar positions as well as proper motions of galactic clusters, associations and variable stars have been carried out at the Shanghai Sheshan Observatory for a long time. A lot of information was obtained.

A few other theoretical problems are under study at the moment, such as the irregular variations of the earth's rotation, long-term polar motion, the non-polar terms of latitude variation, and the computational methods of nutation, etc.

III. Celestial Mechanics

The early works on celestial mechanics in our country were mainly concerned with the motion of asteroids (the computation of preliminary orbits, perturbations, improved orbits and identification of minor planets) and many new asteroids were discovered. Calculations on the perturbations and refinements of many asteroids' orbits were carried out using numerical and perturbative methods at the Zijinshan and Sheshan Observatories. In addition, orbits of many comets were also determined at the Zijinshan Observatory, and three new comets have been discovered.

Work on almanac astronomy, one of the main branches of celestial mechanics, was started just after liberation. This, however, was largely confined to the compilation and publication of astronomical almanacs. The Chinese Astronomical Ephemerides was first published in 1970, based entirely on our own calculations started since 1957. Apart from this, theoretical investigations of the motions of major planets and their moons, solar and lunar eclipses, as well as other relevant computational programmes were also carried out.

The observation of artificial satellites and theoretical studies of their orbits were initiated at the Zijinshan Observatory in 1958, and a forecasting centre was established later. As observational precision gradually improved, theoretical calculations of the statellite orbits were continually refined. In recent years, the second-order perturbation theory including all periodic terms up to second order and all secular terms up to third order, has been formulated independently at the Zijinshan and Nanjing Universities, providing a basis for more accurate predictions of satellite orbits.

There was little research work on the fundamental theory of celestial mechanics before 1970, an example of which is the study of the convergence of perturbative expansions. The quantitative study of the three-body problem has been carried out in the Departments of Astronomy and Mathematics at Nanjing University and the Chinese University of Science and Technology since 1972. The topological structure of the M_8 manifold of the general three-body problem has been investigated, and the necessary and sufficient conditions for the motion of a three-body system in phase and configuration spaces was obtained. These results were also applied in the study of the limit of inclinations and latitude variations with respect to the relative motion of three-body systems. Stability of the numerical methods involved was also studied at the Department of Astronomy at Nanjing University.

IV. Physics of the Solar System

A number of studies were carried out at the Zijingshan Observatory, which included the motions of celestial bodies in the solar system and the location and photoelectric photometry of asteroids (which provided data for the determination of the asteroids' sizes, shapes and rotational axes). Some important data were obtained on Uranus' rings in cooperation with the Beijing Observatory. The radiative transfer theory of planetary rings was also studied through considerations of multiple scattering effects during occultation. In the study of comets, Mr. Kimura and his coworkers investigated the structure of comet tails by means of a 3-dimensional theory. Their results were successfully applied to the comet Arend-Roland, as well as many other cases.

Investigations on the origin and evolution of the solar system were carried out in the Astronomy Dept. of Nanjing University and at Zijinshan Observatory. An original hypothesis was suggested by the late Prof. Dai Wensai on the basis of a careful analysis of the existing hypotheses and a series of theoretical investigations. This hypothesis was able to describe in a comprehensive and self-consistent way, the origins of many of the characteristics of the solar system and formation of the various celestial bodies. It also allowed an analysis of the general process of formation of the solar system. Some new viewpoints were put forward and a number of calculations were made with respect to many problems, especially the structure of nebular discs, the anomalous distribution of angular momenta in the solar system, the Titius-Bode law, the origins of the rings of Jupiter, Saturn, Uranus and their satellites, the origins of asteroids, Pluto and its satellites, etc.

V. Solar Physics

Observational studies of the sun started in the 1950's. Early studies included monochromatic and spectral observations by means of horizontal solar telescopes at the Zijingshan Observatory and the Sheshan station, and sun-spot observations at various stations. Observations of solar activity and the chromosphere were carried out when the chromospheric telescope was installed at the end of the 1950's. In the 1960's, a solar telescope was installed at the Beijing Observatory, and a solar spectrograph with multiple wavebands at the Zijingshan Observatory. Radio telescopes with wavelengths of 3.2 cm and 10 cm were also used for solar observation at these observatories. In 1968, the Beijing observatory conducted the first absolute measurement of solar radiation in the waveband between 0.6 – 2.5 microns at Mount Qomolangma at an altitude of 5000 meters above sea level. This work, together with solar eclipse data, laid the foundation for the study of solar physics in our country.

The numerous daily observations of the sun were useful in the forecasting of solar activities. Based on its own observations as well as those from the Zijingshan and Yunnan Observatories, as well as other stations, the Beijing Observatory published the "Chinese Solar – Geophysical Data" and engaged in predicting the safe periods and periods of major eruptions.

Two major solar eclipse observations were organized in China, both with much success. The first one was conducted in 1958 by the Zijingshan Observatory, the Astronomy Dept of Nanjing University, Beijing Observatory and the Soviet Union, and was carried out at Hainan Island. Data on the optical and radio observations were published in "Acta Astronomica Studies". The second observation was made in 1968 at Xinjiang, by a national expedition consisting of scientists from the Beijing and Zijingshan Observatories, departments of Astronomy at Nanjing University and Beijing Normal University, as well as a number of other institutions. The major accomplishments included the following: (i) corona and coronal condensations were photographed, and much information on the distribution of brightness, electron density and temperature as well as the polarization measurements at a wavelength of 3.2 cm; the various physical parameters of the sun's local radio sources were also obtained.

The Yunnan Observatory accumulated quite a lot of data on the fine structure of sunspot activity regions with high resolution. It was discovered that sunspots with penumbral fibres arranged anomalously and rotating abnormally are particularly significant in the study of solar flares. A statistical correlation between sunspots and solar flares was also obtained, based on the analysis of the physical shapes and magnetic field structure of the sunspots.

In the theoretical area, subjects such as solar turbulence, convection, granulation, flares, nominences, spectra, solar limb darkening, the relationship between luminosity and motion in solar flares and the statistical analysis or solar activity were studied during the 1950's and 1960's. Since the seventies, the main work was on the problem of the stable constraints of electrons in type–I radio bursts studied at Beijing University and the Beijing Observatory; the investigation of the production mechanism underlying the U-type spectrum of type-IV solar radio bursts at Nanjing University and the Zijingshan Observatory — it was found that the spectrum came mainly from synchrocyclotron radiation of non-thermal electrons and the cyclo-resonance absorption of thermal electrons. Scientists at these institutions also studied surge dynamics, and an explanation was obtained for the main dynamical characteristics of the rising accelerative and decelerative phases. The heating problem of the chromosphere–coronal transition region was also discussed on the basis of a theory of interaction between waves and particles. A non-stationary theory for the energy state function of high energy electrons in the solar pulsating X-ray and γ-ray burst sources was put forward by scientists at the Zijingshan Observatory. This theory formed the basis for studying the continuous spectrum in the range of X-ray and γ-rays, and provided some information on the acceleration of particles in the flare region. Attention was also given to the investigation of line formation in magnetic field in the sun, the magnetic gradient in sunspots, as well as the coupling process between the magnetic field and the fluid during sunspot evolution.

VI. Stellar and Nebular Physics

There was much research work on stellar physics after liberation, some of which are summarized below:

Observational Work: A new type of variable stars — thermal super-short period cepheids in the globular clusters M15 and M79, was discovered using the 60 cm reflector at the Zijinshan Observatory. The period of this type of variable stars was less than 0.1 days, with amplitide greater than 0.2 and a spectrum of type A. On the colour-magnitude diagram, they are located on the left-side of the cepheids' unstable belt.

In recent years, the Beijing and Zijinshan Observatories have also used their 40 cm double refractor 60/90 cm Schmidt telescope to observe the variable stars in tbe Oph-Sco. region, and discovered some nebular variables and three flare stars.

The Zijinshan Observatory started work on the photoelectric photometry of the variable stars in 1959, using its 60 cm reflector. Similar work was also carried out at the Beijing Observatory, and since 1975, photoelectric photometry of close binary stars was carried out using the 60 cm reflector at the Xinglong station.

Work on the stellar spectrum began in 1962, and spectral studies of the celestial bodies such as visible binary stars, novae, symbiotic stars, etc. were carried out at the Zijinshan and Beijing Observatories. Many physical parameters of these objects were obtained. Theoretical Work: Before the 1970's, there was little theoretical research work in this area. The main efforts were directed towards the studies on the interval structure and evolution of stars, the period-luminosity relation of cepheids (Zijinshan Observatory); the statistical investigations of several spectral type stars (Department of Astronomy, Nanjing University). Since the 1070's, theoretical work in the field has been given greater attention at the various institutions, and many research results have been obtained.

In the study of cosmic γ-ray bursts, Nanjing University and the Zijinshan Observatory developed the flare model of the magnetic white dwarf, and proposed a model of energy production from the annihilation of neutral sheet magnetic fields in the local activity sources of magnetic white dwarfs. A model of the turbulent accelerating mechanism of plasma was also suggested. The roles played by universe Compton scattering and synchrotron radiation in γ-emission were also explored.

A statistical study of the pulsar was carried out at Nanjing and Beijing Universities. The main results were:
(i) On introducing the idea of magnetic attenuation, there was a better statistical relation between the rate of loss of rotational energy and the time parameter T. The ratio of the moment of inertia I to the square of the vertical component of the magnetic momentum μ_0^2, and the time scale of magnetic attenuation were determined.
(ii) the radio luminosity was independent of the surface magnetic field, but was closely related to the magnetic field at the distance of a light-velocity circle.

Based on the theory of abnormal nuclear state by Professor Li Zhengdao (T. D. Lee), the research scientists at the Chinese University of Science and Technology and Nanjing University, in their studies of neutron stars, suggested that these a new type of compact star might exist — abnormal neutron stars — with their mass reaching 3.2 solar masses (the greatest mass of normal neutron stars is about two solar masses).

At the Zijinshan Observatory, the researchers studied the coupling between low frequency waves and charged particles in rotating magnetic neutron stars as well as its radiation, and explored the possibility of propagation of low frequency waves in relativistic electron gas. The study of X-ray binary stars was also carried out at the Chinese University of Science and Technology and a number of other institutions.

The turbulent-convection theory of pulsating variable stars proposed by the Zijinshan Observatory constituted an improvement over previous theory. Further efforts were also directed towards the study of non-conservative convection in the case of a pulsating velocity field.

Kimura and Lui Caipin, working at the Zijinshan Observatory, studied the polytrope and its stability. When the polytrope's exponent n was any real number not equal to -1, and results obtained were useful for the exploration of nebular evolution. A new method was suggested by scientists at the Beijing Observatory for solving the luminosity curve problem of ellipsing variable stars, which partially overcame the convergence difficulty in Kopal's interative method.

The distribution on the Russell diagram and the luminosity function of white dwarfs were studied at Beijing Normal University. The X-ray radiation of supernova traces was studied at the Beijing Observatory and Beijing University. The spherically symmetric model of the gauge theory of gravity with black holes was pursued at Fudan University and other institutions.

VII. Galaxies and Cosmology

The main work before 1970 involved the determination of the galaxy's rotational parameter A by means of the statistical analysis of cepheids and the 21 cm hydrogen line, as well as the origin of the rotation. In addition, there was some study on galactic dynamics at the Beijing and Shanghai Observatories. The study of galaxies has drawn much of the attention of our workers in astronomy and mechanics since 1972, due in part to the discovery of quasars and the lectures given by Professor C.C. Lin during his visit to China.

Many important results were obtained in the field of density wave theory of spinal galaxies. These included:

(i) The amplitude distribution of density waves was obtained through the uniform effective solution with second order asymptotic approximation, and provided a maintenance mechanism for density waves (Beijing Observatory);

(ii) an understanding of the influence of star formation on density waves was obtained. It was realized that star formation could give rise to instability in density wave modes, and became a mechanism for exciting and maintaining galactic density waves (Chinese University of Science and Technology);

(iii) in the study of the evolution process of galactic density waves from the unstable stage to a quasi-stable stage, scientists at Beijing University pointed out that density waves were produced near the unstable region of the conotation circle, and gradually twined to form tractional short waves which then propagated towards both sides of the conotation circle with a group velocity after reaching the quasi-stable stage;

(iv) the loose rolling density wave on the symmetric plane of a disk with finite thickness was studied at the Institute of Mechanics. It was found that there was a significant difference between loose rolling and tight rolling density waves, and that loose rolling linear density waves with real wave numbers were always unstable. In addition, studies on other stability aspects of density waves and galactic self-justified shock waves were also carried out;

(v) the solution of the Poisson equation of a disk of finite thickness was studied at Nanjing University, and the Zijinshan and Yunnan Observatories. The effect of thickness was investigated, and the spinal structure model of disk galaxies with finite thicknesses was pursued. Furthermore, the velocity of dispersion of spinal galaxies was studied, and a numerical value obtained for the Milky Way which was in accordance with observations.

Statistical research on quasars with radio components in relation to other cosmological subjects has been carried out at the Chinese University of Science and Technology since 1975. The main results obtained include the following: quasars could be classified according to the distance D between their radio components; the relation between their redshift Z and distance r was found to be

$$r = \frac{c}{H}(Z - 0.19Z^2)$$

where c is the velocity of light, H is the Hubble constant; the deceleration parameter q_0 was found to be greater than 0.5, implying a closed universe.

The Zijinshan Observatory carried out morphological analysis and statistical research on two kinds of Seybert galaxies. A common character among them was discovered, namely that the spinal arm was incompatible with the eruption phenomenon It was also found that the Seybert galaxies could be divided into several sub-categories, and the relation between the Seybert galaxies and the other active galactic nuclei on the two-colour diagram was discussed.

Some work was done on normal galaxies with statistical analyses of the form, mass and angular momentum of the galaxies, providing data for further research on the classification of galaxies (Nanjing University); study of the spatial distribution of galaxies and radio sources. Some results pointed to a possible inhomogeneous distribution in space (Shanghai Observatory). Other works include investigations on the relation between supernovae and the type of galaxies (Beijing Observatory); studies in gravitational theory and cosmology (Chinese University of Science and Technology).

VIII. History of Astronomy

Research in this area received much attention soon after liberation, and some efforts were directed towards the editing of ancient Chinese calendars and reports on astrology of particular interest were the ancient records of supernova explosions.

At the end of 1974, under the sponsorship of the Chinese Academy of Sciences, the Editorial and Research Group on Chinese History of Astronomy was established with the joint effort of many institutions in the country, and a systematic programme was set up to compile and edit historical records on astrology and cosmology, astronomical maps and tables, and descriptions of ancient observational instruments. Among the compilations were the two huge volumes, "A General Catalogue of Astrological Records in Ancient China" and "A Collection of Materials pertaining to the History of Astronomy in China," taken from Chinese historical compilations and local chronicles. A volume entitled "An Atlas of Historical Relics of Astronomy in China," which represented a more complete collection of the various astrological atlases in every historical period in China, was also published. A companion volume, "Studies on the Atlas of Historical Relics of Astronomy in China," contained many research papers on the subject.

Most of the articles on special topics on the Chinese history of astronomy were collected in the volumes: "Studies on Chinese History of Astronomy" and "Studies on the History of Science and Technology," and a few were published in "Acta Astronomica Sinica." Some papers on cosmology were collected in "Chinese Ancient Cosmology." The Institute for the History of Science also compiled "History of Astronomy" in China and "A Concise History of Astronomy in China."

There were also some works on the history of astronomy in China using modern scientific methods. For instance, through computer study of the orbital evolution of comet Halley, Professor Y. C. Chang of Zijinshan Observatory was able to fix the year of a punitive war waged by King Wu Wang against King Chou as 1057 B.C. The conclusion that the 11-year solar activity cycle persisted throughout the ages was made through the investigations at the Zijinshan and Yunnan Observatories, after an analysis of ancient historical data.

The brief description given above indicates that research in Astronomy in China has developed in the various areas, and that some outstanding results have been obtained in recent years. Despite the efforts, our progress and achievements are rather limited, when compared with the advanced levels in the world today. In the coming years, we shall devote ourselves to bridging this gap, and strive to achieve the same level of excellence in Astronomy.

CHAPTER 13

ON THE ADVANCES AND DEVELOPMENTS OF WEATHER PREDICTION IN CHINA

by Shu Jiaxin

The main task of meteorology is to improve the accuracy of weather prediction, the advances and developments of which form the foundations for the atmospheric science. After liberation, a powerful contingent was formed in China by combining meteorologists with amateurs. A national meteorological network has been established since 1958 with forecasting offices in every sub-provincial region and weather stations in every county. Their meteorological activities, therefore, have broad mass foundations and rich and unique experiences have been obtained in weather forecasts. A brief summary of these activities and experiences are summarized here.

I. Weather Forecasts with Synoptic Charts

The method for weather forecasts with synoptic charts has been used for more than 100 years. With synoptic charts, the synoptic situation can be physically deduced from the laws of atmospheric motion. Nowadays the application of dynamic meteorology, atmospheric thermal dynamics and soundings from radar and satellite has revived the old synoptic method showing great potential.

1. Short-range forecast

(1) Precipitation area forecast

Precipitation area forecast is a kind of qualitative forecast recently used in China. The physical concepts behind this type of forecast is well understood and the methodology involved is simple, and is widely used in difficult precipitation forecasts. The foundation of the method can be explained in terms of atmospheric energy. The production, development and dissipation of various weather systems in the atmosphere and the whole process of severe weather events are all accompanied by the natural conversion of energy. It has been long in the history of meteorology that many meteorologists studied the synoptic variations in terms of atmospheric energy. However, the Chinese meteorologists were the first to actually adopt these results in practice.

The energy of a single air pocket consists principally of thermal heat energy, latent heat energy, potential energy and kinetic energy. According to the law of energy conservation, it would be presented as

Total energy = thermal heat energy + latent heat energy + potential energy + kinetic energy = constant.

To directly calculate the total energy from conventional observations, a total temperature T_t, equivalent to the total energy, is introduced.

The kinetic energy term can be neglected when wind velocity is less than 30 ms^{-1} but this is rare in nature. The latent heat term can be easily read from a ready table. By knowing the altitude of a station (or the atmospheric levels) as well as its temperature and dew point, the total energy is readily calculated. The total energy or the total temperature T_t of the air is a quasi-conservative quantity in a constant adiabatic process without viscosity and friction, and can be used as a means of tracing the particles in moving air masses.

In the case of heavy rain forecast, when a high energy region is present above the surface of the earth, the air will appear to be less stable than the surroundings and therefore have more accumulated unstable energy. If some circulation systems and favourable situations to produce the heavy rain are also present as synoptic background, the air would release much unstable energy to form an uprush and then to produce the heavy rain. Thus the total temperature T_t can be used as an important parameter for the precipitation-area forecast of heavy rain. Meanwhile, moisture is also a major factor parameterized by K. A small K signifies deep moist layer whereas when K is sufficiently large, then the air will contribute to the formation of heavy rain. The rising motion W is the third parameter which can be computed by electronic computers or obtained using graphic methods. The three parameters are then computed and their isolines analysed for a large area. The superposing area of the three isolines T_t, K, W (Fig. 1) is defined as the precipitation area of heavy rain.

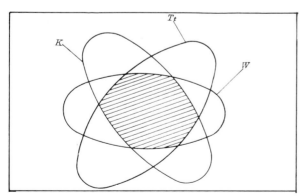

Fig. 1. The precipitation area (shaded part) of heavy rain superposed by three isolines.

Energy as a main parameter can also be applied to the precipitation area forecast of strong convective weather events like hail. From the energy view-point, the precipitation of hail is a process of production with subsequent accumulation and release of vertically unstable energy. In most of the cases, the production process of the vertically unstable energy is a process such that the total energy at lower level increases and that at upper level decreases, or the total energy at both levels increases but with more increments at the lower level. It is a commonly encountered process for cold and dry air flow from north-west at upper level to mix with wet and warm air from south at lower level in a same atmospheric column. This causes the formation of vertically unstable energy, leading to the occurrence of hail. The precipitation-area forecast of hail can be made with the graphic method in which superposing regions of high energy and other quantities are drawn.

(2) Diagnostic forecast

The method of displaying the principal factors effecting weather variations and to deduce qualitatively their trend is named diagnostic forecast. Since the 1960's, particularly with the development of radar meteorology and satellite meteorology, the method has entered into a new stage in China, with various tools and methods of analysis. The forecasting procedure is:

(i) to examine the origin and course of development of the variations of weather systems from successive surface charts (mostly local weather charts),

(ii) to observe the characteristics of pressure, temperature, humidity and cloud variations associated with weather systems from hourly surface reports and station curves, and

(iii) to monitor weather systems with synchronous meteorological satellites and radar, particularly for severe weather events with short life-times.

System genesis, motion, evolution and dissipation are clearly seen with SMS pictures at intervals of 20–30 minutes and with radar tracing or radar pictures. A diagnostic analysis can thus be made with radar or satellite images along with the information described above. For example, a squall line, which is accompanied by strong winds or heavy rain, is often a strong convective line, consisting of thunderstorm complexes. By merging into each other, the thunderstorm cloud clusters often appear as large bright cloudbands on satellite pictures. When the shape of the squall line tends to be enlarged at the head and very narrow at the tail, the latter is most conducive in making new thunderstorms, and the swelling head is only a weakening system. The line activities can also be clearly shown from the successive analysis of radar echoes.

As the diagnostic forecasts are usually effective only for short time ranges, they are often applicable to warn of disastrous weather events, like typhoon, heavy rain, tornado and hail. Its objective basis and operational simplicity are also added advantages. Diagnostic forecasts are widely used in China now.

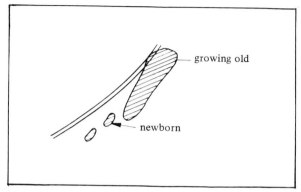

Fig. 2. Schematic diagram illustrating the cloud-band of squall line.

(3) Synoptic dynamic model forecast

In the dozen years or so after the 50's, Chinese forecasters with their rich experiences in weather forecasts developed operational weather models on the basis of physics and synoptic characteristics. This classical method of China is simple and flexible in use but semi-empirical. Therefore it is difficult to formulate with rigorous mathematics. Since the 70's, effort has been made to integrate the theory in dynamic meteorology with synoptic practice to improve the accuracy of weather forecasts. This method is considered as one of the best methods now. Taking heavy rain forecast as an example, equations of atmospheric motion are often used to analyse the causes of heavy rains, while the perfect equation of vertical motion (ω equation) is used to calculate the vertical velocity of the flow, moisture flux and thermal stability of the atmosphere. By analysing the atmospheric motion at different scales and by combining the synoptic analysis with atmospheric dynamics and thermal analysis, a forecast model is thus developed according to the characteristics of weather systems. This is the technique used to predict heavy rains by understanding the physical process of the formation of heavy rains.

2. Mid-range forecast

Mid-range forecast aims at studying the evolution of a weather process for a longer period (3–10 days).

In other countries, the mid-range forecast still lacks effective and fully-developed methods. Since the establishment of New China, the meteorologists have made efforts to develop their unique technique of mid-range forecast, i.e., to understand the evolution of the general circulation and stable weather systems and to make full use of the successive variations of weather elements from historical records as well as from current observations based on synoptic charts which represent the mean and instantaneous large-scale weather processes.

(1) Basic tools

Two tools are mainly used for the mid-range forecast.

(i) Using synoptic charts, including mean and daily charts. Mean weather charts use 5- or 3-day averaging charts or daily space-averaged charts. They filter the negligible troughs, ridges or small weather systems and reveal the essential troughs and ridges that would reflect the mid-range weather variations.

Daily weather charts are used to analyse the evolutions of severe weather systems and along with the mean charts, to understand fully the mid-range and short-range weather variations.

(ii) Using historical climatic data and currently observed evolutions of local elements. The successive variations of parameters of a station often reflect the local weather systems. The annual historical data can reflect the general regularity of local weather variations while the currently observed weather evolutions can describe the fluctuating weather variations. Combining all these data with synoptic analysis, the features of a mid-range weather process of a station can be described on the whole. This technique, widely used in China, provides a unique method for mid-range weather forecast which is quite simple and reliable and is applicable to all weather centres and stations. For instance, in the lower Changjiang (Yangtse) River Valley before the occurrence of prolonged rainy weather, such variations of the parameters are often observed: the continuous increase of temperature and humidity, the decrease of pressure, and the presence of several days of strong south-east winds. Under such conditions, a good mid-range forecast can be easily produced if the characteristics of cloud variations and the large-scale weather background have been analysed in detail.

(2) Basic methods

There are so many mid-range forecast methods that cannot be fully described in detail. Here the author will introduce three of them.

(i) The wide application of the long-wave theory of the atmosphere: Disturbances on various scales are actually present in the atmosphere. A disturbance will be considered as a fluctuation when its life cycle and amplitude remain constant. There is a broad wave-like westerly air mass in upper atmosphere over middle and high latitudes. The wave trough and crest correspond to the low pressure trough and high pressure ridge respectively. Waves of longer length, higher amplitude, slower moving and longer life cycle generally can remain over 3–5 days. In addition, the super-long wave can be regarded as the largest fluctuation in a large-scale synoptic process and its space and time scales are longer than the long wave. In a process of 5–7 days, the manner of the action of the super-long wave dominates the whole mid-range weather process. Long wave and super-long wave, which are shown in t-day mean charts or daily space-averaged charts, interact with each other. When the circulation pattern is adjusting, the super-long wave pattern is often found to have large adjustments first and after 1–2 days the long wave pattern would also change subsequently. The regularity of super-long wave action is widely used for mid-range forecast in meteorological departments in China now. In winter time, a large-scale cold wave event is often a process of formation and development of a super-long wave trough at the east coast of China. If the super-long wave is relatively stable at the coast and the northwest flow is prevailing over the continental part, the occurrence of the cold wave will result in several fine and clear days (equivalent to a mid-range weather cycle).

(ii) Circulation index: The large-scale weather process at middle and high latitudes appears often with zonal and meridional circulations changing alternately. A circulation index is often used to tell the difference between the zonal and meridional circulations.

$$I = \frac{\sum_{1}^{n}(\phi_{45} - \phi_{65})}{n}$$

where I is the westerly index, ϕ_{45} and ϕ_{65} are 500 mb potential heights at 45°N and 65°N respectively, n is the number of sampling points.

When I is large, the zonal circulation is generally dominating smoothly with little exchange between north and south flows. When I is small, the meridional circulation is strong and there are frequent exchanges between north and south flows.

The above-mentioned long wave theory and the circulation index method are used together with the curves of station parameters to develop regular mid-range weather forecasts.

(iii) Weather models: Weather models is one of the most widely used techniques since the 1950s. From the examination and analysis of large numbers of actual weather evolutions and other meteorological data, the models summarize the patterns of the occurrence and development of mid-range weather events and systems to reflect, to a certain extent, the pattern of weather variations. For example, the Changjiang Valley has cold rains in spring time every year. The mid-range weather model of the cold rainy weather is shown in Fig. 3, composing of a south front and a north front (the maximum pressure temperature gradient bands due to the convergence of cold and warm air masses, often used in upper air isobaric charts). The north front extends from Siberia into the Yellow Sea through Mongolia and North China. The south front extends eastwards from the southern Tibetan Plateau to the Lower Changjiang and then into the sea, traversing West China. The joining of the two fronts at the southern Yellow Sea results in a strong frontal region. The northern cold air coming along the north front and western warm and wet air along the south front converge in the middle and lower reaches of the Changjiang. Thus, a quasi-stationary front is consequently formed along the Changjiang, which produces perturbations unceasingly, leading to the maintaining of the continuous low temperature rains in the middle and lower reaches of the Changjiang.

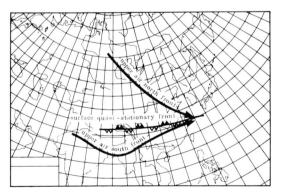

Fig. 3. Illustration of the weather model of spring successive rains in the middle and lower Changjiang Valleys.

3. Long-range forecast

(1) To prepare a good long-range forecast, it is necessary to further master the regularity of the general circulation and long-range climatic variations and to study the factors affecting the long-range weather events in various aspects. Now, the 500 mb

mean circulation charts are commonly used to examine the antecedent variations of weather systems. In some localities, the seasonal variation of the general circulation is analysed based on the natural synoptic period. Some communities study the long-range weather events which have strong influences on farm activities in various regions, such as the autumn cold dew flow in South China and the summer low temperature events in Northeast China. Some institutes study the long-range trend of drought and flood. There are so many methods for long-range forecast that they cannot be described in detail here. Taking the long-range forecast of drought and flood in Shanghai region as an example: by analysing the general circulation and the long-range trend of major weather systems, the forecasters lay special stress on analysing and studying the trends of weather systems that would lead to droughts or floods in the lower Changjiang Valley in summer. A first detailed analytic comparison between the weather systems of the severe flood year of 1954 and that of the great drought in 1967 was made. The extreme events were caused by anomalous circulations. It is the interaction and restriction among the key regions in Europe and Asia that results in the anomalous circulations. From the synoptic analysis and tracing of temporal and spatial variations of the height departures of the key regions in June–July with 500 mb monthly mean charts in the Northern Hemisphere, the related conclusions of the cause-effect activities are made. On the basis of this regularity, the height departures of the correlation fields with higher correlation coefficient are selected from key regions on 500 mb monthly charts as predictors. Therefore, trend forecast of drought and flood in the flood season is developed from the forecasting equations which use data of December and January–April respectively.

(2) Sea temperature has been used as a physical factor for long-range forecast since the early 70's. As ocean and atmosphere form a complex feedback system, the study of sea-air relationship is helpful in understanding the regular annual activities of long-range weather events, such as the subtropical high pressure region in North Pacific (110°E–115°W, north to 10°N) which greatly influences the summer rainfall in China. In particular, the west part of the subtropical high pressure region (110°E–180°E, north to 10°N) has direct influences on the long-range weather pattern. In recent years, the Shanghai Weather Center, cooperating with the Geographical Institute, Academia Sinica, used the Pacific autumn temperatures in forecasting the sub-tropical high pressure region in North Pacific of the following year with satisfactory results. Results show that if the autumn and winter sea temperatures near the equator and off California are higher, the mean ridge line of the North Pacific high pressure region next summer will be located rather southerly and the northern displacement to over 20°N will also occur later, and vice versa. Besides, the main factors which affect the northern boundary of 588 geopotential meter line of the June subtropical high are the autumn kuro-shio, as well as the sea temperature off California. When the equator sea and kuro-shio temperatures are higher, while the sea temperature and pressure in the westerly region are sub-normal, the Pacific subtropical high is likely to extend to the west, and vice-versa for opposite effects. With the results of the above analysis, a set of regression equations has been set up for the long-range forecast of the western parts of the subtropical high pressure-regions.

(3) Soil temperature variation was introduced in long-range forecast in the 1970's as a new predictor. Practically speaking, the main factor which influences the weather change is the thermal condition of the earth's surface. The conflict between cold and warm air masses reflect the thermal condition on the surface. Soil is a heat container which accepts heat from solar radiation and the inner of the earth itself. The antecedent general circulation has its influences on the ground thermal condition which, in turn, affects the later weather. The analysis of soil temperatures in various areas in China indicate that the order of magnitudes of ground thermal variations at various periodicities agrees well with those of energy required by mid-range, long-range and super-long-range weather variations. Therefore it may be assumed that the departure of ground temperature is an important factor which influences the weather and precipitation. Soil is a good filter too. Weather variations of short periods affect the soil temperature in a shallow layer, while only fluctuations of longer periods effect the deep layer. According to the period of soil temperature fluctuations, soil is divided into four layers: surface layer, shallow layer (20–40 cm), middle layer (80–160 cm) and deep layer (below 320 cm). Conditions in middle and deep layers are affected by long-period natural weather, on the average, for half a year or longer. Therefore, the variation of heat storage in middle and deep layers is an energy index in weather forecast for half a year or longer. With this index, we can forecast the temperatures and precipitations for a season, half a year or even a year.

II. Weather forecasts in county weather stations

Weather forecast in a county weather station is called supplementary forecast, i.e., the county station can issue its local forecast which supplements or corrects the weather forecast from weather forecasting offices, if necessary, according to the practical requirements of production. County weather stations can now carry out short-, mid- and long-range forecasts independently and effectively. In these stations many effective forecasting methods developed by combining the mass experiences with new techniques, local parameters with synoptic background and practical experiences with objective information. The methods can be classified into two major types: (1) The method based on the analysis of charts and tables of local parameters has been used since the late 1950's. This type of technique takes into account the weather-observing experiences of the labouring people, together with examining historical records with regard to the scattering figures, curves and composite time cross-sections of meteorological elements as main bases to reflect the physical characteristics of local weather systems, the geographical influences and the small-scale systems that cannot be shown in synoptic charts. From a chart with various weather elements, it is very easy to find out clearly the relationship between the antecedent element variations and the forecasting objects by using the composite time cross-section and thus develop forecasting models of cold successive rains, heavy rain or hail. (2) The technique based on probability statistics. Predictors are selected for the statistics from the experiences of the labouring people and weathermen. With different numerical models, large quantities of data are statistically processed and computed to develop (different forecasting) formulas and criteria for different weather systems. The general model is

$$y = a_0 + a_1 x_1 + a_2 x_2 + \ldots + a_n x_n$$

where y is the predicted quantity such as precipitation and temperature, a_0 is constant, $a_1, a_2 \ldots a_n$ are regression coefficients which can be statistically determined from historical data, $x_1, x_2 \ldots x_n$ are predictors which can either take the initial parameters or the antecedent variates. From the above formula, the predicted quantity can be estimated when x_i values at the forecasting time are known.

The method of statistical forecasting in county stations is characterized by its flexibility, absence of the need to perform complicated calculations, simple operation and repeated uses. There are also many other interesting methods of daily use in these stations. Examples are given below:

1. Cloud observing

China is one of the earliest countries in using observation of the clouds for weather predictions. The evolution of clouds is closely related with weather systems so that clouds can be, to a certain extent, the manifestation of weather change. Now in many weather stations, the evolution of clouds is regarded as part of a composite synoptic chart comprising daily and hourly variations of clouds, meteorological records, synoptic background and mass experiences. For instance, in the middle and lower reaches of Changjiang, the occurrence of *cirrus spissatus (cirro-filum spissatus)* over south, associated with upper air jets between spring and summer, indicates that it is just before the onset of the plum rains (rainy season), and that the upper air jet is just over the middle and lower Changjiang Valley. Once the jet leaps northwards between 35–40°N, it indicates the north boundary of the subtropical high rises to the south of the Changjiang, the frontal system of plum rains will be set up consequently in its lower reaches. It is apparent that cirrus spissatus is an indicator of the onset of plum rains. Another example is in Baoshan County, Shanghai. The clouds of cucumber form (Fig. 4) termed as *altosumulus opacus* within *altosumulus translusidus* are good precursors of heavy rain. They occur in the east-to-southeast sky, 12–24 hours before heavy rain, ranging in a south-north direction.

Fig. 4. Cucumber form clouds As. tr. with As. op.

2. Animal behaviour observing

The observation of animal behaviour has a long history in China. It is demonstrated from practice that the unusual activities of animals like frogs, ants, fishes and snakes can sometimes indicate forthcoming weather variations. In Tianyang, Guangxi, there is a species of toads with activities of ovulation and fertilization in water that are closely related to the later weather conditions. Their cries before jumping into water means that the weather begins to warm. Every time after they jump into the water, successive warm and fine days will follow. In South China, the whistle of cicadas is also an effective indicator for weather forecast. When the moisture in the air undergoes a sharp increase, wings of cicadas get wet and cicadas will find it difficult to shake their wings to make sounds. Hence, on fine but rather sultry days, the sudden stop of cicada sounds implies the potential of an impending storm. And just before the stop of rain, as air humidity decreases, cicada wings again before dry and are capable of vibrating, emitting audible sound. That is why there is a proverb in South China that says "Hearing the cicada whistles in the rain foretells the coming of fine weather".

3. Proverbs and rhythms

The association between two weather phenomena with a certain time interval or between various antecedent weather conditions and some weather phenomena later on is termed a rhythm. The use of rhythms is one of the common methods in weather stations in terms of farming proverbs, such as that "A warm spring is followed by early plum rains and a cold spring indicates late plum rains", "If the peach blooming season is wet, the plum rains season will be dry", "As thunder is heard in the term of Waking of Insects*, flood will come in Grain Full**". The proverbs reflect weather rhythms. After analysis and evaluation with meteorological data, the proverbs are very useful in studying the association between weather conditions beyond a time interval and forecasting the future weather.

III. Statistical probability forecast

The statistical probability forecast or statistical forecast is used to reveal the rules of atmospheric action in terms of statistics. For instance, cyclones move along the isobars (so-called steering current) in warm sectors. This is only a statistical rule at first, and is physically explained by synoptic meteorology

---

* A year is divided into 24 solar terms in Chinese calendar. Waking of Insects is the third term in early March.
** The term of Grain Full is in late April. – tr.

later on. It is apparent that the probability can be used as a basis of formulating the physical essence of the atmosphere. The method is to discover the statistical rule from large quantities of random events; the larger the sample size, the more objective are statistical rules. The success or failure is often related with the selected predictors. Only when the predictors agree well with the physical rules can they reflect the formation and development of weather systems. Predictors of physical and synoptic significance must be selected. The statistical forecast is moving towards a new trend endowed with more physical significance as general circulation, atmospheric stability, sea temperature and other physical parameters are selected as predictors.

Now in China, the techniques to develop the forecasting schemes on the basis of selected predictors, including regression equations, statistical logic, discriminatory analysis, analogue principles, spectral analysis, random processes, etc., are well-developed. The most widely used one is the step-by-step regressive method. The regressive analysis involves determining how the conditional expected values of the dependent variable change with the independent variables. Using this relationship, the conditional expected values of the dependent variables (forecasting phenomena) are derived as estimated values (forecasting values) on the basis of the occurred independent values (antecedent value). For example, the direction of a typhoon movement is highly correlated with the intensity and position of the subtropical high and the direction of the westerly trough, and the observed ridge of the subtropical high and westerly trough can be used to forecast the future typhoon motion. Therefore, the regression equation is set up on the basis of the relationship between the subtropical high, westerly trough and the typhoon motion through the solution of linear equations by computers.

Now, meteorologists in China are applying the statistical method to the study of all kinds of linkages of related factors which affect the atmospheric motion in both time and space to help to explore the regularity of the atmospheric motion, so as to get a better understanding and to improve the reliability and accuracy of weather forecasts.

IV. Numerical weather forecast

Closed equations to describe the laws of atmospheric motion based on the fluid dynamical and thermal dynamical principles are solved by computers to derive the future state of the atmospheric circulation on a large scale and a definite period. This is the so-called numerical forecast. China began to use the operational numerical forecast in the middle and late 1950's with simple graphic methods. It is only since the early 1960's that the quasi-geostrophic barotropic model has been used, i.e.

$$\nabla^2 \frac{\delta\phi}{\delta t} = J(f^{-1}\nabla^2\phi + f, \phi)$$

where ∇^2 is the Laplacian, J the Jacobian, ϕ the potential height, and f the Coriolis parameter. From the above equation the values of $\delta\phi/\delta t$ can be obtained and ϕ can then be evaluated using the difference method.

Because the barotropical model does not include the conversion of internal energy and potential energy into kinetic energy, it cannot describe the formation and development of severe weather systems. In the middle 1960s, the quasi-geostrophic baroclinic models were developed on the basis of the barotrophic models. It takes into account the variation of the wind field with height, and introduces the temperature field and vertical velocity. This kind of model still deviates from the atmospheric motion to a certain extent in many aspects, so the primitive equation model was introduced in the late 1960's. Based on a sequence of coordinate conversions, the primitive equations are

$$\frac{\delta u}{\delta t} + u\frac{\delta u}{\delta x} + v\frac{\delta u}{\delta y} + \sigma\frac{\delta u}{\delta \sigma} + \frac{\delta \phi}{\delta x} - \frac{\sigma}{\pi}\frac{\delta \phi}{\delta \sigma}\frac{\delta \pi}{\delta x} - vf + F_x = 0$$

$$\frac{\delta v}{\delta t} + u\frac{\delta v}{\delta x} + v\frac{\delta v}{\delta y} + \sigma\frac{\delta v}{\delta \sigma} + \frac{\delta \phi}{\delta y} - \frac{\sigma}{\pi}\frac{\delta \phi}{\delta \sigma}\frac{\delta \pi}{\delta y} + uf + F_y = 0$$

$$\frac{\delta \phi}{\delta \sigma} = -\pi\alpha$$

$$\frac{\delta \pi}{\delta t} + \frac{\delta(\pi\sigma)}{\delta \sigma} + \left[\frac{\delta(\pi u)}{\delta x} + \frac{\delta(\pi v)}{\delta y}\right] = 0$$

$$\frac{\delta \theta}{\delta t} + u\frac{\delta \theta}{\delta x} + v\frac{\delta \theta}{\delta y} + \sigma\frac{\delta \theta}{\delta \sigma} = \frac{\theta}{C_p T}Q$$

where $\sigma = P/\pi$, in which π is the pressure at the defined level, α is the heating rate, F_x and F_y the frictional force, θ the potential temperature and C_p the specific heat of dry air at constant pressure.

Since the 1970's, many initial equation models have been developed for the convenience of operation from the above simple primitive equations.

Shanghai began its operational numerical forecast with three-level initial equation models on September 16, 1975. Daily upper air soundings at 0800h (Beijing Time) are used to develop the forecasts of the circulations at 700 mb and 300 mb ranging to 72 hours on a DJS-6 electronic computer. The whole automatic process from input of telegrammed upper air soundings at 1405 h every day to the output of numerical forecast charts achieves very good results. Generally marked variations and adjustments of the large scale synoptic situation at middle latitudes (35–50°N) can be forecast qualitatively and correctly. The results of the operational numerical forecasts have been primarily summarized for a 3-year period from October 1, 1975 to September 30, 1978. It is shown from the statistics that all of the 27 sampling forecasts of the ridge strengthening are good; 45 out of the 56 forecasts of trough deepening are good, 6 mediocre and 5 wrong; and of the 11 forecasts of the maintaining of the smooth circulation east of the Urals Mountains, 9 are good, 1 mediocre and 1 wrong.

V. MOS forecast

The Model Output Statistical forecast, shortened to MOS forecast, involved the direct introduction of the output of a numerical model into a statistical model. It established the statistical relationship between the antecedent (results of) numerical forecast and the observed weather condition. It is then used to develop weather forecasts. From many tests and operational uses in recent years, MOS forecast is regarded as a useful objective method. The forecast is graphically described as:

forecasting model (primitive equation model) → MOS → { precipitation, wind, cloud, severe storm, temperature }

The MOS method is now used to forecast the winter precipitation in Shanghai. The regressive equations of the model which selects predictors from nearly 10 thousand factors are of physical significance. Forecasts from independent data samples achieve satisfactory results and agree well with the actual empirical forecasts. The method can also provide objective quantitative forecasts.

VI. Statistical-dynamical forecast

Since the 1970s, China has begun to use the unique method of statistical-dynamical forecast. The model retains the physical significance of the atmospheric process described by the dynamical equations and, in addition, introduces the statistical treatment in processing the quantities that are difficult to determine dynamically. Successes have been achieved in the operational forecasts of typhoon tracks in Shanghai.

The computed and simplified dynamic equations for typhoons are as follows:

$$\frac{dx}{dt} - U = 0 \qquad \frac{dy}{dt} - V = 0$$

$$\frac{dU}{dt} - f_0 V = F_1 \qquad \frac{dV}{dt} + f_0 U = F_2$$

where x, y describe the position of a typhoon centre, U, V are components of the moving speed of the typhoon center, F_1, F_2 the composite mean effects of all the forces on the typhoon track. These quantities cannot be determined dynamically, and have to be treated statistically. According to the principle of stepwise regression, the automatic screening of predictors, the set of regression equations is established to forecast the typhoon track.

It was shown that the 24 forecasting errors for typhoon tracks in Shanghai were within 100 nautical miles. Forecasts of the trend of typhoon motion are no worse than in other countries.

Summary and conclusions

(1) To increase the accuracy of weather forecast, it is necessary to constantly introduce new techniques from other sciences such as radio, radiosound, radar, satellite and computer technologies into weather forecasting. These new techniques should continue to contribute to the development of weather forecast.

(2) It is very important to strengthen atmospheric sounding and observation for more available data in studying the state of the atmosphere. At present, we should still make full use of telemetric facilities such as satellite, laser and radar.

(3) It is necessary to strengthen the construction of weather networks by combining the data communication system, radar observational system, data transmission system, local forecasting system and information servicing system to provide an automatic, timely and accurate service. This is quite important for short-range severe weather forecasts.

(4) It is also necessary to pursue basic theoretical studies. In the late 50's, the dynamical method was applied to operational weather forecast on a worldwide scale. With the continuous improvement of numerical models and the use of fine-grid models, the dynamical method has become a major tool in short-range forecast for about 20 years. In addition to improving the forecasting models and computing systems, it is also necessary to further understand the physical essence of weather processes, which needs further theoretical studies. Due to the lack of a solid physical foundation in mid-and long-range forecasts, the relationship between the atmosphere and sea, and the effects of surface state and geographic features on the thermal dynamical process of the atmosphere should be broadly studied. We should also constantly study the atmospheric thermal dynamical equations and enhance the mid- and long-range weather forecasts physically, objectively and quantitatively.

(5) The synoptic experiences should still be stressed. Now, a problem has been noticed in operational forecasts in other countries as to how to bring the practical experiences of the forecasters into full play under the condition of the increasing automation of weather forecasts. Otherwise it is hard to improve the forecasting level. In China, we must also give much attention to the synoptic experiences. It is noticed that numerical predictions should be integrated with synoptic experiences and calibrated with the experiences of the forecasters, a "man machine combination". Statistical forecast or dynamical statistical forecast should select predictors with physical significance, which are also related to the synoptic practice. Satellite pictures and radar observations are analysed together with the synoptic charts and practical experiences to improve the weather forecasts.

CHAPTER 14

NEW FEATURES OF THE EARTHQUAKE SCIENCE IN CHINA

by Mei Shirong

China is one of the seismically active countries in the world. During a severe earthquake, direct destruction would occur within only minutes or even seconds, and in an instant, mountains and rivers could be altered. An earthquake that occurred in 1556 in Huaxian, Shanxi Province was described in the annals of China: "In the winter of 1556 A. D. an earthquake catastrophe occurred in the Shanxi and Shaanxi Provinces. In our Hua county, various misfortunes took place. Mountains and rivers changed places and roads were destroyed. In some places the ground suddenly rose up and formed new hills, or it sank in abruptly and became new valleys. In other areas, a stream burst out in an instant, or the ground broke and new gullies appeared. Huts, official houses, temples and city walls collapsed all of a sudden ....." These words vividly depict the shocks, and unlike other natural disasters, are characterized by both drastic and concentrated destructions. Therefore, minimising effects of seismic disasters has become the main goal of earthquake research. In order to realize this goal, seismological research work must include three main aspects: earthquake prediction, seismic regionization and earthquake engineering. Among these three, prediction comes first. We achieved some objectives in earthquake prediction after the Xingtai earthquake, due to the efforts of specialists and the masses, and the employment of modern and indigenous methods. As the work of seismic regionization and earthquake engineering developed actively, we have also gained some useful experience. This paper will devote itself to the earthquake work in China after the Xingtai earthquake. The scientific research on the Xingtai earthquake has indeed played an extremely important role in the development of seismological studies in China. However, progress in any branch of science is a continuous process and cannot be completely detached from its past developments. For this reason, we shall first briefly recollect the history of seismology in China.

I. A Brief Historical Review

I.1 Historical facts about ancient seismological research

The Chinese people had begun to explore the possible clues for earthquake prediction since ancient times. "Shi Ji" (Memoirs of the Historiographer) has a saying: "Earthquake occurs when the morning star appears among the Scorpion". This was the first attempt to relate earthquakes with anomalous patterns in the orbits of planets. In the Eastern Han Dynasty, an outstanding scientist Zhang Heng invented the first seismoscope in the world. "Hou Han Shu" (History of the Later Han Dynasty) describes his invention of 132 A. D.: "It consisted of a vessel of fine cast bronze, resembling a wine-jar. Inside there was a central column capable of lateral displacement along tracks in eight directions, serving as the triggering mechanism. Outside the vessel, there were eight dragon heads, each holding a bronze ball in its mouth, while around the base there sat eight corresponding toads with their mouths open, ready to receive any ball which the dragons might drop." At that time, the theory of elastic waves was far from being developed. However, Zhang Heng already expected that the seismic waves could propagate very far, and used the inertia of an object as a mechanism for recording ground vibrations. What a great invention this was then! His seismoscope was set up in the city of Luoyang, Henan Province. In 138 A. D., a severe earthquake occurred between Lintao, Gansu Province and Minhe, Qinghai Province, some 1000 km away from Luoyang. The bronze ball dropped from one of the dragon's mouths into the corresponding toad's mouth. All the people at Luoyang appreciated the miraculous and mysterious power of this seismoscope.

After that, there appeared periods of strong earthquake activity in Chinese history, and the people suffered greatly. Consequently, the search for the causes of earthquakes was more actively pursued. During the period of 1550-1750 A. D., five seismic shocks with magnitudes over 8 occurred in the continental part of China: the magnitude 8 shock of 1556 in Guanzhong (central part of Shaanxi Province); the magnitude 8.5 shock of 1668 in Tancheng and Juxian, Shandong Province; the magnitude 8 shock of 1679 in Sanhe and Pinggu, Hebei Province and the magnitude 8 shocks of 1695 near Linfen, Shanxi Province and the magnitude 8 shock of 1739 in Yinchuan and Pingluo, Ningxia Province. The tremendous earthquake disasters made thousands of people homeless and caused untold sufferings. It was inevitable that the people should start to look for means of minimizing the destructive effects of earthquakes. According to "On Earthquakes" written by Qin Keda, during the period of aftershock activities of the great Guanzhong earthquake of 1556, the residents of Changan built their houses with wooden walls, so as to prevent collapsing houses from burying people alive. It is obvious that the Chinese people had already realised more than 420 years ago that structures made of soft materials like bamboo and wood are more earthquake-proof. "Local Records of Yinchuan" recorded: "Every year a number of small earthquakes occur in Yinchuan. The people there are used to them. Earthquakes are more frequent in the spring and the winter. When water in wells suddenly becomes turbid, sounds roar like cannon, dogs bark loudly, people should be aware of earthquakes. If in the autumn it rains more than usual, then in the winter, shocks are inevitable." On the basis of pre-seismic sounds, residents of Jinxian county, Liaoning Province, anticipated an earthquake in 1855, thus reducing loss of lives and property. The archives in the former Imperial Palace recorded: "567 huts and houses collapsed from quakes ..... . Before the quake, however, sounds like thunder were heard, and the residents rushed out their huts. Only seven persons were killed." These historical facts still have some value today.

I.2 "On Earthquakes" by Emperor Kangxi of the Qing Dynasty

Emperor Kangxi in his paper, "On Earthquakes", systematically discussed problems such as causes, stages, effects and spatial distributions of earthquakes. He wrote: "All earthquakes are caused by some accumulated 'Qi' (gas). The 'Qi' cannot pass smoothly and after prolonged accumulation underground it must burst out." In ancient China, 'Qi' usually meant a kind of objective material force. When such force is inhomogeneously accumulated underground for a long time, it would burst out causing earthquakes. This paper further asserted: "After earthquakes the accumulated 'Qi' is released, so that no strong shocks would occur again. However, when the 'Qi' is returning along the 'vein' of the Earth, there would be some obstacles which hinder the 'Qi' on its passage back. Only when all 'Qi' are in equilibrium would the earth calm down. So after strong shocks some further tremors would frequently occur, indicating the restoration of 'Qi'." These descriptions are close to the present general understanding of aftershocks. The energy accumulated for a long time is released through main shocks, but in focal regions some residual stress remains. The residual energy, through redistribution and subsequent local accumulations, continues to be released in a series of aftershocks. In this paper the relationship between earthquake effects and focal depths was also described: "An earthquake starts at one place and propagates in all directions; where it comes from can be judged from the east, the west, the south and the north of it ... Yin (negative forces) and Yang (positive forces) meet and quakes take place under the ground. If the shocks are deep underground, then even if they are small, the affected areas are large. On the other hand, if the shocks are close to the surface, the affected areas are limited regardless of their magnitudes. The affected areas may have diameters ranging from about 100 Li (a Chinese Li is equal to 0.5 km) to 1000 Li. Right at the center of the shock, tiles would drop down, walls collapse, the ground fissures and the houses destroyed, the effects varying with distance away from the center." These views are consistent with the modern notions of earthquakes. The earthquakes originate from the interior of the earth and seismic waves spread out from the focus in all directions. If the magnitudes of earthquakes are nearly the same, the deeper the source, the wider the waves spread and the lower the epicentral intensity; the shallower the source, the narrower the waves propagate and the higher the epicentral intensity. Land slides, ground fissures and destruction of buildings may occur in meizoseismic regions. However, with increasing epicentral distance the shock intensity decreases gradually. As for the spatial distribution of earthquakes in our country, "On Earthquakes" states: "In northwestern China, quakes usually occur once in a few decades, but in Jiangsu and Zhejiang no quake is experienced ...". However for regions near the seas, as in the Taiwan Province, earthquakes usually occur several times a month. Today it is known that Taiwan is one of the most active seismic regions in China. There is also a rather large number of shocks in northwest China (Gansu, Ningxia, Qinghai and Xingjiang). We cannot say that there is never any shock in Jiangsu and Zhejiang Provinces for certain. For instance, an earthquake of magnitude 6 occurred in Liyang, Jiangsu Province on July 9, 1979. But, comparatively speaking, earthquakes in these provinces are really rare. Such understanding of earthquakes 200 years ago is truly remarkable. In fact "On Earthquakes" is a summary of the observation and investigation of earthquakes by the Chinese people over a long period. It is a testimony to the level of understanding of seismic phenomena by the ancient Chinese.

I.3 The Seismological Work before Liberation

During the late Qing Dynasty, because of the weak and corrupt ruling class, the Chinese nation suffered from domestic troubles as well as foreign invasions. During the period of the KMT's rule, the national economy was tending towards collapse. Few people were concerned with scientific research, much less with earthquake studies.

The magnitude 8.5 earthquake of 1920 in Haiyuan, Gansu Province, was an enormous disaster for the Chinese people. More than 200,000 people were killed with the disaster covering several provinces. A magnitude 7 earthquake occurred in Dali, Yunnan Province in 1925, while in 1927, a magnitude 8 occurred in Gulang, Gansu Province. Some seismologists and geologists hurried there to carry out field survey and investigation. They had preliminary discussions on the relationship between the spatial distribution of earthquakes in China and their geological conditions. They realized the necessity as well as the difficulty of earthquake prediction in China, and recommended that seismological observatories be set up in China. Thus, in the summer of 1930, seismologist Li Shan Bang began to set up the first seismic station on Jinfeng mountain in the western suburbs of Beijing. It recorded over 2000 shocks up to "The July 7th incident" of 1937. That station was destroyed during the anti-Japanese war. Seimologist Fu Chengyi did some valuable work on the theory of seismic waves. These are the only two praiseworthy seismological works in China over many decades before Liberation.

II. The Development of Earthquake Science in New China

Seismology has been greatly encouraged by the People's government since the early days of Liberation and it has developed rapidly. The previous Institute of Meteorology, by combining the seismologists of the former Central Geological Survey and those who were engaged in geophysical prospecting in the former Beijing Institution, established the Institute of Geophysics, Academia Sinica, in April, 1950. Originally it was sited in Nanjing but was moved to Beijing in 1951. Prof. Zhao Jiuzhang, a meteorologist, was the director.

In 1953, the First Five-year Plan of economic construction began in our country. Before a factory or a mine can be constructed, the seismic intensity of the construction sites must first be known. Therefore, a task of assessment of seismic intensity for constructional departments was assigned to the seismologists. The needs of the country promoted the developments of earthquake science. In order to achieve the task, a committee of seismological affairs was organized under the auspices of the Academia Sinica. Professor Li Siguang, the Deputy President of Academia Sinica, was the director and Professor Zhu Kezhen and Professor Fan Wenlan were the deputy directors. Professor Li Siguang proposed to use the historical records of the Chinese earthquakes to assess the seismic intensity of the construction sites of factories and mines. He appointed Fan Wenlan and Jin Yufu, two historians in the committee to direct the work. It took two years to review the 8000 documents, which included more than 2300 history books, informal historical sources and poems, more than 5600 local annuals, many relics and files, recent journals and newspapers, reports and so forth. Fifteen thousand pieces of earthquake records were selected and verified with these publications. The ancient and present names of localities were checked, lunar time was turned into solar time, two volumes of 'Chronicle of Earthquake Data of China' were completed with divisions according to provinces. It contained valuable information on the seismicity of China in historical times and is thus a very important document for studying earthquakes in China.

On the basis of the above-mentioned data of historical earthquakes, the seismic intensity of construction sites of 156 factories and mines was determined between 1955-1956 and a 'Map of the Chinese Seismic Regions' was made.

The historical earthquakes were catalogued for future research and form a basic part of earthquake sciences in seismic-active countries. There is still much to be done in order to compile related catalogues, such as: 'New Chinese Seismic Intensity Scale' and 'Scale of Magnitude of Historical Earthquakes'. Major historical earthquakes were analyzed by drawing isoseismals, determining epicentral locations, estimating epicentral intensities, coverting quake power into magnitudes, including the epicenters of some recent earthquakes and so on. The material was also compiled according to countries so that the intensities of activities in each country can be estimated. It was not until 1960 that 'The Catalogue of Great Earthquakes' and 'The Catalogue of Earthquakes of Different Counties' were compiled and published, representing another achievement in the 1950's.

When the twelve-year National Scientific and Technological Plan was constructed in 1956, the instructions for the most important scientific and technological tasks included the 'Seismicity and Earthquake Disaster Prevention of China.' The most important aspect of this subject was to establish modern networks of seismic stations. Since 1956 basic stations have been established in Beijing, Guangzhou, Shanghai, Wuhan, etc., This laid a foundation for recording moderate earthquakes occurring in most parts of our country and strong earthquakes all over the world.

In order to further develop the study on regional earthquakes, instruments for observation in regional networks in our country had to be made. The type 581 seismograph for small earthquakes was made by us in 1958. In the observation of microearthquakes, electronic technology was adopted in long-term continuous observational systems instead of the usual galvanometer recording. This was an innovation and was not widely used in other parts of the world at that time. The success of such highly sensitive seismographs played an important role in promoting regional earthquake research in our country.

In order to develop geophysical research in Northwest China, the Lanzhou Institute of Geophysics was formally established in February, 1959. At that time, seismologists had already raised the problem of earthquake prediction as a practical task and organized an investigation team to go to the northwestern part of China to look for precursors of earthquake for prediction purposes. It was found that the most common anomalous phenomena reported by the masses before strong earthquakes were those concerning earth sounds, earth light, underground water, animal behavior, meteorological phenomena and so forth. It is very difficult to ascertain whether these recalled anomalous phenomena were real earthquake precursors. Even if they were true precursors, it is difficult to resolve the technical problems on how to observe them before earthquakes. At the same time, seismologists in most advanced countries also thought it was too early to solve the problem of earthquake prediction. Besides, the experiments on earthquake prediction were quite expensive. Earthquake prediction was thus classified as a long-term research topic. Apart from the tasks of national defence, emphasis was placed on basic seismological research.

In 1959, a swarm of small earthquakes occurred in the Xinfengjiang Reservoir region in the upper reaches of Dongjiang River, Guangdong Province. The largest among them reached a magnitude of 4.3. If the seismic activity had developed further, it would have damaged the dam of the reservoir and caused loss of life and property in the lower reaches of Dongjiang River. Premier Zhou Enlai immediately urged some related government departments to strengthen the reservoir dam so as to be able to withstand large earthquakes. Later, an earthquake with $M = 6.1$ struck the area but did not damage the reservoir. Premier Zhou Enlai then asked Academia Sinica to carry out intensive investigations in that area. Hence, the Xinfengjiang Seismological

Brigade which was a special team for studying the reservoir earthquake began to work in the area and obtained valuable results ten years later. Our reports of those achievements at the First International Symposium of Induced Earthquakes held in Canada in 1975 were highly regarded.

An earthquake with M=6.6 occurred in Urumqi Xinjiang Autonomous Region in 1965 and did some damage there. In 1964, both the Niigate earthquake in Japan (M=7.5) and the great Alaska earthquake in America (M=8.5) did much damage. Thus Japan and America convened a conference on earthquake prediction in Tokyo in 1964. In America, a special conference on earthquake prediction was also held and a ten-year plan of earthquake prediction and earthquake engineering was put forward.

All practical tasks such as: the unsolved problem of Xinfengjiang Reservoir earthquakes; the new construction tasks in Southwest and Northwest China; the occurrence of great earthquakes near big citis, lead to ideas of earthquake prediction and disaster prevention. This together with the interest of the world seismological circle resulted in a decision to place the problem of earthquake prediction on the agenda.

In the spring of 1966 before the Xingtai earthquake, the Institute of Geophysics made an overall scientific program to study earthquake prediction and it became the forerunner of comprehensive research of earthquake prediction in our country.

III. The New Look of Earthquake Prediction

Research Work after the Xingtai Earthquake

1. General Situation

A strong earthquake of magnitude 6.8 occurred suddenly on Mar. 8 1966, in East Longyao County, Xingtai District, Hebei Province, resulting in much damage and many casualties. Then on Mar. 22 an earthquake of magnitude 7.2 again occurred in the Southeast Ningjin County, which also caused great damage and many casualties. This was the first time since Liberation that strong earthquakes occurred in the densely populated areas, causing great destruction. After the Xingtai earthquake the Party Central Committee, Chairman Mao and Premier Zhou showed great concern for the people of the quake-stricken area. Premier Zhou himself went to the stricken area to express sympathy and solicitude for the people. He called on the masses to rebuild their home villages and to rely on their own efforts in rebuilding. At the same time he convened an emergency meeting to consult experts on solving the problem of earthquake prediction. The famous geologist Li Siguang had stated that 'earthquakes are predictable.' At the meeting Premier Zhou instructed that experiments should be carried out immediately to achieve earthquake prediction with the experience gained from the Xingtai earthquake in mind. At the same time he appointed the State Scientific and Technological Commission to mobilize and organize related government departments, such as the Ministry of Geology, Ministry of Petroleum, Survey and Drawing Bureau, Hydroelectric Ministry and the related institutions under the Academia Sinica, including the Institute of Biophysics, the Institute of Acoustics and the Institute of Zoology, Beijing University, the University of Science and Technology of China and others to wage a joint battle at the Xingtai quake-stricken area for earthquake prediction research. After 2 months of intensive work many promising phenomena were discovered. In the last ten-day period of May of the same year, a symposium on the Xingtai earthquake was organized by the State Scientific and Technological Commission in Beijing, during which many fine reports were presented.

Taking various opinions into consideration, Premier Zhou designated the Commission to call on experts from different fields to draw up a comprehensive project for earthquake forecast, prediction, disaster prevention and anti-seismic measures. He also instructed that earthquake monitoring work should be strengthened in the Beijing-Tianjin region. Workers from the Geo-Geophysical Institute went to the Xingtai region for earthquake prediction experiments and very soon completed the establishment of a wired telerecording network of 8 stations for centralized recording in the Beijing-Tianjin region. This allowed earthquake trends in this region to be predicted in time. Next, a special group to study the earthquake prediction problems of the Beijing - Tianjin region was organized in the Geophysical Institute to investigate and monitor seismic activity in this region, thus opening a new chapter in seismological research work in the Beijing-Tianjin region.

An earthquake of magnitude 6.3 occurred on Mar. 27, 1967 in Heijian, Hebei Province, causing slight damage in the vicinity of Tianjin. It seemed that the earthquakes were migrating towards the Beijing-Tianjin region along the north-eastward tectonic line. This was brought to Premier Zhou's attention who then issued instructions: 'It is highly necessary to take a close look at earthquake trends in the Beijing-Tianjin region.' The Beijing-Tianjin earthquake office was set up under the joint leadership of the State Scientific and Technological Commission and Academia Sinica to improve research on quake prediction in this region. At that time the network for observing earthquake precursors was built with stations in successive areas in the Xingtai-Hejian-Beijing-Tianjin region.

Other earthquakes which were recorded included the earthquake of Aug. 30, 1967 with magnitude 6.8 in Ganzi, Sichuan Province; the Bohai shock of magnitude 7.4 in Jul. 18, 1960; the magnitude 6.4 shock of July 26, 1969 in Yangjiang, Guangdong Province; the magnitude 7.7 shock of Jan. 5, 1970 in Tonghai, Yunnan Province; the Feb. 17, 1970 Earthquake with magnitude 6.2 in Puer, Yunnan Province; and the magnitude 6.2 earthquake of Feb. 24, 1970 in Qionglai, Sichuan Province. These earthquakes all occurred in relatively densely-populated areas, causing damage in varying degrees. The occurrences of these earthquakes indicated that the earthquake activity in continental China had reached a new high tide.

With the growing number of earthquakes, it became more urgent to resolve the problems of earthquake forecasting and disaster prevention. After the 1969 Bohai earthquake, the central leading group of earthquake affairs headed by Li Siguang was founded to further strengthen the leadership of earthquake prediction research work in the whole country. Under this group, the Central office of earthquake affairs was set up, and this office became the forerunner of the State Seismological Bureau.

In 1971, the State Council decided to established the Seismological Bureau and put the units which were engaged in earthquake research of other related departments under the State Seismological Bureau. Since then the Institute of Geophysics, the Institute of Geology, the Institute of Engineering Mechanics and the other units engaged in geodynamics field measurements of crustal deformation and deep crustal structure sounding, analogue experiments and instrumentation have been administered under the Bureau. Besides, seismological brigades, corresponding administrative organizations and seismic observation networks consisting of experts and members of the masses have been established in many provinces, municipalities and autonomous regions. At present, China has 17 standard seismic stations and more than 400 regional stations excluding those in the Taiwan Province. Relying on professional teams, comprehensive observation and research work in geophysics, seismogeology, crustal deformation, geochemistry, deep crustal structure sounding, seismic intensity regionalization, earthquake-resistant structures, meteorology, astronomy, oceanography and biology were carried out. In recent years mass movements to observe earthquakes and prevent their disasters have also been in full swing. Many factories and mines, schools, enterprises and rural areas have set up their own amateur organizations for observing and predicting earthquakes. Therefore earthquake forecasting in China has become a mass movement of scientific experiments.

In 1975, a successful prediction was made with respect to the Haicheng earthquake, and in 1976 a series of relatively successful predictions were made in these cases: the Longling earthquake with magnitude 7.6 in the Yunnan Province, the Songpan earthquake with magnitude 7.2 in the Sichuan Province and the Yanyuan earthquake with magnitude 6.7 in the bordering area between Yunnan and Sichuan Provinces. Accurate forecasting had reduced the loss of life and damage in these areas.

However, in 1967, the Tangshan earthquake of magnitude 7.8 was not predicted, causing tremendous losses. There were several reasons for failure to predict in this case according to preliminary analyses.
(1) This quake occurred at a place where no strong shocks ($m \geq 7$) have ever been recorded and before the shock no one had ever pointed out that there were tectonic conditions for strong earthquakes;
(2) The occurrence of this earthquake was entirely unexpected, because it was very close to the Haicheng earthquake both in space and in time;
(3) Before the Tangshan earthquake, a series of earthquakes took place nearby such as the Haicheng earthquake with magnitude 7.3 in the east, the Helingeer earthquake with magnitude 6.3 in the west and the Dacheng earthquake with magnitude 4.8. Thus the precursors of these shocks blended with that of the Tangshan earthquake, and the field of precursors presented a pattern too complicated to indicate whether the precursors predicted a large earthquake or a series of damaging earthquakes. Even after this shock, it was still difficult to define criteria for these occurrences.
(4) No remarkable pre-shock activities took place and impending abnormal phenomena appeared very late before the Tangshan earthquake compared with those of the Haicheng earthquake. On the whole, the Tangshan earthquake fully reflected the difficulty and complexity of earthquake prediction. However, the Tangshan earthquake in China is not a unique

2. Earthquake Precursors

Both positive and negative experiences summarized from work in earthquake prediction for more than a decade have demonstrated that precursors really exist prior to large earthquakes, and the precursors form the objective basis for earthquake prediction. Many kinds of precursory phenomena have already been noted. Examples include seismicity, deformation and ground water. Brief descriptions are given as follows.

Many earthquakes show that various anomalies in regional seismicity often appear at the early stage in a large earthquake-prone area. The seismicity usually increases and the epicentral region of the future earthquake forms a 'gap' (an area with no earthquakes above a certain magnitude) surrounded by medium and small shocks. As time elapses, the seismicity increases and thereafter the frequency of small shocks also decreases with the ratio of large shocks to small ones (usually called the B-value) reduced. At the late stage in earthquake-latent areas, the earthquake epicenters are distributed in belts. Sometimes pre-shock activity occurs in the coming epicentral zone and the B-value in its vicinity swiftly decreases. Main shocks take place while the B-value recovers. Recently, some investigators have discovered anomalies in velocity ratio and in the spectrum of seismic waves prior to some large earthquakes.

All these precursory phenomena are very helpful for identifying earthquake-latent areas. Data have suggested that anomalies of crustal deformations are also common kinds of precursory phenomena in earthquake-prone areas. Based on the results of geodetic surveys, it was discovered that unusual crust deformation in earthquake-prone areas is accelerated within several or even nearly ten years before earthquake occurrence. The range of unusual deformation in areas is generally related to the magnitude of future earthquakes. Surface levellings, short levellings across faults within earthquake-prone areas and their surroundings also indicate apparent anomalies before large shocks.

In addition, various forms of anomalies of ground water appear prior to large earthquakes, such as the sudden falling and rising of water level, and abnormal changes of Radon content. Some characteristic precursory data obtained in recent years for other anomalies include gravity, geomagnetism, earth electricity, animal behaviour and macroscopic phenomena.

It should be emphasized that the above-mentioned probable precursors were summarized from a small number of earthquakes. They must be tested to determine whether they are reliable or can be generalized.

In a strict sense, the precursors must be indications of inevitable earthquakes. Such precursors do exist, but are complex processes. Field surveys, analogous experiments and theoretical investigations have all indicated that large shallow earthquakes are the consequences of sudden ruptures or dislocations of rocks after the earth crust has been acted on by forces for a long period of time. But prior to the sudden failures of crustal rocks, the rocks experience changes with the accumulation of stress and strain. And the more impending an earthquake is, the more drastic these changes become, thus causing a series of variations in the medium within earthquake-prone areas which include complicated variations in seismicity, seismic waves, crustal deformation, ground stress, gravity, geomagnetism, earth electricity, ground water etc.. If a large earthquake is actually causing and controlling the above-mentioned precursory variations, the variations will be correlated intrinsically. If no large earthquake is imminent, these quantities will also undergo changes, but each will vary individually without regularities. Thus, in order to ascertain whether the anomalies are earthquake precursors, every kind of approach should be used, and their joint characteristics should also be taken into consideration. From our own experience, a set of abnormal phenomena as precursors of an earthquake should be relatively concentrated in space, synchronous in time and rational physically. Some years ago we lacked understanding in these problems but through many failures before and after the Tangshan earthquake, we have begun to accumulate information and gained experience.

3. Stages of earthquake prediction

Summing up the past prediction experience, we now divide the prediction of strong earthquakes into five stages: long term, medium term, short term, imminent and aftershock predictions. Each stage consists of different tasks, and evidence of prediction for each stage is different as well.

(1) long term prediction in this stage, the trend of the seismic activity in a given area for several years or more is predicted. Long term predictions are usually made on the basis of stages, periodicity, repetition,

gap-filling and migration of seismic activities in combination with tectonic conditions for the occurrence of earthquake and geodetic measurements of crustal movements. The relationship of earthquakes to earth rotation, solar activities and meteorological conditions are all taken into consideration and methods of statistical extrapolation or statistical correlation are also used. The predicted area and the time scale are comparatively large, and it is difficult to estimate a definite place and time of the occurrence of future earthquakes at this stage.

(2) medium term prediction: in this stage, earthquakes are predicted with a lead time from several months to one or two years. This is mainly based on some precursors appearing in the corresponding period of time prior to earthquakes, besides using the methods of seismo-statistical prediction. Such precursors are: earthquake activities distributed abnormally in belts and zones, accelerations of seismic energy release, decreasing of B-value and seismic velocity ratio, crustal deformation, marked trends in anomalous variations of gravity, resistivity, geomagnetism, ground stress, underground water level and chemical compositions of ground water. Surveys of crustal deformation, gravity, geomagnetism etc. have been carried out extensively to seek anomalies for medium term. The use of these trends in anomalies to discriminate annual variations from secular variations is worth looking into. In addition, in a large region, several areas with medium term anomalies may appear at the same time. This probably reflects several zones of stress-concentration due to the inhomogeneous crustal structure under the strengthened large-scale regional stress field. It is usually difficult to judge at this stage which of these zones will develop into the future hypocentral area. Therefore the task of medium term prediction is to delineate one or several zones with a concentration of medium term anomalies within the dangerous zone which has been predicted by long term prediction, and to roughly estimate their respective time and magnitude in order to facilitate various monitoring work. The plans for large scale monitoring work in many distinctive areas have been decided in this way. Some successful predictions have also been achieved with prerequisites that background and aims of short term imminent prediction be provided by medium term prediction.

(3) short term prediction: main indications for many medium term anomalies entering into the short term stage are evident in the variations in the morphology of observation curves, such as markedly increasing speed and reversal. Within several days or months prior to large earthquakes, apparent abnormal phenomena usually appear around the hypocentral region. Various kinds of anomalies usually concentrate on a certain active tectonic belt and are characterized by synchronism and correlated variations. The presence of these phenomena is helpful in judging the entry into the short term stage. These phenomena may indicate that the regional strain strengthens along a certain belt and enters into a stage when the microfractures link together and the fault is under stable creeps. Special attention should be paid to disturbances from seasonal factors (rainfall, temperature changes etc.) and environmental conditions when short term predictions are based on the above-mentioned anomalies.

(4) imminent predictions: accurate imminent prediction is of paramount importance to be well prepared before the earthquake; otherwise 'everything falls short of success for lack of a final effort.' The failure to predict the Tangshan earthquake is just such an example. In recent years, people have been particularly interested in imminent predictions. Some cases of earthquakes have showed that the imminent earthquakes anomalies occurring several hours or several days prior to earthquakes are characterized by their abruptness. Some short term anomalies which develop into the imminent stage have demonstrated drastic changes in amplitudes and emphasises the presence of the anomalies. The concentration of these anomalies is of significance in order to diagnose the moment of earthquake occurrence. However, practice indicates that many actual difficulties exist in making accurate imminent earthquake predictions immediately on the basis of these anomalies. Such abrupt occurrences can be disturbed by numerous factors (such as abrupt changes of special meteorological, astronomical factors), and their durations are very short, and not easily recognized. Furthermore, several climaxes of anomalies may appear prior to earthquakes, and at present it is still not possible to conclude which of the climaxes is to be followed by an earthquake, so that false alarms easily occur. During this imminent stage, it is also not easy to identify the position of earthquake occurrences. From existing data, although the indications from anomalies are most outstanding in the epicentral region, some examples (the Songpan and the Tangshan earthquakes) have shown that the imminent earthquake anomalies do not at first appear in epicentral regions markedly, but at some points in their vicinities. They appear evidently in epicentral regions only soon before an earthquake. This matter needs further observations and studies before one can judge

whether these past unexpected occurrences are general. This has added difficulty for accurate impending earthquake predictions. Therefore in the case of increasing or decreasing of observed values of complex abrupt anomalies, it is very often easy to hesitate rather than make a prediction promptly and opportunely, especially as this could be influencing many in metropolitan areas. At present several places have to be considered at the same time for further monitoring.

(5) prediction of aftershocks: After a strong earthquake, aftershocks are inevitable. Some aftershocks are strong enough to destroy surviving houses and cause further casualties if people are careless. Therefore the prediction of strong aftershocks is also of great importance. 'Late strong aftershocks' often occur when people are not careful and vigilant, when the earthquake danger appears to have ceased. Such cases are not unprecedented. Much experience has been gained in predicting aftershocks. It seems not too difficult to determine the times, magnitudes of aftershocks with their positions known roughly, but it still remains an uneasy problem. The conventional method for predicting aftershocks is to use the characteristics of aftershock sequences, such as the phenomenon of 'densification quiescence' which is commonly seen prior to strong aftershocks. However, in some regions or at different times for the same region such a phenomenon is sometimes marked and sometimes not. The aftershock attenuation characteristic is another common feature for predicting earthquakes, but many cases have shown that strong earthquakes can suddenly occur during the attenuation of aftershock sequences. Such types of earthquakes which do not follow the usual rule of aftershock attenuation are very difficult to forecast. The earthquake of magnitude 6 occurring on May 18, 1978 in the Haicheng earthquake region is an example.

4. Some major problems in the present stage of earthquake prediction.

Earthquakes are indeed very complicated natural phenomena. Although some possible precursors and prediction methods have been obtained through careful analysis of past experiences, the present techniques can only make predictions for earthquakes in some regions under given conditions. The present predictions are only empirical because the laws governing earthquake occurrences are not yet understood. Therefore an important task in the future is to study in greater depth, the nature of earthquakes and the laws governing an earthquake occurrence and its development. Predictions would then be carried out according to the laws. Only in this way can earthquake prediction be made accurate. The following major problems in particular should be stressed:

The relationship between different types of earthquakes and the multiplicity of precursors merits special attention when precursory phenomena are used to predict earthquakes. Although several studies on the types of earthquakes have been done, few have been conducted, based on the relationship between different precursory characters and the types of earthquakes. The existing data demonstrate that different types of earthquakes cause variations in the characteristics of earthquake precursors, such as presence or lack of preshocks, certain kinds of precursors, processes of development, combination forms, expressions of imminent earthquake anomalies and characteristics of aftershock activity etc. If different types of earthquakes are close to each other, predictions may be made successful by referring to past experiences and applying prediction methods; otherwise failures of predictions are inevitable if one uses the wrong information and methods without analysis to check on the types which may be different. The Tangshan and Haicheng earthquakes are two different types of earthquakes: the short term anomalies increased and decreased respectively three times just before the Haichang earthquake, and preshock activities were extremely outstanding with imminent earthquake anomalies. There were however no such indicators for the Tangshan earthquake. Such earthquakes with no pre-shock activities together with the late appearance of imminent earthquake anomalies make predictions comparatively difficult.

Liyang Province in 1979 had no shocks with shorter term imminent seismic precursors, and the earthquake occurred at a place where there had never been any previously recorded earthquakes of magnitude 6. All this indicated that the prediction of this type of earthquake was more difficult.

Another important problem is to make distinctions between the related precursors in focal regions and general anomalies in regional stress fields. Generally speaking, hypocentral regions that are regions of stress-concentration lead to large earthquakes. During crustal movements, many concentrated stress regions may be formed and anomalies present in many regions will develop to become the earthquake foci. Whether the stress concentration region will lead to earthquakes needs to be studied further. This is usually an important cause of false alarms.

Therefore it is necessary to study not only the precursory characteristics of an individual focus, but also the whole tectonic stress field and the whole precursory field. Research is also needed on the relationship between large-scale regional crustal movement and earthquakes, and methods for discriminating earthquake precursors and abnormal phenomena in usual stress concentration regions from the various anomalies caused by general crustal movements. This is important for detecting earthquake-prone regions and to reduce false alarms in other regions.

The problem of earthquake triggering is an old one in earthquake predicting. Many research reports have been published on the relationships between the time of an earthquake occurrence and other factors (astronomical and meteorological). Some investigators believe that the two are related, while the others have opposed this idea. The cosmic environment should not be ignored but deserve further investigation as they may be the important factors for predicting earthquakes, especially for predicting earthquake occurrence movements during the impending earthquake stage.

IV. Seismic Intensity Regionalization

It is clear that accurate earthquake prediction will enable people to prepare for the event, although buildings cannot be moved. Considerable progress has been made in the past few decades on the reduction of earthquake damage on property. This was achieved by promoting work on earthquake intensity regionalization and earthquake engineering. As mentioned above, the work on earthquake intensity regionization began in the 1950's. The seismic intensity shown in the first seismic zoning map did not consider the time parameter. Problems also appeared in the principles and methods on which the map was based, and the data used in the map were only rough estimates. Thus zoning of earthquake intensity could not meet the needs of the construction department. Early in the 1960's an attempt was made to improve these principles and methods but no progress was made. The seismic zoning map only represents the basic intensity of various areas in the future and it indicates the highest possible intensity of earthquakes which may occur in the future over a certain period and under general site conditions. According to this idea, the basic seismic intensity must be based on earthquake prediction during a certain period to come. Earthquake risk zoning will actually provide a very long term earthquake prediction map in some years ahead. It is difficult to make progress in earthquake zoning before earthquake prediction is applied extensively. In other words, comprehensive application of earthquake prediction will create conditions for the development of seismic zoning.

From 1971 to 1975, the State Seismological Bureau formed a South-West China earthquake intensity team. The team dealt with comprehensive research on seismic intensity zoning, and the main studies concerned the basic principles, methods and evidence of the zoning. They took the boundary areas of the Sichuan and Yunnan Provinces as their experimental sites, where earthquakes occurred frequently and where a foundation in seismic intensity zoning work had been laid. They worked on the following six principal aspects:

(1) seismogeologic tectonic characters and their relation with strong earthquakes;
(2) the relation between gravitational fields and seismic activities;
(3) the relation between recent crustal deformations and earthquakes;
(4) analysis of characteristics of seismic-activity;
(5) determination of seismic risk and the synthetic standard of earthquake intensity;
(6) evidence of effective field.

These research results will undoubtedly be significant to the country's seismic intensity zoning map.

In 1972, under the auspices of the State Seismological Bureau, a seismic zoning map of China was constructed and 'The Seismic Zoning Map of China' (scale 1 : 3,000,000) was published in 1977. This was a significant result in seismic work in the 1970's. There are essential differences between the new map and the one in the 1950's, with great differences especially in the principles, methods and the data used.

In principle, for forecasting the future seismic risk in a certain time interval, the principle of comprehensive analysis and research has been used. This included comprehensive analysis and research in seismic activities, surface and deep structures of the earth, neo-tectonic and recent structure movements as well as geophysical fields. The trends of seismic activities and the probable area of earthquake occurrences in a certain future period were then assessed. In general, the following methods were used:

(1) Delineation of seismic zones and belts: The activity of earthquakes, especially strong earthquakes, is inhomogeneous in the crust geographically. It frequently shows concentration in certain zones and seismic belts. Accurate delineation of seismic zones and belts can provide the necessary conditions

for the research of seismic activities in time, space and intensity. The main bases for the delineation of seismic zones and belts in our country are as follows:
(i) characteristics of seismic activities in time, space and intensity;
(ii) characteristics of structures and thicknesses of earth crust;
(iii) nature and characteristics of development movement and the action and characters of current tectonic stress fields;
(iv) investigation of physical properties of rocks in different seismic zones.

(2) Analysis of seismicity of various seismic zones and belts: its aim is to forecast the developing stages of seismic activities in various zones and belts in a certain future period and the corresponding trend of seismic activities, the maximum magnitude and the possible frequency of occurrence of earthquakes with different magnitudes. This also provides a basis for predicting which places are in danger from earthquakes:

Analyses on seismic activities in principle include: (i) analysis of earthquake active periods; (ii) investigation of the process of strain accumulation and release; (iii) investigation of the recurrence and gap-filling of strong earthquakes; (iv) research on the migration of earthquake activities: (v) analysis of magnitude-frequency relations; (vi) application of methods of mathematical statistics.

(3) Seismotectonic analysis of seismic risk: the purpose of analysing such different magnitudes earthquakes is to study geologic structures within the given zones and belts so as to investigate further the structural conditions of earthquake occurrences, to point out the structural marks of occurrence of earthquakes with various magnitudes where destructive earthquakes may occur within those zones and belts. When analyzing the seismogeological conditions, the main principles should be complied as follows: (i) the analyzing method from the old structure movements to new ones, that is, when making investigations on the characteristics of activities in surface crustal structures in history, special attention should be paid to the characters of tectonic movement from the Neozoic to the recent period and their relation with seismic activities; (ii) the method of investigation from shallow structures to deep ones, that is, besides paying attention to the search for the general features of the surface structures simultaneously, efforts should be made to explore the features of structures at the depth of focus and their relation with seismic activities; (iii) while investigating the structural conditions of earthquake occurrences, the stress should be put on the search for the relation of new active faults (their scale, mode of movement, strength and form of mutual connection and so on) and basins (their scale, type and strength) to seismic activities.

(4) Delineation of earthquake risk zone: through the comprehensive analysis and evaluation of seismic activities, the trends and levels of seismic activities and the frequency of occurrence of strong earthquakes with different magnitudes within each earthquake zone and belt in a certain period in the future are evaluated. Based on this, according to the seismotectonic marks and the features of seismic activities in different earthquake belts, some places and segments where strong earthquakes will probably occur are picked out, and zones with seismic risk of different levels are delineated.

(5) Determination of field of seismic effect and intensity regionalization on the basis of delineating earthquake risk zones, quake power is converted into epicentral intensity according to the empirical relation between earthquake magnitude and epicentral intensity of our country, and the earthquake risk regions are deemed to be regions with epicentral intensity in the future earthquakes. Its field is determined statistically with the attenuation law of seismic intensities in the seismic zones and belts. Statistical data of fields of seismic effects were obtained from isoseisms of historical earthquakes occurring in different zones and belts. It is obtained by applying the methods of equivalent areas and least squares.

The new 'Seismic Zoning Map of China' was obtained by using the above principles and methods. There is an apparent development in comparison with the past, but it cannot be considered as being perfect. As the rules of earthquakes are far from being known, there are still many difficulties and problems against accurate evaluation of seismic risk and the determination of places where strong earthquakes will occur. With further knowledge of the rules of seismic activities and deeper investigation of the structural conditions of earthquake occurrences, the map of seismic zoning will be revised again in the future and will be improved.

V. Earthquake Engineering

For the past ten years or more, earthquake engineering in China has been developed step wise e.g., a new seismic intensity scale of China was made; the seismic intensity of many construction sites was determined; the design theory of antiseismic structures was developed; the antiseismic design standard of industrial and civil constructions was formulated;

the dynamic models of various types of structures in our country were measured in situ; dynamic model tests and mathematical analysis of various types of structures were carried out in conjunction with major construction items; the plan to rebuild home villages was drawn up; houses in some big cities were examined for antiseismic capability and reinforced. At the same time, by combining indigenous and modern methods, some experimental equipment and a moderate number of strong motion networks were gradually set up.

As our economic situation does not allow us to remove or replace the existing non-seismic proof old buildings, since 1967, by relying on the masses, their antiseismic capability has been examined and the strengthening work done in some cities in seismic areas. As for the broad rural area distributed within seismic areas, according to the general antiseismic principles, with experience concerning earthquake damage, local features and the use of local materials, the effective antiseismic measures adopted have been welcomed by peasants and considerable results have been achieved.

On the whole, scientific research of earthquake engineering includes the following aspects: (i) investigation of damaging earthquakes; (ii) observation and research on ground movement due to strong earthquakes; (iii) research on the selection of antiseismic engineering standards; (iv) research on the effects of site conditions on ground motion; (v) research on the earthquake resistance of various structures; (vi) compilation of antiseismic design codes.

VI. Concluding Remarks

From this review and analysis of the history and present state of seismology in China, it can be seen that seismology has undergone quite a long course in history, with its development from correlations and deductions according to superficial phenomena in ancient times to the present observations and analyses with scientific methodology, from individual studies to the present nationwide seismological research, and from very shallow understanding to the present systematic science. Premier Zhou showed the greatest concern for seismological work and his concern has always been a source of inspiration to seismologists. More than a decade prior to the Xingtai earthquake an initial foundation was laid, and the great developments in more than ten years since will surely provide more concrete conditions for future developments. We are full of pride and enthusiasm and look forward to a brilliant future for earthquake science.

CHAPTER 15

A SUMMARY OF MARINE RESEARCH IN CHINA

by Shen Zhendong

It is only more than a hundred years since the emergence of marine science, and the domain of marine scientific research continues to grow. During the last 20 years, new technologies such as electronics, optics, acoustics, remote sensing, satellites and electronic computers have been adopted in all fields of marine science, and it is becoming one of the important components in modern science.

There was hardly any research in marine science in China before liberation. After the founding of New China, institutions for the study of marine science were set up, scientific personnel organized and various programs planned. This led to active investigations and research. Through the ardous efforts in the last 30 years, China now has a contingent of scientific personnel with expertise in almost all areas of marine science. They have achieved good progress in marine research.

I. Investigations

China, bordered by the Pacific Ocean in the east with a long coastline, vast sea areas and many islands, is a country rich in marine resources. Marine investigations have been cooperatively carried out in the Pohai, Yellow Sea, East China Sea and the South China Sea since the founding of New China.

In the period following liberation, agencies from the Chinese Academy of Sciences and institutions of higher education made comprehensive marine investigations with emphasis on fishing grounds. In 1957, the Chinese Academy of Sciences, the Ministry of Fisheries, Shantung University and the Chinese Navy conducted a joint investigation in the Pohai and Pohai Straits. This investigation provided valuable information and experience for organizing comprehensive marine investigations on a large scale. Further investigations from single ship cruises to observations involving many vessels significantly marked the increased marine investigative abilities in China.

In 1958 under the leadership of the Oceanographic Panel of the National Commission of Science and Technology, a comprehensive marine investigation of China's sea areas was made to obtain basic and systematic data. The investigation lasted for 3 years and had the participation of the Chinese Academy of Sciences, the Ministry of Communication, the Central Meteorological Bureau, the Chinese Navy as well as a number of organizations from the coastal cities and provinces. It was the largest investigation of its kind since the founding of New China, as well as the first one conducted according to the regulations of unified marine investigation. It was therefore a great event in the history of marine science in China.

After the completion of this comprehensive investigation, new emphasis was placed on key areas so that further surveys could be carried out. In recent decades, comprehensive marine investigations were carried out systematically on the continental shelves of the Pohai, Yellow Sea, East China Sea and the waters of the Zhongsha-Xisha-Nansha Island. An expedition was also sent to the equatorial areas of the Pacific Ocean. Research vessels were sent to participate in the First GARP Global Experiment in 1978.

The coastline of China which is 18,000 kilometers long is of strategic importance for national economic and defence constructions. In 1960, the National Commission of Sciences and Technology encouraged related agencies from the coastal cities and provinces to conduct general surveys in an effort to obtain basic data of the coastal zones as soon as possible. However, due to various reasons, this general survey was not completed. In 1978, when plans for the division of the national agricultural resources were made, a general survey of the coastal zones was proposed again. Experiments were carried out in Wenchow, Zhejiang Province, to prepare for the undertaking of this general survey.

During the last 30 years, comprehensive and specialized investigations for both productive and defensive purposes were carried out by the agencies concerned.

Large scale intensive investigations have been made for fishing resources and fishing grounds, and investigation areas were being continuously expanded to the open sea. Various kinds of fishes that could be harvested were discovered during investigations of fish inhabiting the upper, middle and lower sea layers of offshore East China Sea. Similar information was also obtained for deep sea fishing resources in the Sansha and the South China Sea. These discoveries were valuable contributions to marine fishery in China.

In the last 20 years, marine geological surveys occupied a prominent position in marine investigation. The marine geological and geophysical surveys aimed at looking for offshore oil were carried out on a fairly large scale in the Pohai, Yellow Sea, East China Sea and the South China Sea. It was through these surveys that the tectonic zones and bottom strata of marine waste products of the Cenozoic era, ideal for the concentration of oil and gas, were discovered in the offshore areas of China. Oil has been found and produced in some of these sites. In 1970, surveys were conducted in the Pohai, Yellow Sea, East China Sea and Zhoushan to look into the feasibility of laying underwater cables to meet the increase in telecommunication developments in China. A survey on a joint China-Japan cable route was conducted in 1974. These surveys were well conducted as evident from the results obtained.

With the increase in marine activities, marine pollution has become a serious problem in need of massive attention. Marine pollution in the waters of the Pohai, Yellow Sea, East China Sea and the mouth of the Pearl River has been jointly investigated by the Environment Protection Office of the State Council and coastal cities and provinces, the Chinese Academy of Sciences, the National Bureau of Oceanography and institutions of higher learning. The investigation revealed that the coastal waters of China were polluted to different degrees. Attention was alerted to those areas where pollution was a serious problem.

Projects on ocean shipping since liberation included hydrographic surveys, estuarine measurements and silt and mud-sand-banks accumulation in harbours and ports like Xingang and Tienjin. Investigations on marine acoustics and surveys of harmful sea organisms were carried out for marine defence purposes. Surveys on mud-banks and marine ecological systems were also made for the fishing industry, while aerial observations of icebergs were carried out for the safety of marine oil platforms and oil-drilling operations.

In addition to cruising and fixed oceanographic observations, permanent tidal observatories, oceanographic stations, estuarine monitoring stations, hydrographic and pollution monitoring stations have been set up in China. This constitutes a preliminary oceanographic monitoring system in China.

China obtained her first refitted specialized research vessel in 1956. Today, China has designed and built dozens of research vessels for marine investigations and comprehensive geophysical and geological explorations. Some of these are ocean research vessels with a tonnage of 1,000 or even 10,000. The oceanographic instruments are extremely important for marine investigations. Before liberation, even common instruments were imported. Today, conventional as well as instruments for specialized observations are manufactured locally; other instruments also include acoustic digital wave and tidal meters, underwater laser televisions, etc. Automatic oceanographic data buoys have also been successfully developed, and have been put to trial at sea.

As the capability for marine investigations increased, the systems for the oceanographic data services and for oceanographic instrument management were set up and strengthened. The Chinese collection of oceanographic data, marine atlases, all kinds of diagrams and tidal tables for the main ports at home and abroad have been published.

II. Researches

In the last 30 years, China has been actively engaged in theoretical research in order to resolve the urgent oceanographic problems. Oceanographic fundamentals were studied and a theory for the shallow seas of China was established.

1. Oceanographic physics

Studies have been made on the formation, development, attenuation process and regularity of the water masses, currents, waves, tidal currents and their spatial distributions and time variations in the Pohai, Yellow Sea, East China Sea and the South China Sea through data analysis, theoretical models and experiments. These studies contributed to a preliminary understanding of the hydrological characteristics of China's offshore areas. The cold water mass in the Yellow Sea has a very important role in the hydrological conditions of offshore areas in China. It has been studied with the emphasis on its formation, properties, circulation patterns and the relationship between its fluctuations and local climate. Another factor which influences the hydrological condition in China's sea areas is the influx of fresh water from the main estuaries such as those of the Yangzi River and the Yellow River. In particular, simple correlation analyses were carried out for the Yangzi River.

Current — Many studies have been made on the nature of the current, in particular, the Kuroshio of the Yellow Sea and the East China Sea. The Kuroshio study is characteristic of the Chinese approach. A Chinese oceanographer presented a paper entitled "Some Study Results of the Abnormality of Kuroshio in the East China Sea" at the Fourth International Kuroshio Symposium held in Japan. The participants were receptive to it as the study filled gaps in various aspects of Kuroshio studies; it also contributed to the research of the Kuroshio and adjacent waters. The methods of residual current analysis and geostrophic flow calculations were inferred in the current studies. The method for the calculation of current under vertical eddy viscosity and sea floor friction effects was also considered.

Tide and Tidal Stream — Research on the tidal theory and methods for tidal predictions have been successfully performed. As a result, these methods have been applied to large area predictions of the tides and tidal streams. The non-linear effects of tidal friction were used to derive the important characteristic values of the tidal energy consumption and the tidal distribution of the Yellow Sea. The formation and the motion of rotational tidal waves were also studied taking into consideration the geostrophic effects and the frictional effects due to the sea floor. A special formula derived from a wave dynamic equation was used to calculate the tidal level gradient. The distributional pattern of the tidal stream was then obtained.

Tidal Storms — China is one of the countries in the world often prone to disastrous tidal storms. During the past 10 years, theoretical research on tidal storms was carried out and an empirical method for the prediction of tidal storms was proposed. The effective timing and accuracy of predictions basically met the requirements for the prevention of tidal disasters.

Waves — At the beginning of the 1960's, the formation, propagation and variation of coastal waves in shallow sea were studied on the basis of energy equilibration. The papers "General Wind-Wave Spectrum and its Applications" and "Swell Spectra" and "The Principles of Sea Waves" were published. The solution of a small amplitude progressive wave on the sea floor with a gentle gradient was derived from the coastal wave research. The wave speed and amplitude derived were in conformity with that obtained under the conditions of uniform depth. All these results enriched the coastal wave theory. Calculational methods were put forward which were comparable in accuracy with other methods used abroad. In addition, multiple models of resonance, shear current and shearing force were developed and the evolution equation of wave surface amplitude spectrum for describing the wind-wave formation was found. The theoretical wave growth rate derived from this equation is more practical than the former one.

Air-sea Interaction — At present, air-sea interaction is an international research subject. Long-term forecast of precipitation in summer has been studied since the early 1950's by using the temperature data from the Pacific Ocean. Studies have also been made on the northwest Pacific Ocean and its influences upon the climate of East Asia, the relationship between the temperature and the air pressure field on the surface of the north Pacific Ocean, the relationship between the ocean circulations in the north Pacific Ocean and Meiyu (plum rains), and the energy exchanges between the atmosphere and the sea during the dry and wet period in summer in the Jianghuai River valley. In recent years, studies have been conducted on large scale air-sea interactions and long-term weather forecasts. Numerical experiments have also been made by the combination of sea and large scale atmospheric circulation. China also participated in the First GARP Global Experiment in 1978.

Marine Acoutics — Marine acoustics is studied with emphasis on the theory of sound propagation, the pattern of sound-speed distribution, the sound-speed variations and the effects of hydrological conditions on sound propagation, and the regularity of sound propagation and marine noises in China's seas. The approximate solution of sound field with the random sound-speed is obtained from the analysis of sound field theory of shallow waters. A general formula was derived for calculating the group velocity and the exponential attenuation, and the fractional amplitude function of simple harmonic waves under the condition of distribution of several typical sound speeds with positive and negative gradients. This provided the theoretical basis for predicting the sonar functional range in the shallow sea. It also succeeded in interpreting the relationship between the interim range and the frequency in the underwater sound experiment and mapped the clear physical images of sound field characteristics. This theoretical analysis possesses a characteristic unique to China.

2. Marine Biology

After the founding of New China, a basis for the theory of marine fauna and flora was established through research on marine animal and plant taxonomy throughout the country. Preliminary analyses were made on the fauna of polychaeta, mollusca, crustacea and echinodermata in the studies of the benthic invertebrates. Algal species of economic importance and their distributions were investigated and a proposal for dividing the algal flora in the northwest Pacific Ocean was forwarded in the research.

Marine Ecological Studies — The basic knowledge for the biomass and the distributions and variations of important species of plankton and benthonic organisms in various sea areas was acquired through the analysis of approximately one hundred thousand biological samples. Upon this analysis, the close relationship between the species constitution and distributions of plankton and the water systems of sea areas was described.

Marine Ichthyology — Investigations on the distribution of marine fish taxonomic fauna were carried out with emphasis on the research of biological fundamentals for fishery, fish habits, hydrology, hydrochemistry and plankton. Basic elements like the biological characteristics and variations of chub mackerel resources within the main factors effecting it, and environmental factors for the formation of fishing grounds as well as the grouping nature were understood. In the studies of fish taxonomic fauna, the treatises *Fish Fauna of the South China Sea, Fish Fauna of the East China Sea* and *The Food Fish Fauna of China — Marine Fishes* were published. In studies of fish population ecology, emphasis was placed on important food fishes such as the small and large yellow croaker and the hairtail. The characteristics of population biology as well as variations within the species were systematically understood. Results obtained included the resolution of the problem of the age determination for the large yellow croakers and the division into three local populations. Also, analytical methods to determine reproductive and growth rates, and population of these fishes in various geographical locations were proposed. In the studies of experimental fish ecology, emphasis was placed on the development of the parent fish sexual gland, mechanism of ovulation and reproductive cycle, hatching of eggs, the development of larvae and the growth of young fish. These studies led to the use of artificial reproductive methods in marine fish cultivation.

Since the 1970's, marine biological investigations have been conducted in the waters of Dongsha-Nansha-Zhongsha-Xisha. In investigations in the waters of Xisha alone, ten thousand samples were collected with 2700 species of organisms identified. 130 of these species were new discoveries in the world. These discoveries provided data for compiling the fauna and flora of China and also provided the necessary basis for the development of biological resources in the South China Sea.

3. Marine Geology

In the past 30 years, geological researches carried out in China's seas include the location of oil resources, the characteristics of geotectonics, the source and composition of sediment, coastal geomorphology, the estuarine evolution and sea floor geomorphology, the geotectonics of the coasts, the continental shelf and important islands of China, and the source and factors involved in sediment formation in the Pohai and the Yellow Sea.

The coastal geomorphology, the development and evolution of typical estuaries were studied with the particular reference to the Yangzi River and Qiantangjiang. Based on analyses of the data from model experiments, the processes of estuarine development and the basic patterns of siltation evolution were discussed, and propositions for harnessing the Yangzi River estuarine waterways were also forwarded.

The coastal dynamic measurements were carried

out at sections selected from the various types of coasts and model experiments of coastal dynamic geomorphology were also performed. The sea level variations and the new tectonic movement since Quaternary in China's sea areas and the evolution and sea floor geomorphology of China's continental shelf were studied.

The study on the problem of sand-siltation, especially in Tang-Gu New Harbour has been carried out for many years. Research work on deep-sea port sites has been done according to the needs of economic and military construction.

4. Oceanographic Chemistry

Data on the variations in the distribution of fundamental chemical elements in the sea have been acquired. The relationship of these elements with marine organisms, hydrology and geology was discovered. The hydro-chemical characteristics for the various sea areas have been summarized and used as a basis for studying all oceanographic phenomena. The measurements of the variations in the distribution of dissolved oxygen, phosphate, silicate and pH in China's seas have been completed. The comprehensive hydro-chemical measurements in the important estuaries such as those of the Yangzi River, the Yellow River and the Pearl River have also been made.

To study marine radioactive elements, the background radiation of China's seas was investigated. The form of radioactive pollution and its transference were also studied. Some oceanographic phenomena were studied by means of radioactive elements. By measuring the dilution of radio-isotopes and the hydrological conditions, the relationship between these factors was also established.

Analytical methods using colorimetric estimation of alkaline organic dyes were proposed to measure trace elements like gallium, iodine, boron and uranium in marine sediments. A new type of uranium absorbent has been developed, which is characterized by large absorption capacity, faster absorption rate, insolubility in water, and the capability of repeated usage and low cost.

Physical and chemical processes in offshore and estuarine areas were studied, with emphasis on the variation of silicate and its coacervation processes. Investigations on the equilibrium and dynamics of marine geo-chemical processes, the basic physical and chemical properties of sea water (for example, the relationship between conductivity, chlorinity and salinity, the relationship between conductivity and density, etc.), and the geo-chemistry of marine trace elements were also carried out. In physical chemistry studies, a mathematical model was used to analyse the distribution and the transference patterns of silicate in the estuaries, and some interesting results were obtained.

III. The Prospects of China's Marine Science

Although great achievements have been made in China's marine science in the last 30 years, China still lags about 10 to 15 years behind when compared with the advanced nations in the world. The present focus of work in the field of marine science should be shifted to modernization with the fundamental goal of realizing the use of automation and computerization of data processing and analysis.

In order to speed up the development of marine science, it is necessary to obtain the cooperation of the various agencies concerned and those of related disciplines to build a strong contingent for marine research, to set up experimental bases accordingly, to extend international cooperation and exchange of oceanographic data, and to participate in well-planned international marine investigations and academic activities.

CHAPTER 16

WINDING ROADS AND A BRIGHT FUTURE — 30 YEARS OF CHINESE PSYCHOLOGY

by Xu Liancang

Psychology is a science that has a long history. There were treatises on psychology in ancient classical Chinese writings on philosophy, morals and education. It has been quite a long time since psychology was introduced from the West to become an independent science. Sixty-two years ago the first psychological laboratory was established in the Philosophy Department at Beijing University. In the old Chinese society, science was neglected by the authorities and thus the progress of psychology was badly affected.

After the founding of the People's Republic of China, the Chinese Communist Party and the Government fostered the development of psychology. In 1951, a psychology research unit was established under the Chinese Academy of Sciences; the Chinese Psychological Society resumed its function, and journals were published. In 1956, the Institute of Psychology was formally founded. In the years between mid-sixties and mid-seventies, psychology was sabotaged by the "Gang of Four". After the downfall of the "Gang of Four", psychology was again in full development.

The development of Chinese psychology can be divided into 4 stages from the time of the founding of the People's Republic of China in 1949.

I

The first stage was the period of learning and reformation which lasted from 1949 to 1958.

Even though the People's Republic of China was founded in 1949 and great changes took place in the political realm, it was not possible to have an overnight change in ideology in the scientific community. Scientists who answered to the call of the Communist Party by learning Marxist-Leninism and Mao Zedong's thoughts, tried to reform their world outlook and to master the ideological weapon of dialectical materialism as a guidance to their research and teaching. A campaign for learning and reformation was also initiated in the field of psychology; the slogan of learning from the Soviet Union was raised, calling on psychologists to study the experience of Soviet psychology, to reform old psychology and to establish a new psychological system.

Soviet psychology at that time had similar problems. In 1950, a meeting on Pavlovian theory was sponsored jointly by the Soviet Academy of Sciences and the Soviet Academy of Medical Sciences. A resolution was passed stressing the importance of Pavlov's theory of higher nervous activity and a trend appeared to replace psychology with the theory of higher nervous activity. This resulted in a controversy that lasted many years in the circle of psychologists. The main issue was the relationship between psychology and the physiology of higher nervous activity with debate on the subject matter, task and methodology of psychology.

Chinese psychologists, in the process of finding a correct path for the development of psychology were much concerned about the discussion of this fundamental issue of psychology in the Soviet Union. Therefore, "fundamental problems related to psychological phenomena" were listed as crucial problems open for discussion in the First Representative Congress of the Chinese Psychological Society in 1955. Opinions differed widely at this meeting. Psychologists at one extreme admitted that "there is no fundamental difference between psychological activity of man and salivation of dog — all being of higher nervous activity"; those at the other extreme stressed that "psychological activity cannot be explained by higher nervous activity as it is a phenomenon of a higher order". However, the majority of psychologists did not agree with both of these extremist views. They felt that to equate psychological phenomena with higher nervous activity, which is to abolish psychology is erroneous, and they also disagreed in separating psychological phenomenon from its material substrate which leads to a form of dualism. They held that psychological phenomenon or human psychological activity is a form of higher nervous activity peculiar to human beings evolved from animals which manifests in a highly complex new quality. They maintained that psychologists should study the brain, the material substrate of the psychological phenomenon, and also stressed the view that psychological activity of man is a reflection of objective reality, and that the decisive function of objective existence (social and natural) should be considered.

This discussion was fruitful and served as a guiding principle to future research work. It should be noted that this kind of knowledge is in accordance with dialectical materialism. However, in August 1958, a campaign was initiated in Beijing Normal University to "criticize bourgeois scholastic thinking". One of the main targets for criticism was psychology. Later on, many other cities also initiated similar campaigns, and the boundary between scientific and political problems became confused. Many research works were accused of being "abstractions" or "biologizing", and many veteran psychologists were accused of "holding white flags". Because of this, the policy of "let a hundred flowers blossom and let a hundred schools contend" was destroyed, and the enthusiasm of many psychologists was thwarted. Although in early 1959, measures were taken to make corrections for this erroneous criticism, ideologically this correction was not thorough. Thus in the period of the interference of the "Gang of Four", when the criticism of 1958 was repeated in many places, psychology was again in chaos. It was only in 1978 at the Second Scientific Meeting of the Chinese Psychological Society that the lesson of the criticism movement was summed up, and in 1979 at the National Planning Meeting of Educational Science, the Ministry of Education formally announced that psychology is a science every educationalist must learn and master. It was agreed that the criticism of psychology in 1958 was scientifically absurd and must be renounced.

During this first period (particularly in 1956), under the direction of Premier Zhou Enlai, the State Council Committee of Scientific Planning convened a meeting with scientists from various parts of the country to help in the planning of a 12-year project for the development of science and technology. Psychology was one of the basic sciences included and was assigned a programme for future development with many research items planned. Although psychology was subjected to much interference during this period, psychologists were still enthusiastic and went on to accomplish much in various fields.

1. Psychological processes and physiological psychology

(1) Psychological processes studies were made on perception of movement, threshold of perception of movement, judgment of speed and phi-phenomenon, the pure tone auditory sensation, thought process and skill learning.

(2) In physiological psychology, studies were made on conditioned reflex. At first Pavlov's classical experiments were repeated, and the discriminative ability of animals was studied. Psychologists studied complex motor chain reflexes in dogs and monkeys.

Studies were made on the types of higher nervous activity in children and animals. In children it was found that even in the establishment of the simplest voluntary motor conditioned reflex, the influence of past life experience must be considered. Thus the difficulty and limitations in the use of the method of conditioned reflex in studying the physiological basis of individual differences were pointed out.

(3) Cortical functional state was studied in dogs under free movement conditions. The amount of

conditioned reflex and the difference in tolerance to drugs were the methods used for this study. Observations were made on the changes in conditioned reflex and unconditioned reflex under sleep and drowsiness conditions.

2. Educational psychology and child psychology

In that period, the psychological test was criticized. In 1936 in the Soviet Union when pedology was criticized, psychological testing was also criticized as a pseudoscience. About 40 tests designed by Chinese psychologists before liberation were rejected without any analytical judgement. Since that time, testing has been a forbidden area in psychology.

At the same time the research works were:

(1) The study of psychological age characteristics of child development. This is a very important problem, as the scientific elucidation of the nature of child psychological development gives rise to correct teaching procedure. Studies were made on the generalization of characteristics of objects, the role of words in children's generalization function, and age characteristics of children's perception, attention, memory, understanding etc.

(2) In educational psychology, studies have been made in the psychological analysis of the solution of arithmetical problems in primary school. It was proposed that the learning of basic knowledge of algebra could be advanced to primary school, and this could help primary children in the solving of arithmetical problems. There were some studies on the teaching of physics, where elucidation in class was joined by life experience to help the students in handling the basic concepts in physics. A quick learning method was devised for adults to learn Chinese characters.

(3) Psychological studies in connection with moral education. Psychologists and teachers collaborated to summarize the work experience in moral education in schools and to study the role of collective behaviour in moral education. Factors influencing the formation of communist consciousness in young students and peasants as well as the training methods were studied.

3. Labour psychology

In response to the policy of "connecting theory with practice", psychologists started the research work in the field of labour psychology in our country.

Chinese labour psychology must be distinguished from industrial psychology in Western countries and the past "psychotechnology" in the Soviet Union. Chinese labour psychology should "come from the masses and go to the masses", with due respect to the wisdom and creativity of the workers.

(1) Studies in operation methods – Psychologists participated in the summary of experiences of advanced workers in various industries, such as the advanced methods used by certain efficient textile workers. A special study was the 1951 standard of textile operation, which was a summation of experiences of advanced workers. Studies were also made on the analysis of operation methods of punch press workers and of the work tempo of the assembly line. Rationalization methods were proposed to the factory authorities to raise efficiency and reduce fatigue.

(2) Accident prevention – Psychologists on the invitation of some government departments helped in the analysis of the causes of accidents in mines. Since human factors were some of the main causes of accidents which also reflected organization and educational problems, proposals were made as to how to prevent such accidents. Psychologists and workers jointly discovered in advance a predication method to raise the level of awareness for the prevention of accidents.

(3) Learning of skills – The application of the psychology of learning to the training of skilled workers obtained apparent effects. Steel workers were trained in visual discrimination of colour, form of flames in the furnace and their relation to furnace temperature, chemical reactions etc. This kind of training showed far better results than allowing the workers to master the skill by trial and error by themselves.

(4) Studies in creativity – Psychologists used the laws of creative thinking, particularly the method of prototype heuristics to promote the inventions and innovations of the workers. This method worked well in the automation of punch press design. Workers under the direction of psychologists created a number of automatic punch models, which helped them to obtain a high production rate with lower accidents on record.

4. Medical psychology

Medical psychologists criticized the psychoanalysis of Freud and tried to establish a new medical psychology under the principles of Marxism.

(1) Psychological studies in the treatment of neurasthenia and the analysis of its causes. Together with personnel in the hospital, psychologists devised

a speedy synthetic therapy for the treatment of neurasthenia which consisted of: (i) imparting to the patient the scientific knowledge of neurasthenia, to relieve the anxiety, (ii) encouraging the patient to struggle against the disease, (iii) helping the patient to take a correct attitude towards the disease. This psychotherapy together with drug and physical therapy gained good results.

(2) Studies of characteristics of higher nervous activity of mental patients. The establishment, inhibition, extinction characteristics of conditioned reflex, with certain types of higher nervous activity of patients and schizophrenia, children's nervousness and its relation to behavior were studied. These served as means for diagnosis.

(3) Studies of psychological literature in ancient Chinese medical books.

II

The second period was 1959—1966. In this period psychology witnessed a steady growth.

Under the course suggested by the Party — "readjust, strengthen, enrich and improve", psychological research advanced smoothly. In 1960, the second conference of the Chinese Psychological Society took place, and in this conference a 3-year plan for psychological development was adopted. In December 1963, the first Annual Meeting of the Chinese Psychological Society (CPS) took place. At this time there were 26 branch societies and a committee of educational psychology was also set up. From 1960 to 1963, some 300 psychological papers were published in various journals. In the 1963 meeting, 203 papers were received and a new scientific research plan was drawn up that further clarified the tasks and the directions for psychology. More advantageous conditions for the development of psychology were created after the 1963 meeting. In 1965 members of the CPS totalled 1087 (in 1958 there were 585 members). The Institute of Psychology in the Academy of Sciences was enlarged to include 170 members and laboratories of a high standard were built. In this period the orientation of psychology was enriched and psychologists in China paid more attention to the achievements of foreign colleagues. Results were obtained in the following fields:

1. In the study of fundamental psychological process

In this field, intensive studies were made on perception, such as of distance, observer's position, brightness, background and their influence on perception, the influence of target size and background cue on distance judgement, convergence studies, size constancy, size-weight illusion, the influence of set-on illusions and the illusion problems in practice, and the way of illusion prevention. Eye-movement during perception of figures, the relationship between information load and response time and information analysis during pattern recognition were studied from the approach of information theory.

There were intensive studies in the memory process, such as visual, tactile learning and recognition, simultaneous learning by vision and hearing and their mutual interference. Also, the learning of individual characters, numbers, and pictorial materials by children, the comparison of different methods of memorising and studies of forgetting and reproduction etc. were also investigated.

There were studies on the recognition of Chinese characters, the influence of central factors in the learning of Chinese characters using the tachistoscope display and the development of learning Chinese characters in children.

2. Child psychology

Emphasis was on the characteristics of children of different ages. There was a series of studies on the development of the formation of concepts, such as the concept of kind, causality, mathematical concepts, and concept of space. The development of memory ability (figural memory and verbal memory), development of verbal language, verbal and written studies were carried out. There were also studies on the characteristics of higher nervous activity of normal and abnormal children.

3. Educational psychology

Educational psychological studies were carried out in combination with our reformation of the teaching systems, such as explorations on the possibility of starting primary school at the age of six and not seven. Experiments on teaching Chinese characters and calculations were made in kindergarten; and algebra was taught in the primary schools, with lower grades in primary schools being taught to use English. The aim was to look at the possibility of certain middle school subjects being taught in primary schools from the psychological point of view.

Psychology of learning in schools involved studies including analysis of mental activity in solving applied arithmetical problems, thinking process in solving geometry problems, the problem of understanding subjects taught at schools and the formation of skill in athletic training, abacus operations, etc.

Foreign studies on programmed instruction were also introduced into China and tried in schools. Linear and branching programmes were adopted. A number of texts for programmed teaching were compiled, and simple teaching machines were developed; all with some success. It was also found that the time spent in experimental classes was shorter than in the control classes and the results were also better in experimental classes. Other problems were uncovered, such as the more extensive the textual reference material the more difficult it was for students to form whole concepts. Later, in keeping with the practical conditions in our country, programmed instruction was compressed and simplified and used in students' self-study exercises.

4. Labour psychology and engineering psychology

Research work in this field developed, but studies on rationalization of work were not as popular as in the 1950s. Nevertheless, psychologists participated in the study of the prediction of aviation ability of aircraft pilots, and studied advanced workers' skill characteristics in factories. During this period, the emphasis in research was switched on to engineering psychology. The discrimination of flash light signals was studied systematically to determine perceptual thresholds, equal discrimination scale and speed of flash light signals, semantic noise in signal coding and the effect of interference evoked by the differences in the frequency of different flash light signals.

In addition, information theory was used as an approach to study the influence of the spatial arrangement of signals on a panel board on information transmission rate between man and machines. This could be advanced by the arrangement of signals compatibility with reaction keys. There was also research on the effects of verbal and motor responses on information transmission.

Other studies were made in relation to the establishment of various kinds of standards, such as visual evaluation of sports stadium, illumination standards for use in classrooms etc. Engineering psychologists appraised the layout of instrument displays and aircraft control panels by examining their shape, scale, pointers, markings, design of numbers and lighting conditions.

An extensive investigation in the field of engineering psychology was made on the design of signal display in a control room for a large hydro-electric power station using a new control system (with low electric voltage). The items under study included the arrangement of the control room, fatigue evaluation of the monitoring workers, design of signal panel and ways of presenting signals, and the colour of panel and coding system. Reactions of workers when accident signals occur have been analyzed in a simulation laboratory imitating actual task environments. The result of this study has been put into use in the construction of a large scale hydroelectric station.

5. Medical psychology

Extensive investigations have been carried out in this field. Besides examining the speedy synthetic treatment of neurasthenia, psychotherapy was applied in the treatment of long-term schizophrenia. Several treatment methods were adopted to suit changes (activation, change and presence of delusions) in the illness. The therapy included giving lectures to the patients and allowing the patients to discuss and organize themselves in group activities. This was found to be beneficial in enabling patients to regain their health. At the same time, the characteristics of higher nervous activity of schizophrenia, patients' EEGs, and their ability to form abstract concepts were studied. The physiological and psychological basis of schizophrenia were theoretically explored and some criteria for its psychological diagnosis were proposed. There was also research on psychological factors in patients with such chronic diseases as high blood pressure and peptic ulcer, and experiments with a combined treatment like "qigong" and hypnosis yielded positive results.

6. Physiological psychology

In this field, EEGs had been studied systematically. EEGs were studied for different age levels to show the stages in brain development. The EEGs for orienting reflexes were investigated. EEGs were also used as a tool for the investigation of some pathological phenomena such as cortical disturbances of schizophrenia, and mental retardation.

In animal psychology, studies have advanced from classical conditioning studies to more complex behavioral researches, like dogs' relational conditioned reflexes to complex stimuli and monkeys' recognition of repeatedly appearing phenomena in the environment.

III

The third period was from 1966 to 1976. During this period when psychology came under the influence of Lin Biao and the "Gang of Four", progress in psychology came to a halt.

In October 28 1965, the Guang Ming Daily published a paper entitled "Is this the scientific method and correct direction in psychological research?" by Gu Ming Ren , which was actually a *nom de guerre* used by one of the members of the "Gang of Four" — Yao Wenyuan. This paper criticized a publication in the Acta Psychologica Sinica entitled "Differences in Colour and Form Preferences", but actually it was meant to reject the claim that there were common psychological laws. It maintained that psychological research should be replaced by class analysis, and that a psychology studying the common laws of human beings is a pseudoscience. When this paper was published there were different responses. Many psychologists disagreed with the views expressed in this paper. They maintained that: (1) the main aim of the paper "Differences in Colour and Form Preferences" was not in preference itself but to study the process of abstraction of children (from form abstraction to colour abstraction), (2) one should not take the attitude that all psychological problems are inaccessible to experimental verification, but rather some factors may be singled out for individual examination, (3) to avoid the influence of subjective conditions, the more abstraction in the study of the general laws of form and colour preference development, the more one can avoid the interference of other factors, and the general laws could be obtained in this way, (4) there are different preferences inclinations for different individuals, but this may not always be understood in a social context. Many comrades pointed out then that to replace psychology with class analysis was wrong, for if this were done it could destroy psychology.

During the cultural revolution, Yao's paper was regarded highly, for criticizing psychology, by some people and they gave credence to this paper as a dogma so that psychologists who were against the viewpoint expressed in Yao's paper were subjected to persecution. Due to this misdirection most parts of psychology were refuted, the remaining small part was removed from psychology and looked upon as physiology or medicine. Research work and teaching of psychology stopped and experts in psychology were forced to change their profession. The loss of library facilities and equipment in laboratories was also great. Institutions of psychology had been abandoned. Psychology as a science, which had begun to flourish after Liberation under the protection of Party and Government, suffered a lot during this period.

IV

The fourth period was from 1976 till the present time. During this period the study of psychology regained importance and progress was made.

Psychologists actively participated in the criticism of the "Gang of Four" and discussions were held on the reestablishment of research and teaching in psychology. In 1977, under the guidance of the Chinese Academy of Sciences, new plans were made for all spheres of science. In August, psychologists from various universities and institutions met in Beijing to discuss plans for the development of psychology. A draft for a long-term project was made, and psychologists were assigned various projects. Psychologists from various districts of China discussed this project and many who had left the field returned to work in psychology. Study and research groups in psychology were resumed. New research projects were started. An enlightening picture appeared in psychology and this meeting was considered by Chinese psychologists as a major turning point in the development of Chinese psychology.

In May 1978 another scientific meeting was convened in Hangzhou and was attended by representatives of 35 teaching and research institutes from 24 provinces, municipalities and autonomous regions. At this meeting 3 problems were discussed: (1) the development of psychology, (2) the problems of educational psychology, and (3) evaluation of the historical personality in psychology, Wilhelm Wundt.

After a suspension of 15 years, the Second Scientific Meeting of the Chinese Psychological Society finally resumed in December 1978. Delegates from 28 provinces, municipalities and autonomous regions and guests from various research and teaching institutes were present. At this meeting the lessons and experiences of 15 years were summarized, the developmental trends of international psychology were introduced, and a decision was made to hold a Third Scientific Meeting in 1979. The Acta Psychologica Sinica was resumed and special committees were established; the leading organizations of the Chinese Psychological Society were improved.

Chinese psychologists then worked with renewed vigour and accomplished some preliminary results in the last two years. These works may be summarized as follows:

1. Developmental psychology and educational psychology

A nationwide collaborative study was organized in this area. Topics included: development of number concepts in children from 3–7 years of age, survey and investigation of supernormal children, and the development of language ability. These collaborative studies are producing good results. A variety of methods have been used in these studies which include experimental methods, surveys, summing up of teaching experiences of advanced teachers. "Intelligence testing", a forbidden topic ever since Liberation, has been used, and foreign psychological tests have been analysed with consideration given to conditions in China.

Presently, valuable results have been obtained in developmental psychology (development of concepts, language, personality in supernormal and mentally retarded children) and educational psychology (moral education, psychology of teaching and learning various subjects, sport psychology).

2. General, experimental and engineering psychology

The main research topics concern visual perception and hearing. Many of these researches are related to standardization, luminous efficiency function of Chinese eyes, standards for Chinese skin colour for use in colour television and colour photography, with tolerances of memory skin colour, tolerances for most commonly used colours. The results obtained from the studies of visual performance of Chinese eyes provide the basis for the prescription of illumination standards for factories and mines. To set standards for noise control, research has been undertaken on the measurement of loss of hearing. The acuity of depth perception among the Chinese was also tested, providing a basis for the design of optical instruments. The results obtained from this research have been adopted by the relevant governmental departments.

3. Medical psychology and physiological psychology

Psychologists have also conducted research on the role played by psychological factors in acupuncture, anaesthesia and analgesia. These include the influence of such psychological factors as suggestion, attention and emotional states. Psychologists also collaborated with medical doctors to study emotional changes before, during and after surgical operations as well as the relationships between emotional changes and physiological and bio-chemical changes. Because the operation is done under acupuncture analgesia, the patient is fully conscious and is able to relate his experiences. It is not possible to obtain data of this nature with drug anaesthesia. To develop a basic theory of pain, research has been done on biorhythmic changes in pain thresholds and the effects of acupuncture on the functioning of the hippocampus.

In physiological psychology, some studies have been made on the neural and biochemical basis of learning and memory. With electrical damage and effects on different areas of the hippocampus and injections of protein and enzymes, their relation to the dark avoidance reflex pattern was studied. It was proved that during the early phase of learning and memory the hippocampus played an active role and was related to the synthesis of proteins.

In pathological psychology, psychologists studied psychodelic drugs, especially Chinese herbal psychodelic drugs, and tested them with animal behaviour experiments. In psychiatry, hypnosis and psychotherapy were studied. These treatments exert a positive effect on the patients after hospitalization and reduces the re-admissions to the hospital. Chinese psychologists have also experimented with the new techniques in psychotherapy from other countries, such as bio-feedback, behaviour modification. Neuropsychology flourishes in many topics: asymmetry of cerebral function especially in vision and hearing, the transfer function of corpus callosum, deficiencies in speech and memory in brain damage patients.

4. Theoretical studies in psychology

This is a field of systematic psychology. Recently, to commemorate the centenary of the first psychological laboratory founded in Leipzig by Wundt, Chinese psychologists appraised Wundt's contributions and 29 of his papers were read at the 1978 annual meeting. Many participants announced that the attitude of complete negation of Wundt in the past was wrong, and due acknowledgement to the contributions of Wundt must be given. Other prominent psychologists in the world have also been reviewed, such as Soviet psychologist Leotiev, concerning relationships between activity and consciousness.

Several experiences and lessons should be remembered when we reflect on the course of the development of Chinese psychology over the last 30 years.

(1) Marxism is the guiding principle for psychological science, but psychology cannot be replaced by philosophy. We must use dialectical materialism to guide our psychological studies, and not by simply referring to quotations as proofs. We must adhere to

the principle that the only standard for the verification of truth is practice.

(2) Never leave the principle, "let a hundred flowers blossom, let a hundred schools of thought contend". This is a prerequisite for developing psychology. In the past we had tried to treat different scientific viewpoints as political problems and subjected them to criticism. This was wrong and stupid. It should not be repeated in the future.

(3) We must treat foreign psychology in an objective manner, using what is good and useful to us and rejecting what is not. The blind anti-foreign attitude, with only criticism without learning can only harm ourselves. We should translate more foreign books and introduce psychology from overseas, promote scientific exchange and broaden our horizon.

(4) Psychology is in a stage of development and borders on other sciences, such as the social sciences, natural sciences and the technical sciences. Studies of the frontier sciences — neuropsychology, social psychology, ergonomics, artifical intelligence and biofeedback should be encouraged. In the past, psychology was limited to a narrow field and its development was restricted, so we should not limit the study of psychology by traditional boundaries.

(5) At the moment, the meaning of psychology is not well-known among the masses due to the lack of effort by the relevant departments. Therefore, psychologists should now make contributions to serve the "Four Modernizations" as well as popularise this science so as to create better conditions for the development of psychology.

CHAPTER 17

A BRIEF INTRODUCTIONN TO TRADITIONAL CHINESE MEDICINE

by Tsai Chinfong

Traditional Chinese Medicine (TCM) has a history of several thousand years and was fundamentally formed and developed in the feudal Chinese society. Progress was retarded this century before liberation due to the cultural invasion by imperialism. In the short period of 30 years after liberation, tremendous progress has been achieved and this may be summarized in the following three categories.

I. Fundamental Theory

Generally speaking, the various schools of thought in TCM may trace their origins to ancient medical classics like "Huang Di Nei Jing", "Shen Nong Ben Cao Jing", "Nan Jing", "Shang Han Za Bing Lun", "Zhen Jiu Jia Yi Jing", "Mai Jing", etc. A lot of work (revisions, checking, compilation of notes and comments, and translation into modern Chinese) has been done on these medical classics since liberation. In the past 30 years, roughly about a thousand papers on TCM (including those combined works of traditional Chinese and Western medicine) have been published. These included teaching materials, comprehensive introductions for TCM, and symposia of various subjects (e.g. the Yin-Yang Principles, the Five Revolvers and the Six Climatic Factors, Acupuncture and Moxibustion, the visceral organs and meridians, pathogenesis and pathology, clinical medicine, Chinese materia medica, etc.). These publications encompassed new ideas and developments, as well as the traditional approach to TCM. For instance, "An Outline of the Methodology of Ophthalmology in TCM" utilized "treatment based on the differential diagnosis of signs and symptoms" according to the method of the six meridians in ophthalmology. It combined the theory of TCM, anatomy of the interior of the globe, "Fire Wheels" and the eight parts of the external eye with the differential diagnosis of viscera, thus pushing traditional Chinese opthalmology a step forward in the old medical academy. Various collections of veteran physicians' clinical experiences, case reports and medical lectures have been summarized and published. A new dictionary, "A Concise Dictionary of TCM", was the first of such ever published since liberation. Several important ancient medical classics, such as those mentioned above, (i.e., "Huang Di Nei Jing", "Shen Nong Ben Cao Jing", etc) have been translated into Modern Chinese with detailed explanations. Reviews, comments and annotations on "Pi-Wei" (spleen and stomach), "Wen-Yi" (Epidemic Febrile Diseases without chills), and "Corrections on Medical

Problems" were published. Discussions on theoretical problems included summaries on the Febrile Diseases School, Four Academic Schools of the Jin Yuan period, School on Warm Tonics, School on Febrile Diseases without chills, "San Jiao" (Three "Burnt" places), the treating principles of antipyresis by sweating and with warm Nature drugs, and the strengthening of the five elements to ameliorate the earth elements. All these have contributed much to the development of traditional Chinese Medicine. Practical discussions involve the question as to whether the theory of five elements and "Zi Wu Liu Zhu" (circulation of vital energy and blood in the meridians determined by the regular turning of 5 revolving and 6 climatic factors for the acupuncture-moxibustion treatment) is worth retaining. A trial has been done on the unification of the traditional programme for differentiating febrile diseases. It was clinically applied for instance, to obtain a structure for the differentiation of diseases according to the six meridians, "San Jiao", and the "Guarding-vital energy-nutrient blood doctrine". The latter was extended to the treatment of non-infectious diseases and the rules concerning its application were further summarized.

All of the above promoted the development of TCM. Undoubtedly, the continuing development and contributions of pure TCM will further the development of the integration of two systems of medicine, TCM and western medicine, thus creating a new medicine in China.

It should be mentioned that much attention has recently been given to a continued development of medicine practised by the National minorities. An important Tibetan medical classic, "The Four Tantras (rGyud–bzhi)" has been translated into Chinese and will be published when revised and finalized. Moreover, work on the inherited medicine of the National minorities (which includes Tibetan, Mongolian, Uighur Yi, etc) with the aim of integration with western medicine is still going on. The symposia on the treatment and application of Materia Medica have been published. These are of great significance to the continued inheritance and development of the ethnological medicine of National minorities.

Research on the fundamental theory of TCM is of utmost importance to its development. At present, this research cannot be strictly limited to discussions on the theoretical problems. To develop TCM further, the Party calls on us to address ourselves to the theoretical problems of TCM using scientific theories and methods. This is in accordance with the historical evolution of medical science in general. Western Medical science, for example, evolved gradually after the Renaissance from ancient Roman and Greek medicine when the modern approach and theories were adopted. Without the adoption of modern scientific concepts and methods, TCM will not change fundamentally and its modernization will be impossible.

Recently, armed with modern scientific and technological measures research work on the fundamental theories of TCM has progressed and this progress may be summarized under the following topics:

1. Diagnosis.

Diagnosis in TCM possesses excellent features. In antiquity, there was no scientific apparatus for performing relatively objective examinations. Physicians had to make efforts to diagnose disease by direct sensation. Of the four diagnostic methods (i.e. inspection, auscultation, inquiry and palpation), the methods of pulse feeling and the examination of the tongue have been greatly developed. The pulsation method, especially, had an extensive and profound effect domestically and internationally. Ancient Islamic pulse feeling was influenced by the Chinese method. More than 20 kinds of pulses were distinguished in ancient China through a subjective feeling method. "The Pulse Classic" (canon of pulsation) by Wang Shuhe of the Jin Dynasty mentioned 24 kinds of pulses, while in the "Pinhu Mai-Xue" (pulsology of Li Pinhu), written by the famous Ming naturalist Li Shizhen, 27 were mentioned. To determine whether these pulses were merely subjective sensations, or whether they were truly objective, pulsilograms have been used for detection and checking. The 15 kinds of pulsilographs of various conditions were "fu" (floating), "chen" (sinking), "chi" (slow), "shu" (rapid), "cu" (short), "xi" (tiny), "ruo" (feeble), "ru" (soft), "wei" (minute), "jie" (slow-irregular), "dai" (rapid- irregular) and "huan" (delayed).

Certain instruments recorded the waves of "unsmooth" and the "xuan" (tense) pulses, while others recorded on graphs the concurrent pulses such as "tense-smooth", "tense-minute", "minute-smooth", etc. These graphs form the objective proofs of traditional Chinese pulse-feelings. Objective indices and parameters were proposed and used so as to enable the analysis of these pulsilographs on a large scale. With these indices and parameters, and by comparing them with the standard pulsilograph, a patient's pulse wave can be determined and classified for diagnostic purposes. The analysis of pulsilographs of 20 different kinds of pulses was scientifically based and an equation for the hemodynamic index was formulated and stored in a computer for further automation of pulse diagnosis.

Pulsilography is also applicable for grouping diseases according to the Chinese traditions and is a useful reference for treatment and prognosis. For instance, it is known that the appearance of the "xuan" (tense) pulse is related to arteriosclerosis and increased peripheral resistance of local vessels. It is observed that there is a related and important reference value for determining the sequelae and prognosis of the shock condition in the pulsilogram of patients suffering from septicaemic shock. It was observed that pulsilogram changes occurred prior to sphygmomanometric changes, and there is a correlation between the intensity of the pulse and blood pressure manifested in the pulsilogram. The development of the pulsilogram of a patient reveals changes in the following sequence from a "shallow inch pulse" to a "juncture-foot pulse" to a "foot pulse", a "subfoot" pulse until finally, the pulse is absent. Hence, prophylactic measures may be taken in advance according to changes in the pulsilogram. There is also a close relationship between pulse manifestation and diseases, e.g. "wei" (tiny), and "ruo" (feeble) pulses indicate a weakness syndrome in the Yang principle, while "xuan" (tense) pulse is often seen in liver diseases and liver-stomach incompatibilities, etc. Recently, the use of modern biomechanics and pulse analysis techniques in pulse research has also made a good start.

Bio-microscopy, lamp examination, physiological biochemical, microbiological and physiological biochemical, microbiological and pathological examinations, including biopsies, pH determinations and bacterial cultures have been employed for advanced studies of the tongue. The coating on the tongue proper was analyzed physio-pathologically. It was found that changes of the coating and tongue proper are closely related to the blood circulation (including capillary changes and haemoglobin), digestive function (including liver, pancreas, enzyme systems, metabolism and nutritional condition), the nervous system and the endocrine system. Any changes in these systems (eg. edema, cellular droppings, hyperplasia and infection) manifest themselves as various kinds of changes in the tongue texture and coating. The lingual picture is important in diagnosis and prognosis. For example, in liver carcinomas, purplish stripes or irregular spots can be seen on both sides of the tongue. This is not seen in any other disease. The course of the disease also manifests itself by changes of the tongue coating. As the disease becomes severe, the coating may change its colour from yellow to grey to brown and finally black. As the patient's condition improves, eg. in hepatitis and acute myocardial infarctions, greasy coatings may become non-greasy, the coating area and thickness decrease and the texture of the tongue returns to normal. These textural changes of the tongue form the scientific basis of "lingual diagnosis" in TCM. However, more research would further establish the knowledge of lingual diagnosis on a more solid scientific basis.

2. Theory of Visceral Organs

This is also an important component of the fundamental doctrine in TCM. Research on the theory of viscera has been continuing since liberation, concentrating mainly on individual viscera like the "shen" (kidney) and "pi" (spleen).

Advances have been made with investigations of the effects on the kidney due to diseases (dysfunctional uterine bleeding, bronchial asthma, lupus erythematosus disseminata, sclerosis of coronary artery and neurasthenia) which reveal common manifestations with the weakness of vital energy (pneuma, Yang principle) in the kidney. It was discovered that the concentrations of 17-OH corticosteroids in the urine of these patients were low. Studies monitoring the daily levels of 17-OH corticosteroid, Su-1885 and radioactive immune plasmal ACTH, discovered a malfunction in the hypothalamus pituitary-adrenal-cortical system decreasing the activity of the adrenal cortex. Observations on the respiratory quotient, glycogenolysis in erythrocytes and oxidation reactions of phosphate pentoses in basal metabolism, also disclosed abnormal metabolic functions in such patients. These patients were treated with drugs which "warm the vital energy" (Yang) of the kidney. After treatment, the conditions improved and were checked, and the 17-OH corticosteroid levels in the urine returned to normal. This indicated that the kidney in TCM is a regulator of the neurohormonal system. Other observations also revealed that there are definite correlations between the kidney and the immune system, and the metabolism of mineral salts like potassium and sodium chlorides. These investigations provide a basis for the study of the nature of the kidney as well as rules to be followed in the treatment of diseases with "weaknesses of vital energy" in the kidney.

And there is the question of the nature of the "spleen". The definition of "spleen" in TCM is vague. Studies on the digestive, respiratory and immune systems in patients suffering from weakness of "pi", demonstrated a relation with ingested food that was not well digested and stools with fat granules, and muscle fibres. Results obtain from iodine excretion by the parotid gland and salivary amylase content examination suggested that the activity of the digestive glands were below normal. This mechanism needs further investigation. Excessive coughing and sputum in patients with a weakness of the spleen were very common and this gave a scientific basis for the old saying in TCM that "spleen" is the origin of "sputum".

Apart from the spleen and kidney, preliminary work on the nature of the liver, "San Jiao", has also been carried out with certain success. Recently, in the study of the nature of "pneuma" (vital energy, "Qi" in Chinese), it was claimed that pneuma was a kind of message of the human body that can be transmitted, stored, and exchanged. Due to the original definition of pneuma being so extensive, investigations on the physio-pathology of some diseases can reflect its nature in not only one aspect. The studies of cardiovascular diseases, led to preliminary conclusions that cardio-vascularly and haematologically, pneuma alludes to the function of the left heart, the haemolytic and coagulation mechanisms etc., and is closely related to haemodynamics and haemo-rheology. The treatment principles which involve the activation of the blood circulation and stagnation, benefits the pneuma or vital energy and ameliorates the physiological function. They probe into the nature of pneuma and preliminary results form a basis for further investigations. Laboratory and clinical studies were made on the main aspects of pneuma and the "spiritual pneuma" ("Jing Qi" in Chinese) from the immunological, physiological, biochemical, and histological standpoints. It is believed that "spiritual pneuma" is related to energy metabolism and immune function. Based on ancient literature, the ideas of constitution, differential diagnosis constitution and constitutional pathology were classified according to the fundamental concept involving the Yin-Yang principle. The blood and pneuma, sputum and dampness, and the normal human body were classified into six groups, namely: "normal", "gloomy", "greasy", "dried-red", "stagnant-coldness" and "fatigue-paleness". The common features of such a classification in the two different systems of medicine were investigated. Indeed, it provided a good start for the study of the theory of constitution by the method of combined medicine.

It should be noted that there have been preliminary trials in attempts to apply the theory of cybernetics to explain and investigate that theoretical aspects of TCM which holds that the body is a black-box. Treatment based on the differential diagnosis of signs and symptoms in TCM shows the process of input and output regulation and adjustment of messages. The concept of the regulation-adjustment system of the body resulted from the theory of "pneuma" with the principle of equilibrium from the Yin-Yang idea and an inter-reliance of the "Five Elements".

3. Acupuncture, Moxibustion and Channels

Studies on the theory of acupuncture, moxibustion and channels included two components, acu-anaesthesia and the mechanism of anaesthesia. The technique of anaesthesia has a basis derived from ancient classical acupuncture-moxibustion, and analgesia is a great achievement representing a combination of western and traditional Chinese medicine. The advantages of this technique are:

(1) The patient is in a conscious state and is able to cooperate with the surgeon to ensure efficacy;
(2) High safety;
(3) Equipment and installation are simple and easy to handle,;
(4) It increases the resistance of the patient against pathogenic organisms, inflammation and shock; and
(5) Quick convalescence after surgery.

Although there are shortcomings, acu-anaesthesia is a safe, effective and simple process. To date, more than two million cases have been operated under acu-anaesthesia. It has been used in all kinds of minor and major operations, complicated brain and heart surgery which require artificial heart-lung preparations, in surgery involving the thyroid, maxillary sinus, glaucoma, and ligation of Fallopian tubes through the abdominal route.

It has also been routinely used in Caesarean sections, subtotal gastrectomy, splenectomy, and laryngotomy. Acu-anaesthesia is the first choice in brain surgery, prostectomy, excision of meniscus and lobotomy. The recent search for effective acu-points, modifications in manipulation and variations of the stimulating index, application of rational adjuvant drugs, and the improvement of surgical instruments, have led to the formulation of clinical rules in acupuncture for preoperative prediction, adjuvant administration of drugs and manipulation. Guided by these advances, relevant developments in surgery and physiology also occurred. Acupuncture is currently used in 30 other countries in the world.

Studies on the mechanism of acu-anaesthesia showed that pain sensation is a kind of endogenous reception. It is affected, to a certain degree by psychological factors, so that evaluations of results obtained from animal experiments are limited. Nevertheless, investigations on the mechanism of acu-anaesthesia are very important. When stimulation is applied to the proximal segments of experimental animals, acu-point is most satisfactory at a high frequency of 80/sec; and on distal segments, acupoint is most satisfactory at a low frequency of 2/sec, showing a difference in mechanism between the two segments. In analgesia, different types of nerve fibres reveal different kinds of analgesia effects. The role of the central nervous system in analgesia is suggested in experiments which showed that the stimulating signal is first transmitted to the spinal cord then ascends through the lateral spinal tracts of the external thalamic system. The analgesic action is accomplished by effects on the external thalamic nucleus, on both its ascending and descending tracts and its afferent sensational signals at different levels of the central nervous system. It has also been found that the cerebral cortex plays an important role in the conformation of pain sensation signals. It reflects, to some degree, the process of acu-anaesthesia and pain sensation. The analgesic action may be related to its inhibitory action on the specific and non-specific projection system.

Much research work has been done on the action of nervous media in acu-analgesia. A method using a flourescent photometer has been developed for locating the transmitting media such that within the same nerve tissue, 3 kinds of mono-amines are transmitted simultaneously, namely the 5-OH serotoninergic system, noradrenaline and dopa, and its related metabolites. The results have revealed that the 5-OH serotoninergic system is a necessary factor in acupuncture for which its analgesic effect is proportional to its ratio with prednisolone. The level of noradrenaline decreases when 5-OH serotonine increases in a negative-feedback relationship. In the caudate nucleus, acetyl-choline is another important analgesic transmitter. The catecholamine in the central nervous system may have some antagonistic action on the analgesic process. The use of signal detecting devices first proved that naloxone can partially antagonise the analgesic effect of acupuncture in the normal body. The liberation of endogenous opiates in the central nervous system during acupuncture has already been established. Moreover, much work has also been done on the humoral aspect in acu-analgesia.

The ancient meridian theory is one of the important components of the fundamental theory in TCM. One may ask what is the material basis for this. The clarification of such problems is important for the creation of a new medical system in China. There is current research on the nature of meridian centers in the phenomenon of sensation propagation along the meridians (PSC). This sensation is produced when acu-points are stimulated, and the sensation is transmitted along those tracts described as meridians in ancient medical literature. Intensive research on this subject has been going on since 1958. This included a study where 12-24% of a randomly selected part of the population (100,000) revealed PSC of different degrees. Moreover, there were 60-70% with latent neural PSC, showing the PSC is a phenomenon commonly occurring in the human body.

Other investigations on PSC demonstrated the following features:

(1) it can be aroused by physical stimulation such as puncture, pressure, temperature (moxibustion), low frequency pulsed electric stimulants or even electro-magnetic waves of high frequency;
(2) the transmission of PSC is generally in the direction of the meridian. In the presence of lesions near the meridian, PSC will propagate towards these foci, coinciding with the idea that "vital energy goes toward the lesions" in ancient literature. Finally, the propagation can be affected or even interrupted by various objective factors. Under normal conditions, the speed of the propagation is about one hundredth (several cm/sec) that in the nerves.

The transmitting speed is proportional to the local temperature, i.e., the higher the local temperature, the greater the transmitting speed, and vice versa. The transmission of PSC is interrupted when the local temperature is lowered to 20°C, when pressure on a local point is 500g/cm^2 or when local anaesthetic or normal saline is injected. PSC will continue to propagate after the disturbance factors are withdrawn.

It has been shown earlier that meridians have a marked effect on their corresponding visceral organs (e.g. stomach meridians on stomach, pericardial meridian on the heart, etc.). For instance, an interruption in the spleen meridian will arouse PSC in the tongue and make the patient's taste more sensitive.

On the basis of the above findings, the "wave transmission of channel" hypothesis was proposed stating the PSC is an electromagnetic wave charged with information propagating in the direction of low resistance.

Most experiments on the conduction of acupuncture-moxibustion proved that when the related neural-humoral factor is interrupted, suppressed or excised, the effect of acupuncture will be diminished or removed. This indicated that the conduction of acupuncture stimulation is closely related to the nervous system, humoral factors, and possibly others. Histological and anatomical studies on the acupuncture patients have shown that there are definite relationships between the points with their linking lines and the surrounding nerves and vessels. Some 20 acu-instruments and techniques have been developed since Liberation, which include electro-acupuncture needles, scalp-needles and other physical techniques, chemical acupuncture point therapy etc. They have been applied in the treatment of 300 different diseases, a hundred of which succeeded with good results.

4. Studies on the Differential Diagnosis of Signs and Symptoms

Diagnosis in TCM is made by differentiating a patient's signs and symptoms. Is there any scientific rationale? What is it based on?

Investigations on the " Eight 'Gang' " (Key Link) revealed that the "exterior" ("Biao" in Chinese) syndrome manifested itself mainly as an inflammatory reaction of the mucous membrane of the respiratory and the five sense organs, edema in the muscle, and congestion or ischaemia in the central nervous system. Most of these are due to the defensive convulsion of arterioles. The "Li" (interior) syndrome is mainly a disturbance of body function or energy metabolism influenced by pathogens, resulting in either excess or insufficient heat production, or the uneven distribution of heat throughout the whole body. It may be manifested locally or through the whole body with symptoms such as anaemia, ischaemia, stagnation of blood and edema, etc. The two key links of "cold" and "hot" are related to the autonomous nervous system and endocrine system, and particularly to the disturbance of thyroid functions. The physiological function of the sympathetic nervous system of "hot" patients is increased and excretions of dopamine are markedly increased. Conversely, there are degenerative changes in the cells (e.g., atrophy, fatty degeneration of solid organs and endocrine glands) in the "cold" and "weakness" patients. As a result, these organs increase in size, volume, pigmentation and there is a proliferation of connective tissues leading to sclerosis and fibrosis. The "xu" (weakness) syndrome usually presents itself in chronic diseases and had been intensively studied in the past. It was shown that a lack of "Yang" is accompanied with a decrease in cGMP levels, and the ratio of cAMP/cGMP is markedly lowered. A lack of "Yin" is accompanied by a marked increase in cAMP and slight increases in the cAMP/cGMP ratio. These changes affect the functions of the nervous, circulatory and endocrine systems, due to a mechanism involving the manifestation of "weakness" in both "Yin" and "Yang". A weak immune response in "weakness" in "Yin" with "roaring fire" caused by excessive increases in the activity of solid organs in "Shi" (excessive) syndromes resulting in the stagnation of blood, tumours and constipation.

Investigations on the key links of "Wei-Qi-Ying-Xue" (guardian-pneuma-nutrient-blood) were carried out on "Wun" diseases (Febrile disease without chills). It was discovered that disease of the "guardian" syndrome is due to the reaction of superficial nerves and vessels. The lesions are mainly centered in the upper respiratory tract. The "Pneuma" syndrome is the result of hyperpyrexia which produces disturbances in humora and mineral metabolism with toxemia in these patients. In the "nutrient" syndrome, there is pathological degeneration and necrosis, disturbances of the coagulation system, and infiltration and damage due to intoxification of the capillaries in the central nervous system. The "Blood" syndrome is the critical phase of acute infection where pathological changes such as bleeding in the central nervous system and internal organs are extremely severe.

5. Treating Principles

These form major links in the system of treatment based on differentiating signs and symptoms in TCM. Most of these principles were proposed in ancient China. What forms the scientific basis for these principles? How are we going to check them with objective indices so as to enable these principles to become even more reliable and applicable in order to increase their therapeutic effects? Recently, investigations were conducted with emphasis on the principles of activating circulation and removing stasis, supporting body resistance and consolidating the base.

The principle of activating circulation and removing stasis was applied to the treatment of blood stagnating diseases including cardio-vascular, digestive, urino-genital diseases, nervous system diseases, collagenous diseases, connective tissue diseases and diseases of the immune system. It was discovered that treatments based on these principles improved the rheology, blood circulation, anoxia and ischaemic conditions of patients. Observations on the rheological indices including haematocrits, viscosity of whole blood and plasma, electrophoretic time of RBC and fibrinogen, resulted in the discovery that the blood of such patients become more viscous, sticky and packed than normal. Drugs which activate circulation and remove stasis are capable of improving these pathological conditions, i.e. they ameliorate the rheological indices, viscosity, etc. Breaking of the stagnant nodules, activating circulation and removing stasis can lead directly to the dilating of peripheral vessels. By observing the circulation of the capillary beds under the nails, aggregation of blood cells, blood emulsion and looping condition, the ischaemic and anoxic conditions can be improved. Experiments on animals demonstrated that treatments based on these principles resulted in the removal of aggregation and adhesion of thrombocytes, dilating arteries, and softening connective tissues. They can also be used for the prediction and prevention of ischaemic epilepsy. There has also been progress in the treatment of keloids, thus indicating that such treatments have a hopeful future.

Another principle of treatment, which supports body resistance and consolidates its base has been much studied in recent years. It is mainly used in debility and chronic consumption and it was discovered through the treatment of many chronic diseases such as asthma, chronic bronchitis, malignant tumours, etc., that treatments based on this principle can ameliorate the disturbed function of the thalamo-pituitary-adrenal-cortex system. The phagocytic activities of leukocytes were also markedly increased.

Comparisons of the immune indices of patients with malignant tumours were made before and after treatment by monitoring the level of plasma cortisol, phagocytic activities, lymphocyte transformation and rosette forming tests. It was discovered that this principle of treatment improved the function of the adrenal cortex and increased immune indices. The humoral and cellular activities were increased. It was also discovered that these changes were fulfilled through antibodies produced by B cells and transformation of T cells by the action of nucleic acid and cAMP. Animal experiments showed that this treating principle has an effect in strengthening tolerance to anoxia and "coldness" of the organism. In short, it has been proved objectively that the principle of supporting body resistance and consolidating its base leads to the strengthening of body resistance.

A good start in the investigation of other principles including antipyrexia, cathartics and antitoxic penetration has already been made. Finally, it should mentioned that traditional breathing exercises have also been studied. EEG and infra-red thermograph analyses showed that the point temperature was largely elevated, indicating that the blood circulation and metabolism of the organism during exercise were significantly changed. This provides an important basis for further study on the mechanism of traditional breathing exercises.

II. Clinical Studies of Combined Medicine

The use of TCM has achieved clinical successes in the treatment of various diseases, some of which could not be adequately handled by western medicine. Just after liberation, published summaries on B. encephalitis and piles revealed successful treatments by TCM. However, it must be admitted that all the clinical studies then were actually based on combined medicine. The cases were diagnosed and observed according to the rules of western medicine but treated by TCM. Over the past years, treatment based solely on TCM, which was applied to certain diseases such as epidemic haemorrhagic fever and optic atrophy, was found to be effective for children. Chronic hepatitis was also effectively treated by the TCM. Satisfactory results were achieved in most cases of vitiligo treated by the national medicine of Uighur with complete success for 8.4 per cent. B. encephalitis, thrombo-angitis obliterans, cholelithiasis, certain types of lupus erythematosus disseminata, functional uterine haemorrhage in menopause, aplastic anaemia and chronic nephritis treated with TCM or combined medicine, malaria and bacillary dysentery treated with acupuncture; all achieved satisfactory results and are worthy of further study.

Methods were designed for clinical research in combined medicine. These include standards for the identification of clinical effects, methods for classifying diseases by combining the differential diagnosis of syndromes in TCM with that of western medicine, objective indices for observing the duration of diseases and its sequelae, the combined use of Chinese herbal drugs with western drugs and the design of the treatment programme etc. Successful experients were accomplished in those aspects, Currently, combined methods are widely used in all clinical departments with some success.

1. Cardio-Vascular Diseases

The studies concentrated mainly on coronary diseases. According to the theory of TCM, angina pectoris is a result of the "retention of pneuma and stasis of blood" and should be treated by a programme designed on the principle of pneuma activation and blood stasis removal. For example, an agreed recipe, the "Guan Xin (Coronary heart disease) No. 2", (Composed of *Salivia Miltiorrhiza, Ligustica Wallichii, Carocus Sativus, Paeonia Veitchii, dalbergia hancei*) was used to treat 164 cases of angina pectoris in CHD in the Beijing region. The follow up examinations for 1, 2, 3 and 4 years, demonstrated the curative effect in EKG as 37.4 per cent, 46.7 per cent, 55.6, and 66.7 respectively; and the rates of alleviation of angina pectoris were 89.6, 89.9, 92.9, 93.9 percent respectively. It is believed that the mechanism of such effects may involve the inhibition of thrombus, formation and haemolysis of RBC, increased activity of fibrinolysinase, decreased activity of Factor XIII, and the inhibition of aggregation of thrombocytes in the peripheral vessels and within the myocardium. The Ligustrazin extracted from *ligustica wallichii* was very effective for the treatment of acute obliterated cerebral vascular diseases. Its effective rate which can exceed 90 per cent is better than that of morphine and nicotinic acid. Most of the patients can take care of themselves without difficulty. Mortalities due to acute myocardial infarctions have been markedly lowered, when the treatment principles of activating circulation, eliminating stasis, supporting body resistance and nourishing essence were adopted. For example, between 1976-1977, mortality of the disease was decreased to between 13-14.6 per cent in Beijing. Satisfactory results were achieved when CHD was treated by acupuncture and EKG improved by 53 per cent. Studies in electro-cardio-oscillography, ultrasonic electro-cardiography, cardiac output and rheography demonstrated that acupuncture is capable of promoting coronary and cerebral circulation, and the function of the left heart.

2. Acute Abdominal Problems

One of the greatest advantages in the treatment of acute abdominal ailments by the combined method of TCM and western medicine is the avoidance of surgery in most patients. The rate of non-operation for acute appendicitis is 70-80 per cent, the effective rate being as high as 85 per cent. About 70 per cent of acute perforations of stomach and duodenal ulcers can also receive non-operative treatment with better results, both short and long term, than simple operative sutures. The rate of expulsion of stones in hepatic and biliary ducts is 60 per cent with 25 per cent complex expulsion. Total expulsion rate in cholelithiasis might reach 78 per cent when treated by electro-acupuncture with magnesium sulphate. The expulsion rate within 1-5 days is 69 per cent. "Hungry-thirst therapy" was basically abolished and replaced by combined TCM and western methods. Up to 60-70 per cent of acute intestinal obstructions were indicated for non-operative treatments. Surgery was avoided in more than 90 per cent of intussusception of intestine in children treated by airenema, and satisfactory results were also achieved in adults. Surgery was also avoided in the majority of ectopic pregnancies so that some may even retain their reproductive abilities. The effective rate achieved in checking the abnormal positions of foetuses (29-40 weeks) treated by moxibustion applied at "Zhi Ying" points, was more than 80 per cent.

A new branch of medical science, concerning acute abdominal ailments, thus formed under the direction of dialectical materialism characterized by the advantages of both TCM and western medicine.

For instance, physiopathologically, it breaks the old view that cathartics are contraindicated in acute abdominal problems. Traditional theory maintained that the 6 hollow viscera must be "functioning unobstructed"; the new theory maintains that the hollow viscera, with decreasing peristalsis will lead to a bad prognosis. That cathartics will aid in the cure of acute abdominal problems was supported by experimental studies. The new idea forms the basis of the new science of acute abdominal medicine.

3. Orthopaedics

Fractures healed by the method of small splint fixation in combined medicine possesses several strong points. They include restoration of function, shorter duration of recovery, rapid cure, less suffering and low cost. There are indications that such therapy may be applied to different kinds of fractures, including new fractures of the four limbs (forearm, shafts of humerus, femur, tibia and fibula), old fractures (including malformed fractures, and delayed fractures), fractures of the vertebrae (including cases with intact and partially injured spinal cord), open fractures, complicated and intrajoint fractures, (rotation-dislocation of the medial condyle and medial epicondyle of humerus, intercondylar fracture of the humerus and carpal-scaphoid fractures), fractures around the joints (e.g., separation of epiphysis of the proximal end of the humerus and fractures of the humerus proximal to the condyles). Formerly in a rotation fracture of the lateral condyle of the humerus, surgery was necessary, but the result was not satisfactory. Bone union can occur if treated by the method of small splint fixation, and the function of the elbow joint can be restored satisfactorily. The duration of treatment might be shortened by a third of the time required if surgically treated. Another example is in the intrajoint fracture of the carpal-scaphoid bone. The average course for complete bone union by small splint treatment with combined methods will last only 94 weeks. Fractures are treated by the combined methods of TCM and Western medicine according to the principles of the "four combinations", i.e., combination of part and whole, combination of fixation and movement, combination of treating the soft tissues and the bones, and combination of drug therapy and fixation operation. This is a new system of fracture treatment. The duration of this treatment might be shortened by a third to a half of that treated solely by TCM.

Orthopaedics in modern medicine is a set of treatments developed on the basis of traditional experience combined with the knowledge of modern medicine. Satisfactory results were obtained in treating diseases of cervical vertebrae, protrusion of vertebral discs, disturbances of the posterior joint of thoraco-lumbar vertebrae, fissures of crista of lumber vertebrae, piriform muscle injury, pain in the shoulder and other soft tissue lesions. A new hypothesis on the pathological mechanism of soft tissue injury was proposed, which maintained that injury and degenerative lesions of the local tissues would result in the lost of equilibrium or stability of the vertebrae column as a whole. In addition, shifting or inclining displacements of the spinal processes of the body of vertebrae and laceration of soft tissues can also cause minute changes in anatomical position. Based on experience in the Beijing region, the application of this new hypothesis has guided the treatment of soft tissue injury with satisfactory results.

4. Burns

This is another area of notable achievements in the clinical field of combined medicine. In the emergency treatment of burns of large areas, more than 10 per cent of 3rd-degree burns ended in recovery. In 1958, the combined method was successfully used to treat a patient with a total burn surface of 89 per cent of which 28 per cent was 3rd degree. After that until 1960, the rate of recovery for large area burns was raised to 70.2 per cent and the duration of treatment shortened by half. In 1969, another patient but with 98 per cent burn area (of which 80 per cent was 3rd degree) was successfully treated. Local medicine made from herbal drugs was used with satisfactory effects, and electro-acupuncture was applied for the prevention and treatment of shock in burns. Heterogenous skin grafts which were treated with combined Chinese herbal and western drugs can be kept for several months. Preparations of Chinese herbal drugs such as "Fukang 131", "injectio-pulse recovery", bletilla plasma substitute and cinnamonin have been used for injections, and theories concerning the treatment of burns with the combined methods of TCM and Western medicine are being proposed. An example is the proposition of methods to obtain scientific data that can give indications of the depth of burn lesions by electro-acupuncture. Chinese herbal drugs and electro-acupuncture are capable of adjusting the functions of the organisms, promoting micro-circulation and decreasing the infiltration of body fluid, although the understanding of the mechanism needs further study. To render the treatment of deep burns more effective and safe, herbal drugs have been used for controlling the formation of scars while skin grafts were carried out. The immunity of burn patients can be strengthened by the treating principles of supporting the resistance and abolishing pathogens. Moreover, progress has been achieved in the treatment of keloid and the division of stages in burns.

5. Urinary diseases

Satisfactory results have been obtained in the treatment of kidney stones by the combined methods of TCM and western medicine. It has changed the traditional view that stones with diameters greater than 0.8 cm should be removed by surgery. Based on the traditional Chinese treating principle, different methods have been administered for different types of kidney stones. These aimed at eliminating stones and dispelling dysuria, activating pneuma and removing blood stasis, eliminating and dispelling "hot and dampness", activating stagnant blood and dispelling nodules or hard mass with adjuvant "warming" tonic therapy for both spleen and kidney, managing vital energy and getting rid of stagnation. Concrete therapeutic measures involving the elimination, snaring and dissolution of stones have been used. The success rate of stone eliminations ranges from 48 per cent to 90 per cent, depending on their size and volume; the largest stone could have a cross-section as large as 1.8×1.1 cm^2. It has been reported at a conference that an estimated 53.6 per cent of examined cases successfully eliminated their kidney stones.

The principle of treatment for chronic prostatitis involves the removal of stagnant blood, dispelling stagnancy and diuresis, detoxification, "warming" the kidney and dispelling "coldness" adjuvants with analgesics, and activating vital energy with additives. The treatment course lasts 1-2 months as a rule. It was estimated that 66.2 per cent of examined cases were completely cured and 11 per cent just basically cured. This treatment is better than simple massage, drug or physical therapy.

6. Cataracts

There are many advantages in the treatment of cataracts with the combined methods of TCM and western medicine, such as the small incisions, quick healing, minimal reaction and simple manipulation. Eye vision is rather safe from damage owing to the incision being situated far from the cornea. The results are better than those achieved by other surgical methods reported abroad. According to reports from the Beijing region, the vision of 66.9 per cent of the patients who had undergone this treatment was found to be as good as 1.0–1.5. There are presently 4 kinds of such operations, namely, "couching," "snaring-couching", "sucking-couching" and "grasping couching". They are performed on senile cataracts with large kernels, hard consistency and small volume; matured and unmatured cataracts of old, middle-aged and ablebodied patients, traumatic juvenile or congenital cataracts in infants and secondary cataracts after surgery.

7. Procto-anal Diseases

Good results have been obtained by long existing traditional Chinese methods including mecnecrotic ligation and "thread-hanging" for the treatment of procto-anal diseases. Many improvements were achieved over the past 20 years through the study of local anatomy and therapeutic agents. For instance, prolapse of rectum treated according to the severity of lesions with a local injection of 8 per cent alum, had a success rate of 75 per cent for 3rd degree lesions. An even higher success rate could be achieved for 1st degree lesions. Incisive thread-hanging therapy for the treatment of complex fistulae was adopted. Fistulae situated at a low position could be incised gradually by a successive series of incisions, cleansing the infected tissues and "hanging threads". Adhesion of muscles to the surrounding tissues result. Hence, sequelae such as anal incontinence are avoided. Successfully treated cases based on several thousand examined cases exceeded 90 per cent. External incisions combined with internal ligation and injection have been adopted for ring-form mixed haemorrhoids.

By this method, external haemorrhoids were treated with incisions of separated regions, isolation of the haemorrhoid venous plexus with retention of dermal bridges among the incisions in the opened incised area. For internal haemorrhoids, treatment involved the ligation of a separate region, with the mucosa of the ligated area retained. The injection of 80 per cent alum into the haemorrhoid artery and small internal haemorrhoids prevented sequelae or complications such as incontinence and stricture of anus, stricture of rectum, massive haemorrhage and relapse. The results were very satisfactory in 200 cases. Injection of co-alummelaphis chinensis for the treatment of advanced internal haemorrhoids was also reported as rather satisfactory.

The above mentioned clinical approaches — using combined TCM and western medicine, were examples of the more important cases. Combined TCM and western medicine have also been widely used in the treatment of other kinds of diseases with satisfactory results.

III. Studies on Chinese Herbal Drugs

Tremendous progress has been achieved since liberation in the study of Chinese herbal drugs. A lot of work has been done in sieving, pharmacology, chemistry, pharmacognosy, forms of drugs, processing, recipes, cultivation of Chinese herbal drugs and others.

During the past years, researchers of Chinese materia medica have searched for and studied Chinese herbal drugs with the result that many individual herbal drugs and recipes, effective for certain diseases, were discovered thus contributing a lot to clinical therapy and the study of Chinese herbal drugs. An example is a new antimalarial remedy, "Qing Hao Su", (sieved and extracted from the *artemisia annua* from among 200 kinds of herbal drugs) was discovered through a thorough study of the ancient Chinese medical literature in "Remedies for Emergencies", a work written 1600 years ago. In this text, this original crude drug was recorded for the treatment of malaria. Clinically, "Qing Hao Su" has been successfully used for treating tertiary and malignant malaria. In malignant cases, oral administration of the drug has revealed that the time for antipyresis is on the average 35.6 hours (compared to 50 hours with chloroquine); a negative plasmodial count resulted after 27.8 hours compared to 39 hours with chloroquine. The composition of the drug was studied by X-ray diffraction. It was discovered that the effective fraction is a sequiterpene with a peroxide bridge, a chemical structure which is completely different from those of antimalarial drugs already known. A new method in the treatment of malaria was thus obtained. "Qing Hao Su" is a highly effective drug with fast effects and low toxicity and is especially satisfactory for emergency treatments of cerebral types of malaria. The pharmacological action is mainly at the intra-erythrocytic stage of the life cycle of the malarial protozoa. Tremendous research efforts in searching for effective anti-tumour drugs have been accomplished in the laboratory and clinic. The sieving and studies of hundreds of herbal drugs led to the discoveries of some 40 kinds of effective drugs which include OH-cantharidin amine, sodium-cantharindinate, *cephalotaxus fortune* and *camptotheca acuminata*. These drugs were effective not only in treating tumours but also for strengthening the effects of chemical and radiological therapy as well as eliminating its side reactions. In animal experiments, extracts from *polyporus umbellatus* markedly inhibited the growth of transplanted tumour and the effect was even more remarkable if chemotherapy was adjuvantly administered. Clinically, it has been demonstrated that the extracts of *polyporus umbellatus* had certain effects on lung tumours and tumours of the gastro-intestinal tract. Symptoms and pain were diminished and life prolonged. Approximately 53.2 per cent of patients with primary carcinomas of lung survived 1 year after treatment. The discovery of new antimicrobial herbal drugs, especially antiviral drugs, opens a new way for the treatment of infectious diseases. Moreover, cardial-tonics, anaesthetics, drugs for decreasing blood lipids and lowering blood pressure have also been found. Still others are found to be effective for immunological and allergic diseases.

Anaesthesia with herbal drugs is an important achievement in the field of anaesthesiology. This is a new kind of anaesthetic obtained from the records of ancient Chinese medical literature. The effective agent of this drug is scopolamine (Hyoscine) extracted from *datura metel* plants. The drug is itself a compound recipe, combining the alkaloids of *datura metel*, anisodine and *daphne giraldii*. On the whole, about twenty kinds of herbal drugs with anaesthetic potency have been sieved. At present, Chinese herbal extracts have been chosen for anaesthesia in more than 100 kinds of major and minor operations.

These anaesthetic herbal drugs possess advantages of anti-shock action, high safety, various routes of administration (oral, muscular injection, intravenous injection, acupuncture points injection, etc.), simple manipulation, and the patients regain consciousness on their own. Undoubtedly, it bears boundless prospects.

It has been demonstrated that certain recipes and drugs in TCM produce different effects in organisms in different physical states and of various constitutions. For instance, Ginseng can promote either glycogenolysis or glycogenesis; *citrus aurantium, rheum palmatum, croton tiglium* possess the same dual action on the intestinal tract. However, it was also discovered that some recipes and herbal drugs has their therapeutic action only when used in pathological conditions.

Although further studies are needed, a preliminary interpretation of the scientific basis of treatment can be given, based on the differential diagnosis of signs and symptoms in TCM.

About 100 kinds of crude herbal drugs and more than 200 recipes have been studied in the laboratory. The scientific basis of these ancient drugs was studied and identified by pharmacological, biochemical, biophysical and biomolecular means and techniques. On the other hand, drugs have been sieved through several indices in order to understand their activities and the relation between their primary and secondary actions. Pharmacologically, thorough studies were made on the treating principles such as activating blood circulation, eliminating blood stasis, supporting body resistance and consolidating foundation, benefiting agents and tonics, antipyretics and detoxified agents, managing pneuma, eliminating stagnation and cathartics. More emphasis was placed on the first two and the last principles.

Drugs with composite recipes are complex subjects for study. During the past few years, research has been done on the composition of recipes, the compatibility of individual drugs and dosage proportions. It was demonstrated that the dosage and compatibility of individual components of ancient composite recipes had been established on scientific bases. It was found that the presence of certain individual crude drugs in a recipe may act as a catalytic agent to speed up the therapeutic action of other components. It may have synergistic action to increase the concentrations of the effective components. Interestingly, such critical activities never manifested when the crude materials were adminstered individually but only in a composite recipe. The total action of the composite recipe will be much inferior if the critical component of the crude drug is absent. An abvious example is the neutralizing activity of *srophularia ningpoensis* on *diphtheria bacilli* toxins in the recipe, "Cleansing lung and benefiting essence (Yang Yin Qing Fei) Decoction". The various individual components of the recipe are supplemented and restricted by each other. They not only strengthen their original effects, but also adjust and diminish their toxicities and side effects so as to achieve stronger and more effective action.

It may be noted that the application of a couple of fixed herbal drugs in the composite recipes is another subject for study. This kind of compatibility was obtained after long term practice, e.g., *rheum palmatus* and *mirabilite, ephedra sinica* and *pranus armeniaca* (kernel), *glycerrhize uralensis* and *bupeurum chinensis.*

Herbal drugs used with high frequencies and in large dosages in composite recipes in ancient literature were selected for study. Correlations were made between the composition of the drug and its relation to the clinical indications. The purpose of this study, which bears practical significance for the study of pharmacology of crude drugs and the identification of experimental design and indices, is to determine the role played by the chosen drugs. For example, experiments on the application of gold foil revealed that the metal did not play any conspicuous role in the composite recipe, and therefore led to a suggestion that such usage should be abolished, thus saving several thousand "lian" of gold annually.

Although the mechanisms of the actions of composite recipes warrant further study, the above mentioned work has provided some scientific basis for treatments based on the differential diagnosis of signs and symptoms in TCM. The studies already made, provide a good foundation for further studies of the mechanism of such composite recipes and the creation of a new medicine and new pharmacology. The studies on the preparation and form of drug administration have been done to overcome shortages in pharmaceutics, and to improve methods of processing herbal drugs. For example, routine programmes and methods for preparing decoctions have been worked out through large scale experiments with determinations by physical constants, bacterial inhibition tests, content of decocted extracts, etc. The types of vessels used for decoctions were also important. Results based on the examinations of the pH of decoction extracts, qualitative and quantitative tests of ions, affirmed that the use of earthenware pots in ancient times was of scientific value. Aluminal, enamel and glass vessels are all satisfactory for simmering decoctions and may be used for preparing decoctions in large amounts on a major scale. Five different methods for preparing decoctions, i.e. flow steaming, earthenware, water baths, double boiler steaming, direct boiling and warm water immersions, have been experimented.

The main contents of extracts in the solution, rate of evaporation and bacteria inhibition tests, were criteria used for comparisons of these methods. It is certain that the two former measures are the most satisfactory methods, providing another scientific basis for preparing decoctions on a large scale. Other experiments showed that when coarse powder is used for preparing decoctions, the amount of crude drug needed for producing the same treating effect was only 1/2 – 1/3 that with crude fragments. This saves herbal drugs from unwanted wastage.

New methods of preparation have been designed to improve on the old ones. Traditional methods of preparation of decoctions have been studied. Granules dissolved by boiling water, concentrated pills, drop pills dissolved in the intestines, capsules, membranous and oral ampoules have been made. New forms such as spray injection have also been invented. Not only has the art of preparation (clearance, solubility, stability and standards of drugs) been studied on a large scale, work on the standardization of herbal drugs has proceeded to ensure the safe use of herbal drugs. It is worth mentioning that the combined use of Chinese herbal and western drugs has also achieved great progress. Effective compositions of herbal drugs have been prepared as solid pills and micro-capsules. Studies on the techniques of refining the preparation of traditional plasters for local applications according to different seasons led to the formulation of rules for the refining of such plasters. Certain traditional pills with distinguished therapeutic effects have been prepared instead for injections. Artificial preparations have been found to substitute for precious drugs such as pearls, bezoar and rhinoceros horn.

The knowledge of the processing of herbal drugs for the elimination of toxicity and strengthening of therapeutic effects, has been accumulated by the Chinese people from long term experiences. One can ask whether the knowledge, experience and technique have been worth retaining. Is it necessary to improve and simplify them? These questions have long been the subjects of study since the founding of New China and much work has been carried out on the systematization of Chinese pharmacology and the knowledge of processing. Cleansing and processing methods for the crude drugs are believed to be necessary and scientific. For example, the reason why the thorns but not the old stems and branches of *uncaria rhynchophylla* were used for treatment is based on the fact that the former has a more effective composition for lowering blood pressure. On the other hand, some cleansing and cutting processes appear unnecessary, like the cleansing of cilia of loquat leaves and the cutting of *phellodendron amurense* into fine filaments after being immersed in water. These procedures cause loss of effective composition and waste of crude drugs. Studies of this nature provide for a more rational, simple and concrete cleansing programme. The processes for detoxification and attenuation of toxicity have also been investigated. Traditionally, *aconitine carmichaeli*, kernel of apricot and *strychnos nux-vomica* should be processed. Studies on such crude herbal drugs have revealed that they become toxic or have decreased toxicity after processing. Their effective actions were retained, indicating that the traditional processing was necessary. Subsidiary materials for processing, such as vinegar, wine, salt, swill from washing grain and stir-fry processing have proved to be scientific. Advances in the cultivation of herbal plants and domestication of animals have contributed a lot to the production of our country. These advances included the realization of the value of cultivating *gastrodia* and *ganoderma japonicum* by fermentation, and the researches on bezoar, rhinoceros horn and pearl substitutes.

Pharmacognosically, the differentiation and identification of certain species of crude drugs that were easily confused have been achieved. One example is the species *daphne genkwa* which exist in different forms including "Chuzhou" *daphne genkwa*, "Yellow" *daphne genkwa*, and "Jinzhou" *daphne genkwa*. A histological study of the features of the original crude material proved that merchandized *daphne genkwa* is the same as the "Chuzhou" and "Jinzhou" *daphne genkwa* described in the "Zheng Lei Ben Cao". (Classified Chinese Materia Medica). Successful identification of substitutes and pseudodrugs *achyranthes bidentata*, bones of *panthera tigris*, *arisaema consanguineum*, *glechoma longituba* and *trichosanthes kirilowii* has been made.

Lastly, systematization and verification of ancient pharmacological works in TCM have also been advanced. A reference dictionary, "The Complete Dictionary of Chinese Materia Medica", recording 5767 kinds of herbal drugs, 4773 plants and 740 animals, has been completed. "Compilation of Herbal Drugs in China" and other works on processing, pharmacognosy and prepared drugs have been published. All these contribute to the creation of a new Chinese pharmacology.

IV. Summary

Through the tireless efforts of all members of the

medical field in the past thirty years, tremendous strides have been made in the evolution and development of traditional Chinese medicine. It should also be noted that a new system of medicine, one that combines all the strong points and advantages of both TCM and western medicine is being established.

There are various ways of achieving such a combination. The theory of TCM may be used as a directory with modern scientific knowledge, methods and techniques as its means; combination of "syndrome" ("Zhen" in Chinese) in TCM and "disease" ("Bing" in Chinese) in western medicine; treating different diseases with the same method or treating the same disease with different methods; combination of the theory and practice of western pharmacology with Chinese traditional pharmacology. All of these are exploring the possibility of the creation of a new theory and practice based on a new system of medicine and pharmacology. The theory of the new medical system is being organised. For instance, by studying the nature of "kidney" (in a TCM sense), a "hypothesis of constant Yin-Yang threshold" has been proposed and applied preliminarily to direct medical practice. Another example is the suggestion that the old and traditional fundamental medical system be rearranged and reclassified. They assume a breakthrough of the old traditional medical department on anatomical and functional systems, and propose to rearrange and reclassify the contents of medicine according to new ideas. Although these are all ideas, they are nevertheless signs of a new medical trend and trials that are worth encouraging for the development of medicine.

CHAPTER 18

COMMEMORATING THE CENTENARY OF THE BIRTH OF THE GREAT SCIENTIST ALBERT EINSTEIN

by Zhou Peiyuan

In the wake of the Third Plenary Session of the Eleventh Central Committee of the Communist Party of China and the Working Conference of the Party Central Committee and at a time when the focus of the work of the entire Party and of the whole country has been shifted on to socialist modernization, it is of particular significance for scientific workers in the capital to have this get-together here in solemn commemoration of the centenary of the birth of the great scientist Albert Einstein. We are having this grand commemoration of Einstein not only because the scientific contributions of his whole life had profound impact on the development of modern science but also because he displayed great courage in his exploring and pioneering efforts and his dedication to truth and social justice. All this has set an example for us to learn from and constitutes a force inspiring us to strive for the speeding up of the realization of the Four Modernizations. However, the great name of such a universally-esteemed scientist was subjected to humiliation and slander in our country some time ago. It is with a view to restoring the glorious image of this great scientist that this ceremonious meeting is held in his honour today.

In the following paragraphs I will deal mainly with Einstein's life story, his scientific contributions, his philosophical thinking and his noble qualities.

I

Albert Einstein made great pioneering contributions to modern physics and was the most influential natural scientist of the 20th century. He was born to a Jewish family at Ulm, an ancient town in Southwest Germany on March 14, 1879. His father was the owner of a small electrical appliance workshop. In 1880 the whole family moved to Munich and again in 1894, it shifted further to Milan, Italy. His childhood and youth were spent in Munich. In that period, he did not show unusual promise. On the contrary, he was slow in growth and did not learn to speak until the age of three and was therefore regarded as an obtuse child. In primary and secondary schools, his teachers believed that he would not have a brilliant future when he grew up. However, beginning from the age of four or five, he began to develop a strong curiosity in some natural phenomena. When he read a set of popular science series at eleven, he began to get interested in science. The next year he studied with joy, the Euclidean geometry by himself. These events had great influence on his subsequent development.

He was so disgusted with the militaristic education

in German schools that he left Germany all by himself in 1894, gave up German nationality and abandoned Judaism. In 1895 when he sat for the entrance examination for the Federal Polytechnic Institute in Zürich but failed, he had to attend the Cantonal School in Aarau to take up a preparatory course. The next year he was admitted into the Federal Polytechnic Institute and majored in physics in the pedagogical department. He loved to have direct contact with reality during his four years at college, where he did not achieve outstanding results as a large part of his energy was directed to his own study of the works by well-known authors instead of the regular courses. At the same time, his stress on independent thinking in studies and his disinterest in small matters of his daily life gave rise to dissatisfaction on the part of the professors, with the result that the three other graduates became assistants at the Institute with Einstein alone left out without a job. After two years of continual efforts he was finally employed by the Swiss Patent Office in Berne in June 1902. In the year preceding he was naturalized into Swiss nationality. During his unemployment from 1900 to 1902, Einstein wrote three theses in succession of which the one published in June 1902 was on the statistical theory of thermodynamics independent from Gibbs, the American theoretical physicist, thus further developing statistical mechanics and based upon which the kinetic theory of gases of thermodynamics.

Having a fixed income from the Patent Office, he proceeded much faster and on a broader scale in his scientific research. In the six-months from March to September 1905, he succeeded in accomplishing major break-throughs in three different fields — the quantum theory of radiation, the theory of Brownian motion and the special theory of relativity. His first dissertation on the special theory of relativity was entitled "On the Electro-dynamics of Moving Bodies", written when he was only twenty-six years old. The history of science had seen no precedence in a scientist, who conducted all researches in his spare time and without the guidance of a well-known professor but who could nevertheless accomplish historic successes in three fields in the short span of half a year.

Einstein stayed in the Patent Office for seven years running. Rocking the physics circle with his special theory of relativity, he was successively appointed professor at the Zürich University, the Prague University and at his alma mater, the Federal Polytechnic Institute at Zürich. In 1913 Max Planck and W.H. Nernst on behalf of the Prussian Academy of Sciences invited him to return to work in Germany. In 1914 he went to Berlin to serve as director of the Institute of Physics in the Kaiser Wilhelm Society for the Development of the Sciences (renamed the Max Planck Institute after World War II) and professor at the Berlin University, both being highly esteemed academic positions on the European continent. Fearing Nazi persecution he moved to the United States in 1933 and served as professor at the Institute for Advanced Study at Princeton. He obtained US nationality in 1940 and died of illness in Princeton in 1955.

II

After the establishment of classical mechanics by Isaac Newton based on the discovery of Galileo Galilei and on Kepler's summary of astronomical observations of planetary motions, physics continued to develop for almost two hundred years. Down to the mid-19th century, brilliant achievements included the discovery of the law of the conservation and the transformation of energy, the founding of thermodynamics and of statistical physics, and the establishment of the electromagnetic theory by James Clerk Maxwell based on earlier works of Michael Faraday. By the end of the 19th and the beginning of the 20th century many physicists believed that the theoretical problems in the realm of physics had all been solved in principle, based on the space and time concepts of the Newtonian mechanics. In other words, physics was thought to have reached its peak leaving the coming generations nothing to deal with except some minor amendments. However, contrary to expectations, many new phenomena which could hardly be explained by the then existing theories were discovered in rapid succession, e.g. the negative result in the ether drift experiment, the radioactivity of chemical elements, the motion of electrons, black body radiation, photo-electric effect, etc. In order to resolve the contradiction between old theories and new experiments, the average physicists proposed necessary amendments within the framework of the old theories, but could hardly justify themselves.

The quantum theory of Max Planck was the first to break through the confinements of old theories. Because the theory of black-body radiation in the region of short wave length gave rise to the so-called "ultraviolet catastrophe", Planck put forward in 1900 a daring hypothesis that the energy of radiation emitted from matter was not continuous. This concept of discontinuity of energy was like a square peg in a round hole in physics of the time. Constantly worried by this inconsistency, Planck dared not to move a step further from the discontinuity of energy. Instead he took a step backwards by attempting to use the continuous concept in the explanation of the discontinuity of the emitted energy. A diametrically different approach was taken by young Einstein, who, not tied down by old traditions, developed in March 1905 the quantum concept of Planck by putting forth the hypothesis that the energy of light was not continuously distributed in space and that the energy of light beam possessed the particle property, i.e., the quantum property in the entire course of propagation, absorption and production. This theory satisfactorily explained the photo-electric effect and won him the 1921 Noble Prize in physics. We know that by the beginning of the 19th century, the wave theory of light had already been universally accepted in physics. The quantum theory of light now caused the corpuscular theory of light to appear in a new form. This wave-particle duality of light also formed the foundation for the subsequent development of L. de Broglie's wave theory of matter in 1923-1924 and the wave mechanics established by E. Schrödinger in 1926. In 1906, Einstein applied the quantum concept to the question of specific heat of solid and later to photochemical phenomena. In 1916, using N. Bohr's concept of quantum transitions introduced three years earlier, he put forward the theory of induced radiation and derived Planck's black body radiation formula. In this paper, he laid the theoretical foundation of laser technology that was to come forward in the sixties. In 1924, when he read the work on statistical physics by S. N. Bose, a young Indian physicist, he gave warm support and established the so-called Bose-Einstein Statistics. Therefore, in the first twenty years of the twentieth century, as Max Born put it, "He was a pioneer in the struggle for conquering the wilderness of quantum phenomena" and was "our leader and standard-bearer".

The second piece of work that Einstein completed in 1905 was in the field of the kinetic theory of fluid which was a follow-up to his work in 1902. In his two articles in April and May 1905, he studied Brownian motion by a combined method of statistics and mechanics and worked out the relationship between the root-mean-square value of the displacement of the Brownian particle and the number of fluid molecules per unit volume. This theoretical prediction was verified three years later by the experiments of the French physicist, J. B. Perrin. In those days, whether the atom was in existence was a question under heated debate and the work by Einstein and Perrin provided such evidence and arguement for the atomic theory that even Ostwald, the most stubborn opponent of the atomic theory, had to publicly acknowledge its existence.

At the turn of the 20th century, physics came to a period of transition from the macroscopic motion of matter into the region of microscopic motion. Mechanics, electro-magnetic theory, thermodynamics are all theories of macroscopic motion. However, the discoveries of X-rays, radioactivity of chemical elements and the electron all led physicists inevitably into the realm of microscopic physics. While the laws of macroscopic physics could no longer fully explain the phenomena of microscopic motions of matter, Einstein's theories of light quantum and Brownian motion were creative contributions to microscopic physics.

Einstein published the third piece of his work in 1905, the special theory of relativity, which was also the most important. While Maxwell's theory of electro-magnetism was highly successful, it brought forth a new problem: the question of whether ether, the medium transmitting the electro-magnetic waves, could be detected. Various electro-magnetic and optical experiments failed to detect the relative "drift" of ether with respect to the earth. Theoretically, this reflected the contradiction between the electro-magnetic theory and Newton's space and time concepts and kinematics. Einstein named the relativity principle of Newtonian mechanics and the relativity principle of the electro-magnetic theory as manifest in the Lorentz transformation of coordinates as the principle of relativity in general. By taking the principle of relativity and the principle of the constancy of the velocity of light as the basis, the special theory of relativity was established and the Lorentz transformation of coordinates derived. According to the kinematics of the special theory of relativity, whether or not "ether" or absolute space exists is not a relevant question, because even if "ether" does exist, in an electro-magnetic experiment performed in the system of reference moving uniformly with respect to "ether", the phenomenon of the "ether drift" cannot be detected.

The special theory of relativity changes the space-time concept in Newtonian mechanics in essence and brings to light a four-dimensional space-time uniting space and time, thus unifying the kinematics of mechanics and electro-magnetism. The dynamics based upon the kinematics of the special theory of relativity is applicable to the high velocity motions of matter, with Newtonian mechanics as a specific case for low velocity motions. The dynamics of the special theory of relativity leads to the equivalence of mass and energy, thus laying the theoretical foundation for the use of atomic energy which began to emerge in the forties.

After establishing the special theory of relativity, Einstein set out to further explore the relativity in the field of accelerated motions. Based on the experimental fact that all objects in the gravitational field have the same acceleration, he put forward in 1970, the principle of equivalence for the uniform gravitational field and uniform acceleration. With another eight years of hard work and with the help of his old schoolmate, mathematician Marcel Grossmann, and by applying the geometry of curved surfaces established by B. Riemann half a century ago, he finally established in November 1915 the general theory of relativity, including the principle of equivalence, the principle of general relativity, the theory of gravitation and the dynamics of a particle. The principle of general relativity requires that physical laws under coordinate transformations should obey the mathematical condition of covariance and therefore should be expressed by tensor equations. The gravitational field equations and the equations of motion of a particle are both tensor equations.

The gravitational theory or the general theory of relativity further reveals the relationship between the four-dimensional space-time and matter. It also points out that the nature of space-time is closely related to the characteristics of matter and that the structure and nature of the space-time are dependent upon the distribution of matter. In reality, space-time is usually not the flat Euclidean space-time but the curved Riemannian space-time. In accordance with the gravitational theory based on the general theory of relativity and its equations of motion, he obtained the result that light propagates along a curved path in a gravitational field. This prediction was confirmed by British astronomers in 1919 in the observations of a solar eclipse. When the result was published, the whole world was shaken. J. J. Thomson, the then President of the Royal Society of London, at that time praised Einstein's theory as "one of the greatest achievements in the history of human thought". On his lecture tour in China in 1920, the British philosopher Bertrand Russell referred to Lenin and Einstein on a number of occasions as the "two great men" of our time in social revolution and scientific revolution respectively. Such appraisals as seen today, sixty years after these statements were made, still appear to be pertinent.

Though he was not like Newton who invented infinitesimal calculus in mathematics in order to solve the dynamical problem of planetary motions, his assertion that tensor equations should be used to express the laws of physics according to the principle of general relativity has had great impact in such fields as physics, mathematics and engineering. Since the publication of the general theory of relativity, the often neglected tensor analysis has gained wide application.

In 1938, Einstein made another big stride forward in the theory of motion in the general theory of relativity when he derived the equations of motion from the gravitational field equations. That is to say, the equations of motion become a part of the solution of the gravitational field equations instead of being an independent hypothesis, thus revealing in greater depth the unity between spacetime, matter, motion and gravitation. Thanks to the publication of the theory of motion, the "many-body problem", a mathematical difficulty unsolved for many years in the gravitational theory, was also resolved. Since the sixties, as a result of big advances in laboratory technology (e.g., the Mössbauer effect) and in astronomy, the research works in general theory of relativity and the gravitational theory have won ever greater attention. For instance, Einstein's prediction in 1918 of the existence of the gravitational waves has been indirectly proved quantitatively in recent astronomical observations of the motion of the binary stars.

Again, following the general theory of relativity, Einstein began to explore two other fields. As early as 1916, he began to apply the theory of gravitation in the general theory of relativity to study the structure of universe, and he published his first article in 1917, laying the foundation of modern cosmology. In this paper he added a cosmological term in the gravitational field equations in accordance with the requirement that the universe is filled with static matter and arrived at a static, finite and closed model of universe without boundary. In the same period, the Dutch astronomer W. de Sitter put forward a model with motion but without matter, in which it was predicted that there existed motions of expansion of the galaxies, with the relative velocity of expansion directly proportional to the relative distance. In the twenties, the Soviet physicist A. Friedmann set forth a model of universe with both matter and motion, which actually included the two previous models as its limits. The discovery of the law of the redshift of the spectral lines emitted by other galaxies by the American astronomer Hubble in 1929 verified to a certain extent the prediction of relativistic cosmology. This science is yet to be further developed both in theory and in observation.

After the twenties, Einstein devoted most of his attention to research in the unified field theory, attempting to establish a theory combining gravitation and electro-magnetism that is capable of explaining the quantum phenomena and the structure of matter. In the early stage when the gravitational theory based on the general theory of relativity was published, the propagation of light affected by gravitation was verified by observation during a solar eclipse. Therefore the question necessarily follows whether the gravitational field could be unified with the electro-magnetic field, which in Einstein's belief, should be the third phase in the development of the theory of relativity. However, this exploration, which almost took up the whole of the later half of his life, did not produce specific results. This alienated him from quantum mechanics, an area of physics then in rigorous development. In recent years, as the gauge field theory unifying the weak and the electro-magnetic interactions has gained much support from new experiments, Einstein's concept of a unified field theory has again shown its vitality in a new way.

III

Einstein's remarkable successes in science were partly due to his perseverance in practice and must partly be attributed to his critical philosophical approach. Fond of philosophical cogitation when he was young, he began to read Kant at the age of thirteen. In the first three years of his stay in Berne (1902-1905), he usually spent his evenings with two young friends, studying together, discussing philosophical works and various philosophical and scientific problems. These academic activities, which they jokingly named "the Olympic Academy", enabled Einstein to broaden his vision, activate his thinking, judge things at a greater height, see into a greater distance and think in greater depth than his comtemporary physicists, thus scoring a series of fundamental break-throughs by advancing still further on the basis of accomplishments made by his predecessors. In those days, only a few scientists possessed philosophical minds, as he remarked just two weeks prior to his death.

Obviously, his study of philosophy was primarily for resolving contradictions in physics. He said: "The present difficulties of his science force the physicist to come to grips with philosophical problems to a greater degree than was the case with earlier generations." He held that during revolutionary periods in the theory of physics it was essential for physicists to use philosophical inference by themselves. In spite of the rather great influence on him by Spinoza, Leibnitz, Hume, Kant and Mach, the main thing was that he was capable of adhering to the materialist stand on natural science. He explicitly declared, "The belief in an external world independent of the perceiving subject is the basis of all natural science." In his early period he was much closer to Hume and Mach's sceptical empiricism, but after the establishment of his general theory of relativity he fully realized the limits of empiricism and thereby markedly inclined to the rationalism of Spinoza. His guiding philosophical thinking could be regarded as rationalist materialism. And it was precisely because of this thinking that he did not conform with the Copenhagen School headed by Bohr in their philosophical interpretation of quantum mechanics but engaged in protracted polemics against them.

Einstein had made a number of penetrating analyses on the theory of knowledge, the theory of scientific methodology, space and time as well as on the knowability and unity of the universe. Approaching from the materialist stand of natural science, he believed that practice was the only source of knowledge. He said, "all knowledge of reality starts from experience and ends in it. Experience alone can decide on truth." He sharply repudiated apriorism of Plato and Kant, saying "the philosophers have had a harmful effect upon the progress of scientific thinking in removing certain fundamental concepts from the domain of empiricism, where they are under our control, to the intangible heights of the *a priori*". In the history of philosophy, the most forceful criticism of such apriorism came first from Hume and then from Mach. Young Einstein primarily had their influence in this respect and the influence was basically positive and therefore should be affirmed. On the other hand, however, he also realized that the critical philosophy of Hume and Mach "could only destroy perilous insects" and "could produce no living thing."

His constant mention of his belief in the God of Spinoza should not lead us to think that he had been taken captive by theology. As a matter of fact, Karl Marx had earlier explained that the "God" as referred to by Spinoza was none other than "Nature". Therefore, Feuerback referred to Spinoza as Moses for contemporary atheists and materialists.

Unreconciled to the force of old habits in his new independent thinking, Spinoza was excommunicated from Judaism at the age of twenty-four and was banished to the countryside where he eked his livelihood by grinding optical glass and led a dull and secluded existence throughout the remainder of his life. This had great influence on Einstein who took Spinoza as his example in his attitude towards life. Taking Spinoza's creed of *"amor dei intellectualis"*, which was to seek the understanding of the unity and law of nature as his supreme goal in life, Einstein despised vanity and the pursuit of material enjoyment of the average bourgeoisie. He placed Spinoza and Karl Marx side by side, saying that both of them embodied what was traditional in the Jewish nation — "A love of justice and reason", and both "lived and sacrificed themselves for the ideal of social justice." Obviously, this was precisely a portrait of the spiritual aspect of Einstein himself in his whole life.

IV

Einstein was not only a great scientist and an outstanding thinker full of spirit in philosophical exploration but also a righteous man, with a high sense of social responsibility. He had a profound feeling for the impact the work of a pursuing scientist could produce, and the responsibility that a far-sighted intellectual should assume for society. Hoping with all his heart that science would benefit mankind instead of leading to disasters, he was persistently opposed to wars of aggression, militarism and fascism, national oppression and racial discrimination, waging resolute struggles for the progress of mankind.

He pointed out to students of science and technology that: "It is not enough that you understand applied science as such. Concern for man himself must always constitute the chief objective of all technological effort". He said, "Man can find meaning in life, short and perilous as it is, only through devoting himself to society." So he said, and he acted likewise. He believed in the importance of the role of specific examples to ennoble the human race, and he felt he was duty bound to assume such a role. Consequently he would make open statements of his attitudes towards all major political events that he had experienced himself. He would also publicly condemn any social evils and political persecution that came to his knowledge. Failing to do so would make him feel that he was being "involved as an accomplice". This approach was uniquely manifested in his struggles against German Nazism from the twenties to the thirties as well as in his fight against McCarthyism in the U.S. in the fifties. It was unprecedented in the annals of mankind that a man who has made historic contributions to natural science in its creation could have been so serious and warm in his dealings with social problems.

In 1914, less than four months after his return to Germany, World War I broke out. It was a time when he was pretty close to the completion of his general theory of relativity and was thus in a new period of great scientific productivity. Nevertheless, he actively threw himself into the struggle against the War by issuing open anti-war statements and taking part in underground anti-war activities. He warmly upheld both the 1917 Russian October Revolution and the German November Revolution in 1918. In particular, he showed extremely great vigour towards the November Revolution in which he personally took part.

After the First World War, he visited many countries with a view of restoring and enhancing the mutual understanding among peoples. In November 1922 he went to Japan on a lecture tour and on the round trip his boat anchored in Shanghai for three days. The May 4th Movement had taken place not long ago. The sight of the labouring people in the abyss of suffering, aroused his deep sympathy and strong indignation. He had this to say in his travel diary: "The city (Shanghai) showed the difference in the social position of Europeans and Chinese which makes the later revolutionary events partially comprehensible. In Shanghai, the Europeans form a class of masters, while the Chinese are their servants ... (the Chinese are) the poorest people of the earth, cruelly abused and treated worse than cattle ... This is a working, groaning, yet stolid people." The fact that he could have such a profound passion for the sufferings of the Chinese people in his brief sojourn of three days in Shanghai amply demonstrates his strong sense of justice and of the distinction between good and evil. How passionate and sincere his feelings for all the oppressed nations and the enslaved peoples were!

Prior to the Second World War, he had witnessed the atrocities of German fascists and realized clearly that fascism was the mortal enemy of mankind. Heedless of his personal safety and coming out boldly, he made a series of public statements and talks denouncing Nazi atrocities and calling the slumbering European and American peoples to struggle against Hitler's Germany. Giving up the stand of absolute pacifism, he raised a cry of warning to the whole world that fascism meant war and that peace must be defended with arms. He was hunted by German Nazis. His house was ransacked, his property confiscated, his works burned and his theory of relativity declared a Jewish science against Germany. Nazi organs offered a reward of 20,000 Marks to anyone who could kill him. Much worried about him, his bosom friend von Laue, wrote him a letter advising him to be worldly wise and play safe on political matters. To this, he immediately gave a categorical reply: "Where would we be, had men like Giordano Bruno, Spinoza, Voltaire and Humboldt thought and behaved in such a fashion.

I do not regret a word of what I have said and am of the belief that my actions have served mankind."

In the years of cold war after World War II, Einstein rushed around campaigning for the prevention of a nuclear war and against nuclear tests in the service of policies of aggression and war, actively working for world peace and the defence of democratic rights, with the result that he was accused by the McCarthy-regime as "an enemy of the U.S.", "subverter", and "communist". He was not in the least intimidated but openly exposed the plots of McCarthy and his like and called upon the American intellectual to be prepared "for the sacrifice of his personal welfare in the interest of the cultural welfare of his country."

On April 18, 1955, Albert Einstein left the world with grave concern over the dangers of fascism and of war. He was very much aloof in his attitude towards his own life or death. In accordance with his will, his ashes were secretly kept, no obituary notice was issued, no public funeral ceremony held, no tomb built and no monument erected. But he will live in people's hearts for ever.

V

Einstein's lifelong scientific achievements were most brilliant. His position in the history of science could only be matched by Copernicus, Newton and Darwin. Yet he was never complacent and conservative, never satisfied with the results accomplished and always ready to search for unknown truth very modestly and with an open mind. The words he frequently quoted to urge himself forward were the German enlightener Lessing's famous dictum that the search for truth is more precious than its possession. He sharply ridiculed those who equate their power with the possession of truth by saying that "whoever undertakes to set himself up as judge in the field of Truth and Knowledge is shipwrecked by the laughter of the gods."

At the age of forty, when he had won the highest reputation the world over and his theory of relativity became the central subject of the world's news coverage, the peoples of different countries gave him warm praise and high respect. But he felt quite uneasy about this, for it did not fit into his democratic ideals. He said: "My political ideal is democracy. Let every man be respected as an individual and no man idolized. It is an irony of fate that I myself have been the recipient of excessive admiration and reverence from my fellow-beings, through no fault, and no merit, of my own."

Einstein called himself a socialist from his youth. At the time of the German November revolution in 1918, he was regarded "as a kind of arch-socialist". He thought he had the obligation to do his best to "help bridge the gulf between workers of brain and brawn". Though some of his views on socialism, were not in conformity with Marxism, yet, among them there were quite a lot of ideas which hit at the contemporary evils and therefore were of great value. For instance, in an article entitled "Why Socialism" published in 1949, he asked some deep questions in the conclusion: "A planned economy as such may be accompanied by the complete enslavement of the individual. The achievement of socialism requires the solution of some extremely difficult socio-political problems: How is it possible, in view of the far-reaching centralization of political and economic power, to prevent bureaucracy from becoming all-powerful and overweening? How can the rights of the individual be protected and therewith a democratic counterweight to the power of bureaucracy be assured?" As we have gone through the disastrous destruction brought about by Lin Biao and the "Gang of Four", we keenly and painfully feel the seriousness and urgency of this issue, which is certainly a question of fundamental importance to all the countries eager to build socialism, and this is precisely why Marx and Lenin both brought forth and stressed the "Principles of the Paris Commune".

Einstein was a friend of the Chinese people and, to the mass of Chinese intellectuals, his name is both familiar and intimate. We shall never forget his repeated appeals to all countries after the September 18th incident in 1931 for a united economic boycott to stop Japanese military aggression on China. Neither shall we forget his just support in 1937 on the "Seven Gentlemen Incident". I feel personally fortunate to have had the experience of working for a year before the outbreak of the Anti-Japanese War in the Institute for Advanced Study at Princeton where Einstein was professor and I took part in the seminar under his personal auspices. During the discussion periods, he warmly urged us young people to work and put forward our ideas with genuine sincerity, and at the same time, he solicited with modesty comments of young scientists on his work. He was modest, honest, kind towards others, while leading a simple and frugal life himself. When he was travelling by train in Europe, he used to go third class instead of the first or second. When he talked to me alone, he showed deep sympathy to the Chinese working people for their tragic experience and entertained great expectations for us, a nation with a long history of civilization.

It is now almost a quarter of a century since Albert

Einstein left us, but the fruits of his scientific research, the brilliance of his thinking, the strength of his moral integrity still live among the people. The far-reaching historic significance of the new epoch he opened up in science to man's life has been increasingly recognized. The whole purpose of this solemn commemoration is to carry on further the cause for which he had fought throughout his life, to learn from his noble qualities of fearing no hardship or danger, defying any tyrannical power, and his readiness to fight for truth and to sacrifice himself for a just cause. We want to learn from him the scientific spirit of the absence of blind faith in authorities or old customs but the adherence to truth, seeking truth from facts and daring to think independently and with pioneering efforts. We want to learn from his indomitable spirit of exploration with which he was never self-conceited or complacent along the road of science, to learn from his democratic spirit against arbitrary practice and worship of idols but respecting reasoning, caring for and respecting others. Finally, his open-heartedness and his thinking and acting in one and in the same way, and his consistent approach towards life in search of truth and for the betterment of mankind should always be our common ideal.

References

Contents

I	A CHRONICLE OF EVENTS IN SCIENCE AND TECHNOLOGY	(151)
II	BRIEF INTRODUCTION TO PERIODICALS AND NEWSPAPERS OF THE NATURAL SCIENCES	(229)
III	NAME LIST OF MEMBERS OF ACADEMIA SINICA DEPARTMENTS	(271)
IV	LIST OF PAST SCIENTISTS	(274)
V	PRIZES AND CERTIFICATES OF MERIT IN SCIENCE	(302)

SECTION I

A CHRONICLE OF EVENTS IN SCIENCE AND TECHNOLOGY

1949

October 1
The People's Republic of China was established.

Article 43 of the Common Program adopted at the Chinese People's Political Consultative Conference stipulates that efforts are to be made to develop the natural sciences with a view to facilitate industry, agriculture and national defence by disseminating scientific knowledge, encouraging and issuing awards for scientific discovery and invention.

Article 18 of the Organization Rules for the Central People's Government of the People's Republic of China stipulates that the Academia Sinica (the Chinese Academy of Sciences) is to be established under the Government Administration Council.

Front view of the Academia Sinica at its founding

Gate of the Academia Sinica

October 19

The third session of the Central People's Government Commission approved the following appointments: Guo Moruo as President of the Academia Sinica and Chen Boda, Li Siguang, Tao Menghe, Zhu Kezhen as Vice Presidents.

November 1

The Academia Sinica was founded.

The Bureau of Science Dissemination under the Ministry of Culture was established.

December

More than 200 persons participated in seven discussions held under the auspices of the Bureau of Science Dissemination. The problems discussed were the objects, methods, organization, and manpower mobilization involved in the work of science dissemination. A decision was made to publish popular science journals.

1950

January 13

The Physics Institute of the former Beijing Research Academy, now taken over by the Academia Sinica, successfully designed and manufactured the Panama Telescopic Sight.

February 11

Twelve natural science societies in the Beijing area convened at their annual conference.

February 17—28

The Bureau of Science Dissemination of the Ministry of Culture sponsored the Beijing Spring Festival Science Seminar, the first effort to popularize science since the founding of New China.

March 16
Hua Luogeng, the renowned Chinese mathematician, returned home from the United States, arriving in Beijing.

May 6
Li Siquang, the renowned Chinese geologist, returned home from England, arriving at Beijing.

May 15
Kexue Tongbao (Science Bulletin) published its first issue.

June 12
The Geophysics Research Institute, co-operating with the aviational and meteorological sections under the command of the East China Military Zone, carried out meteorological analysis in terms of 7-day weather forecasting.

June 17
In his speech on cultural and educational work made at the second session of the National Committee of the Political Consultative Conference, Academia Sinica President Guo Moruo pointed out: "The Academia Sinica has taken over 22 units including research institutes of the former Central Science Academy and Beijing Science Academy, and found some of them overlapping in nature. After reorganization and readjustment, the Academia Sinica now boasts of one astronomical observatory, one industrial technology laboratory and 14 research institutes. A recent preliminary survey indicates that apart from the Academia Sinica, China has ninety three scientific research institutions (science, engineering, agrotechniques, medicine, etc.), with, in addition, seventeen local science museums and twenty nine local scientific instruments manufacturing factories."

June 20–28
Policies and programs were decided at the enlarged executive meeting of the Academia Sinica, at which state leaders Zhu De, Zhou Enlai and Lu Dingyi delivered their speeches.

June 24
A delegation from the Academia Sinica was dispatched to the DDR to attend the 250th anniversary of the founding of its science academy.

August 18–24
The All-China Natural Scientists Congress was convened in Beijing. The 468-person participation demonstrated an unprecedented solidarity among the Chinese scientists. Premier Zhou Enlai made an important speech at the congress. Thereupon two bodies came into being: The All-China Federation of Scientific Societies, with Li Siguang as its chairman and 51 scientists including Li Siguang, Hou Debang, Mao Yisheng on its national committee, and The All-China Association for the Dissemination of Scientific and Technological Knowledge, with Liang Xi as its chairman

and 50 scientists including Liang Xi, Ding Xielin, Zhu Kezhen on its national committee. And Wu Yuzhang was acclaimed as the honorary chairman of both bodies.

August 25

Qian Xuesen, the renowned Chinese aerodynamics expert, on the eve of his departure from the United States for China, was detained by the U.S. government.

August 27

An Academia Sinica delegation headed by Wu Youxun and Hua Luogeng, attended Hungary's first mathematics conference.

August 31

Zhongguo Kexue (Scientia Sinica) published its first issue.

September 12

On their way back from the United States, a party of three Chinese, the renowned atomic physicist Zhao Zhongyao and two science-engineering students Luo Shijun and Shen Shanjiong, were detained by the American occupation army in Yokohama, Japan.

September 24

The All-China Federation of Scientific Societies released a protest against the illegal detention by the U.S. government and army, of Qian Xuesen, Zhao Zhongyao, and the two Chinese students.

September 30

189 scientists and professors in Beijing signed a protest against the illegal detention by the U.S. Government of the four Chinese including Qian Xusen and Zhao Zhongyao.

October 9

Detailed Rules of the Regulations for the Protection of Inventions and Patent Rights and Rules of the Invention Examination Committee were promulgated by the Financial and Economic Commission of the Government Administration Council.

October 31

The foreign language edition of *Kexue Jilu* (Science Proceedings) resumed publication.

November 1—7

The Supervisory Committee for China's geological plan held an extended meeting. A 3-year priority-oriented plan was thereupon decided in which China's geological work would concentrate on prospecting coal, iron, petroleum and non-ferrous metals in order to meet the needs of the anticipated upsurge in economic reconstruction.

November 28

Zhao Zhongyao, Luo Shijun and Shen Shanjiong returned home, arriving at Guangzhou.

December 5

The All-China Federation of Scientific Societies and the All-China Association for the Dissemination of Scientific and Technological Knowledge jointly sponsored a forum, at which Li Siguang, Ding Xi and other prominent scientists spoke, urging China's scientists to take an active part in the "Resist U.S. Aggression and Aid Korea" campaign.

December 26

Wu Youxun was appointed Vice Chairman of the Academia Sinica at the 10th session of the Central People's Government Commission.

1951

January 3

Wu Zhengyi, Chen Huanyong, Hou Xueyu, Xu Ren and other Chinese scientists from the Academia Sinica attended the symposium on the origin and distribution of commercial plants in southern Asia, sponsored by India's Genetics Breeding Association.

February 2

Academia Sinica President Guo Moruo delivered a report, entitled "A Summary of Academia Sinica's Progress for 1950 and Major Plans for 1951", at the 70th executive meeting of the Government Administration Council, and approval was given.

February 3

The Academia Sinica Library was established.

February 22

The Academia Sinica sponsored the first discussion for the Chinese Science History Editing Committee.

March 5

The Government Administration Council issued a directive on scientific research work.

April 10–12

The International Scientists Union held, in Paris and Prague, its second convention, at which Li Siguang was elected Deputy Chairman of its executive committee.

May

The Academia Sinica decided to edit and print a series, on the scientific writings of modern China. The series, arranged in chronological order and subject classification, was to cover all the important scientific papers from the time modern science was first introduced in China to the year 1949.

June 29

Zhu Kezhen attended Poland's first science conference.

July 11

The Academia Sinica Industrial Laboratory succeeded in extracting pure cobalt from earthy cobalt

July 31

The Academia Sinica Industrial Laboratory successfully developed a nodular cast iron.

October 31

The Pharmaceutical Research Group of the Organic Chemistry Research Institute and the East China Pharmaceutical Corporation trial-produced penicillin G, sodium and potassium, and penicillin procain.

November 5—11

The All-China Association for the Dissemination of Scientific and Technological Knowledge held its first working conference in Beijing.

1952

February 23

The All-China Federation of Scientific Societies and the All-China Association for the Dissemination of Scientific And Technological Knowledge jointly announced their protest against the American invasion forces for their crime in conducting bacteriological warfare in Korea.

March 10

The All-China Federation of Scientific Societies and the All-China Association for the Dissemination of Scientific and Technological Knowledge announced a joint statement protesting against the United States for conducting bacteriological warfare in North-east China.

May 8

The Internal Affairs Ministry and the General Publication Bureau of the Central People's Government released a joint circular for the correction of two geographical names: Mount Everest was corrected to Mount Qomolangma and Trans-Himalaya to Gandisi Shan.

July 1

The International Scientists Committee for Investigating Germ Warfare Conducted in Korea and China was established in Beijing. Invited by Guo Moruo in the capacity of Academia Sinica President and Chairman of the Chinese People's Committee for World Peace, Joseph Needham and other distinguished scientists from Sweden, France, Italy, Brazil, the Soviet

A chronicle of events in science and technology

Union either sat on the Committee or acted as advisors. The Committee liaison was Professor Qian Sanqiang, director of the Modern Physics Research Institute of the Academia Sinica.

August 2 – December 23
The Yellow River Huanghe Water Conservancy Committee, in co-operation with the Ministry of Fuels Industry, organized an expedition team. It went into the grassland in Qinghai Province and continued along the Yellow River Valley, investigating the source of the Yellow River.

August 6
Sponsored by the Academia Sinica, the Ministry of Agriculture and the North China Agrotechniques Research Institute, a central agrotechniques investigation team was formed and sent to various parts of China to conduct on-the-spot surveys.

August 31
Academia Sinica President Guo Moruo and four Vice Presidents, Li Siguang, Tao Menghe, Zhu Kezhen and Wu Youxun, attended the signing ceremony of a report presented by the International Scientists Committee for Investigating Germ Warfare Conducted in China and Korea.

September 30
The Northeast China Scientific Research Institute Dalian Branch (Industrial Chemistry Research Institute) successfully extracted methylebezene, which was followed by trial production.

October 31
The foreign language edition of *Zhongguo Kexue* (Scientia Sinica), having incorporated *Kexue Jilu* (Science Proceedings), published its first issue.

The Northeast China Scientific Research Institute (Changchun Complex Research Insitute) accomplished nodule bacteria isolation and selection through vaccination with field soya beans. This led to an average increase of 10% in soya bean output.

1953

January 5
The expedition team from the Yellow River Water Conservancy Committee completed its investigation work on the source of the Yellow River.

January 13
The 12th session of the Central People's Government Committee approved the appointment of Zhang Jiafu as Vice President of the Academia Sinica.

February 24 – June 17
A visiting delegation, headed by Qian Sanqiang, from the Academia Sinica, toured the Soviet Union.

May 3

The investigation group, co-organized by the Academia Sinica and the Ministries of Agriculture, Forestry and Water Conservancy, started for Northwest China to examine water losses and soil erosion in that area.

May 5

The Academia Sinica, together with the Yellow River Water Conservancy Committee, the Ministries of Agriculture and Forestry, co-organized 10 groups to investigate water losses and soil erosion in the Yellow River basin.

May 20

A delegation from the Academia Sinica was sent to Budapest to attend the 1953 annual conference of the Hungarian Academy of Sciences.

June 19

At Budapest, Academia Sinica President Guo Moruo was conferred an honorary academian degree by the Hungarian Academy.

September 21

The Zhoukoudian Exhibition, organised by the Paleovertebrates Research Institute of the Academia Sinica, of Peking-Man relics, was inaugurated.

September 26

The Academia Sinica, the All-China Federation of Scientific Societies and the Central Health Research Academy jointly celebrated the 104th anniversary of the birth of Pavlov. Guo Moruo delivered a commemorative speech on the occasion.

September 27

The All-China Federation of Scientific Societies, the All-China Association for the Dissemination of Scientific and Technological Knowledge, the Chinese People's Committee for World Peace, the All-China Federation of Literary and Art Circles and the All-China Union of Literary Workers conducted joint activities commemorating four world prominent cultural figures: the 2230th anniversary of the death of the Chinese patriotic poet Qu Yuan, the 410th anniversary of the death of the Polish astronomer Nicolaus Corpenicus, the 400th anniversary of the death of the French writer Francois Rabelais and the 100th anniversary of the birth of the Cuban writer and national independence movement leader Jose Marti.

October 14 — November 7

The directors of the research institutes under the Academia Sinica gathered to discuss the projects program for the next few years and the priorities for 1954 in the light of the general line and general tasks during the socialist transitional period.

November 1

Guo Moruo was conferred, in Beijing, the title of honorary academician by the Bulgarian Academy of Sciences.

A chronicle of events in science and technology

1954

January 21
The Government Administration Council held its 204th executive meeting at which Teng Daiyuan, Minister of Railways submitted a proposal for the construction of a Changjiang bridge at Wuhan, which was accepted.

January 28
The Government Administration Council held its 204th executive meeting at which two reports were approved: the Present Situation and Future Work of the Academia Sinica by Guo Moruo, and an account of the visit to the Soviet Union by Qian Sanjiang, head of the Academia Sinica visiting delegation. Also passed at the meeting was the Central People's Government directive on the intensification of the work on forecasting, warning and prevention of disastrous weather.

February 27 — March 29
The Academia Sinica sponsored a conference on seismic intensities in places like Fulaerji, Jiamusi, Zhuzhou, Chengdu, Haerbin, Beijing and Baotou.

March 6
The Government Administration Council announced the directive on intensifying the work on forecasting, warning and prevention of disastrous weather.

May 17
An 8-person party, including Hua Luogeng and Zhang Yuzhe, went to Moscow to attend a scientific conference sponsored by the Mathematics-Physics Department of the Soviet Academy of Sciences. They also attended the inauguration of the restored Pulkov Astronomical Observatory, and participated in the June 30 astro-physical observation of a total solar eclipse in the Soviet Union.

May 19 — June 17
Academia Sinica Vice President Zhu Kezhen went on an investigative tour to study water losses and soil erosion in Northwest China.

July 1
The All-China Federation of Trade Unions and the All-China Association for the Dissemination of Scientific and Technological Knowledge announced a joint directive on the popularization of science and technology.

July 16—28
The Academia Sinica and the Ministry of Agriculture jointly sponsored a Conference on soil and fertilizers.

August 1
The Science Publishing House was founded.

August 16 — 20

The Academia Sinica held a Chinese zoological atlas conference.

August 31

The Applied Chemistry Research Institute succeeded in its research on carbon black production.

The Ministry of Geology promulgated the provisional regulations about the discovery and reporting of minerals by non-professionals.

September 20

The first session of the first National People's Congress passed the Constitution of the People's Republic of China. The 95th article of its Third Chapter, "Basic Rights and Duties of the Citizen" stipulated that the PRC guarantees for its citizens, freedom in scientific research, literary and artistic creation and other cultural activities, and that the state encourages and helps its citizens in their creative work in science, education, literature, arts and other cultural undertakings.

The exhibition "Natural Environment and Mineral Resources of our Motherland" was opened in Beijing.

October 13—19

The Academia Sinica and the Chinese Chemistry Society jointly sponsored in Beijing, the first national conference on high molecular chemistry.

October 22

The Chinese Medical Society Headquarters issued a directive on strengthening traditional Chinese medicine.

October 28

The President of the Academia Sinica Guo Moruo cabled to the President of Columbia University, his congratulations on the bicentennial of the University's founding.

November 2—29

The Yellow River Water Conservancy Committee held a water and soil conservation conference in Zhengzhou.

December 22

The Geological Research Institute of the Academia Sinica, in co-operation with the Ministry of Geology, developed a measuring method for the determination of ferrous, ferric and chromium proportions in Chromite ore, using phosphoric acid as solvent.

The Geological Research Institute of the Academia Sinica successfully constructed China's first "Double-Sample Fully-Automatic Thermo-Differential Analyser".

A chronicle of events in science and technology

President Guo Moruo speaking on the 25th anniversary of the discovery of the Peking-Man's first cramium piece.

December 24

At the invitation of the Indian government and the Indian Union of Science Conventions, the Chinese scientists delegation, headed by Qian Duansheng, went to India to attend the 41st annual conference of the latter.

December 27

The 25th anniversary of the discovery of the first cranium of Peking-Man was marked under the auspices of the Academia Sinica. On this occasion President Guo Moruo elaborated on its discovery.

December 28

Academia Sinica President Guo Moruo was conferred in Beijing the title of Academician of the Polish Academy of Sciences.

Under the auspices of the Academia Sinica, discussions about the unearthing of Peking-Man fossils and an exhibition were both held.

1955

January 20

Oculists Xu Yanan and Tang Jidao from the Municipal People's Hospital in Liuzhou, Guangxi, succeeded in an alien cornea transplant with a monkey's cornea.

January 21

Qian Duansheng and Xue Yu attended the 7th conference of the Pakistani Association for the Advancement of Science.

January 28 — February 6

Agriculture and Forestry Science Societies under the All-China Federation of Scientific Societies separately held their 1955 symposiums.

February 2

The Academia Sinica sponsored the "Peaceful Utilization of Atomic Energy" forum. President Guo Moruo informed the participants about Premier Zhou's assessment of the international situation then and his instructions on the peaceful utilization of atomic energy. An organizing committee for a popular lecture series on atomic energy was thereupon formed.

February 7—10

The Academia Sinica held discussions on the control of caterpillars in Beijing.

February 10—17

The Ministry of Geology held the first national conference on regional hydrographical geology in Beijing.

February 12—19

The National Health Science Research Committee held the fourth session of its first congress and decided to rename itself the Medical Science Research Committee under the Ministry of Public Health of the PRC.

March 7

The All-China Association for the Dissemination of Scientific and Technological Knowledge, the Central Committee of the New Democratic Youth League and the All-China Federation of Trade Unions released a joint circular on the dissemination of scientific and technological knowledge.

April 4—12

The Ministry of Water Conservancy held a national conference on scientific experiments in hydrological research.

May 5

The Beijing Antibiotics Research Committee was founded.

May 15

A comprehensive survey for water and soil conservation in the middle reaches of the Yellow River started in the western Shanxi province.

May 23

Deng Shuqun representing the Academia Sinica attended the annual conference of the Hungarian Academy of Sciences.

May 31

The 10th session of the State Council plenary meeting approved a report presented by the Academia Sinica on the organizational preparation for setting up its academic departments and their inaugural conference, and confirmed the nomination of the departmental committee members.

June 1—10

The Physics-Mathematics-Chemistry Department, the Biology-Geoscience Department, the Technical Sciences Department and the Philosophy-Social Sciences Department of the Academia Sinica held their joint founding conference in Beijing. The conference heard and discussed the progress reports by President Guo Moruo and chairmen of the four departments. The draft of a five-year program for the academy and its 1955 projects program was discussed and passed. The standing committees were formed and their members elected. Premier Zhou Enlai, Vice Premier Chen Yi, head of the Party's Propaganda Department Lu Dingyi and other leading comrades also addressed the conference.

June 3

The State Council officially announced the name-list of the departmental committee members of the Academia Sinica. (see elsewhere in this year book)

View of the inaugural conference of the academic departments of the Academia Sinica

June 11

The Academia Sinica held its first inter-departmental committee conference.

July 24

The Academia Sinica worked out its projects program for the first five-year plan period, which was based on the major tasks in the state five-year plan.

July 30

The first five-year plan for the development of national economy of the PRC was passed at the second session of the First National People's Congress. The third section of the Eighth Chapter of the plan prescribes the requirements for scientific research. Over these five years, the research priority was to be on the following 11 items: (1) peaceful utilization of atomic energy, (2) new steel and iron complexes, (3) petroleum,

Presidnet Guo Moruo addressing the inaugural conference

(4) earthquakes, (5) investigation in terms of river basin planning and reclamation, (6) investigation of South China tropical plant resources, (7) natural and economic regionalizations, (8) antibiotics, (9) high molecular polymers, (10) theoretical problems related to China's reconstruction during the transitional period, (11) Chinese modern history, contemporary history, modern ideological history and contemporary ideological history. In addition to the Academia Sinica, other units concerned were asked to establish corresponding scientific research institutions. The state was to make regulations to encourage discoveries and inventions by scientists and engineers.

August 4
The Central Committee of the New Democratic Youth League and the All-China Association for the Dissemination of Scientific and Technological Knowledge released a joint directive for the planned regular lectures on natural sciences to be conducted in large and medium-sized cities.

August 5
The State Council held its 17th plenary meeting which approved the Academia Sinica's Interim Regulations on scientific awards and research students and confirmed the nomination of the chairmen and deputy chairmen of the academic departments.

August 10
Wang You, Wang Yinglai and Xue Gongzhuo attended the Third Biochemistry Conference in Belgium.

August 17
A Chinese-character teleprinter that China first developed met with success in the preliminary tests in Beijing.

August 29
Zhang Yuzhe, Dai Wensai, Wu Xinmou and Ye Shihui attended the Seventh Annual Conference of the International Astronomical Union held in Dublin, Ireland.

August 31
The State Council officially announced the Academia Sinica Regulations for Scientific Awards and the Academia Sinica Interim Regulations on Research Students.

September 15
The 39th executive meeting of the Academia Sinica Standing Committee approved the instructions of the Academy on the making of its long-range (15 years) development program.

September 20—24
The Academia Sinica held a conference on analytic chemistry in Beijing. The Academia Sinica Standing Committee held an executive meeting that decided for the first time to issue scientific awards in 1956.

A chronicle of events in science and technology

October 10

The Academia Sinica and the Ministries of Agriculture, Forestry, and Water Conservancy jointly sponsored a national conference on water and soil conservation.

October 17

Dai Fanglan and Ding Ying attended the second anniversary observance ceremonies of the DDR's Agricultural Science Academy and were conferred the title of its corresponding academician.

October 25

The National Antibiotics Research Committee was established.

October 28—31

The Academia Sinica and the All-China Federation of Societies jointly sponsored the centennial of the birth of Michurin, a Soviet botanist.

Qian Xuesen, the renowned Chinese fluid-mechanics expert, returned home from the United States, arriving in Beijing.

November 2

A scientific delegation headed by Academia Sinica President Guo Moruo visited Japan.

November 3

The Technical Sciences Department of the Academia Sinica, together with 14 units including the Higher Education Ministry, the Heavy Industries Ministry, the Constructional Engineering Ministry, the Water Conservancy

Arriving home from the U.S., Qian Xuesen was warmly greeted

Ministry, the Railways Ministry, organized a project planning group for civil engineering and water conservancy construction research, co-ordinating the related 1956 projects of the whole country.

November 4
China's first planetarium began its construction in Beijing.

November 7
The Agricultural Ministry held a national working conference in Beijing on plants protection and quarantine.

November 8
The combined delegation from China's technological and engineering societies, headed by Yan Jici, started for Yugoslavia to attend the fourth congress of its technological and engineering association.

November 15 — December 8
The Agricultural Ministry held a national working conference on agricultural scientific research, and a 13-year agricultural science research program was worked out.

December 1—6
The Academia Sinica held the 1955 antibiotics symposium in Beijing.

December 4
Terramycin was successfully produced for the first time in China, with improvements in the process of penicillin making and aureomycin lab-produced.

December 5
The Agricultural Science Research Co-ordinating Committee was established in Beijing.

December 19
A traditional Chinese medicine research academy was established in Beijing under the Ministry of Public Health.

1956

January 2
Liu Chongle, representing the Academia Sinica, attended India's 34th Science Conventions annual conference.

January 14—20
The Central Committee of the Chinese Communist Party held a meeting to discuss the problems of intellectuals. Comrade Zhou Enlai delivered a speech. Comrade Mao Zedong addressed the closing ceremony, calling for the whole Party to endeavour to study sciences and unite with the non-Party intellectuals in order to catch up with the advanced scientific developments elsewhere in the world.

January 16
Liu Chongle, representing the Academia Sinica, attended the 8th annual conference of the Pakistani Association for Science Promotion.

January 18
The Forestry Ministry put forth its preliminary plan for national afforestation in 12 years.

January 30
The Chinese Physics Society held a semi-conductor symposium.

January 31
At the second plenary session of the second national committee of the Chinese People's Political Consultative Conference, Academia Sinica President Guo Moruo delivered a speech entitled "A Mission for Intellectuals in the Upsurge of Socialist Revolution."

February 1—7
The Central Physical Culture College held, in Beijing, a scientific symposium on physical culture, the first of its kind in China's history.

February 16
Natural and technological scientists gathered in Beijing to plan a programme for scientific work.

February 20 — March 3
The national hydrographical conference was convened in Beijing.

February 22
Duan Xuefu and Gong Sheng attended the South Asia mathematics pedagogical conference in India.

February 29
The Academia Sinica held a conference for the Earthquakes Committee, discussing China's Seismic Intensity Scales (draft) and Beijing Seismic Intensity Minor-regionalization.

The Chinese Zonal Stratigraphical Charts (draft), jointly compiled by the Chinese Geological Society Editing Committee and the Geological Research Institute, went into publication.

March 3
The All-China Association for the Dissemination of Scientific and Technological Knowledge held discussions in Beijing about the further stepping up of science dissemination in national autonomous regions.

March 14
The Science Planning Committee under the State Council was officially formed.

March 17

Zhao Zhongyao went to the Soviet Union in preparation for the Institut Unifié de Recherches Nucléaires.

March 20–26

Representatives from 11 countries, i.e., China, Albania, Bulgaria, Hungary. the DDR, Korea, Mongolia, Poland, Rumania, the Soviet Union and Czechoslovakia, met in Moscow to hold an international conference on the setting up of the Institut Unifié de Recherches Nucléaires and signed an agreement on the matter.

March 28

The Central Meteorological Bureau convened a national meteorological conference in Beijing.

April 1–4

The World Scientists Union held its executive council meeting in Beijing.

April 3

The World Scientists Union celebrated the tenth anniversary of its founding in Beijing with the participation of more than 1,400 scientists from China, France, India, Czechoslovakia, Denmark, Japan, Pakistan, the Soviet Union, England, etc.

April 4

Wang Ganchang went to the Soviet Union to participate in planning the long-range development of the peaceful utilization of atomic energy.

April 23

Chinese scientists Qian Sanqiang, Wang Ganchang, Peng Huanwu, Zhao Zhongyao, He Zehui, Yang Chengzong took part in planning the long-range development of atomic science in the Soviet Union.

May 1

The Beijing Ancient Astronomical Instruments Museum, an affiliate of the All-China Association for the Dissemination of Scientific and Technological Knowledge, was established.

May 4

At the national advanced workers conference Guo Moruo delivered a speech entitled *March of Science and Technology.*

May 14

Zhao Guangzeng, He Yizhen and Wang Chuanyu attended the Sixth International Spectroscopy Conference in Holland.

May 21–26

The Ministry of Culture convened a national museums conference in Beijing.

A chronicle of events in science and technology

May 26
Lu Dingyi delivered a speech entitled "Let a Hundred Flowers Bloom and a Hundred Schools Contend".

Premier Zhou Enlai presided over a grand party entertaining scientists engaged in national science planning.

May 30
Natural Regionalizations of China (draft) compiled by the Chinese Geography editing board went into publication.

June 6
Academia Sinica President Guo Moruo spoke to a Xinhua News Agency correspondent about the 20-year long-range plan for scientific and technological development and the qualified personnel needed.

June 12
The Four-country (China, the Soviet Union, Korea and Vietnam) Fishery, Oceanography and Limnology Research Conference was held in Beijing.

June 14
Chairman Mao Zedong and Party Politburo members Zhou Enlai, Zhu De, Chen Yun, Kang Sheng, Lin Boqu and Deng Xiaoping received scientists engaged in planning national long-range scientific work programmes.

June 15
China, the Soviet Union, Korea and Vietnam formed the Western Pacific Fishery Research Committee in Beijing whose first meeting was immediately held.

June 17
Huang Yuxian and eight other Chinese scientists attended the World Power Conference held in Vienna, Austria.

June 25
Mao Yisheng, Li Guohau and Zhang Wei attended the fifth annual conference of the International Bridge Structural Engineering Association held in Lisbon, Portugal.

July 9—13
The Chinese Natural Science History Research Committee of the Academia Sinica held its first academic symposium.

July 17
Zhou Ka and five other Chinese scientists attended the Eighth International Photogrammetry Conference held in Stockholm, Sweden.

July 27

Wu Xianwen, Bao Qingzhi and Liu Yuyi attended the Thirteenth International Conference on Theoretical and Applied Limnology held in Helsinki, Finland.

July 29

Feng Depei and six other Chinese scientists attended the Twentieth Conference of the International Physiology Association held in Brussels, Belgium.

August 1

The Eighth International Plants Quarantine and Protection Conference was held in Beijing.

August 3

The Chinese Chemistry Society held a research paper reading in Beijing.

August 10

The Academia Sinica and the Higher Education Ministry jointly held a genetics symposium in Qingdao.

August 13

The Chinese Mathematics Society held a mathematics research paper reading in Beijing.

August 21

The draft of a long-range (1956—1967) plan for scientific and technological development came into initial form.

A national railway scientific research conference was held in Beijing

August 24

The All-China Association for the Dissemination of Scientific and Technological Knowledge celebrated the sixth anniversary of its founding in Beijing. By now it had 26 local branches at the provincial (and autonomous region and municipality) level and more than 500 local branches at the county level, with a total membership of more than 176,000.

August 29

Chen Zongqi, Zhu Gangkun, Lu Baowei, Chen Honge and Hu Qing attended the International Geo-physical Year Ionosphere Committee Conference held in Brussels, Belgium.

Ma Rongzhi and five other Chinese scientists attended the Sixth Conference of the International Society of Soil Science held in Paris, France.

September 3
Zhu Kezhen, Liu Xianzhou, Li Yan and Tian Dewang attended the International Science History Association Conference held in Milan, Italy.

September 5
Zhou Peiyuan, Qian Weichang, Zheng Zhemin and Zhu Zhaoxiang attended the International Applied Mechanics Conference held in Brussels, Belgium.

September 15
Li Yi, Wang Ganchang and Hu Ning went to the Soviet Union to attend the inaugural conference of the Institut Unifié de Recherches Nucléaires.

September 29
An Academia Sinica delegation headed by Yin Zanxun left for Poland for a visit.

October 15
The Institute of Scientific and Technological Information of the Academia Sinica was established.

October 15 – November 1
The Seventh National Forestry Conference was held in Beijing.

October 24
The Academia Sinica held a national conference on tropical botany resources in South China.

October 29 – November 3
The All-China Association for the Dissemination of Scientific and Technological Knowledge and the All-China Federation of Trade Unions jointly held the first conference of activists for the dissemination of science and technology among the workers and staff in state enterprises.

November 5
The All-China Association for the Dissemination of Scientific and Technological Knowledge held the second plenary session of its first national committee.

December 2
The Botanic Garden of the South China Botanic Research Institute was officially established.

December 4–12
The second session of the National Schistosomiasis Prophylaxis and Therapy Committee was held in Shanghai.

December 12
The Academia Sinica, the All-China Federation of Scientific Societies,

the Chinese People's Committee for World Peace, the Chinese People's Association for Cultural Relations with Foreign Countries and the All-China Federation of Literary and Art Circles jointly sponsored a commemoration meeting in Beijing for three world famous cultural figures: the American scientist Benjamin Franklin, the French scientist Pierre Curie and the Polish scientist Marie Curie.

December 28
The Chinese National Committee of the World Power Conference was officially formed in Beijing.

December 31
The Chronological Records of Earthquakes in China, edited by the The history group of the Earthquakes Committee of the Academia Sinica, was published.

1957

The Academia Sinica and some interested parties from the First Machine Building Ministry co-produced Ge semi-conductor components, which marked the beginning of transistorization of China's electronic technology.

The Academia Sinica organized a sand control team that went for an all-round survey of the drought-stricken areas in North China.

January 24
The Academia Sinica announced its 1956 scientific awards (for natural sciences only). Among the 34 winners, 3 received the first award, 5 the second award and 24 the third award. The honoured scientists were given gold medals, merit certificates and money (10,000 yuan RMB for the first award, 5,000 yuan RMB for the second award, 2,000 yuan RMB for the third award). (see the awards list elsewhere in this year book).

February 5—10
The Mathematics-Physics-Chemistry Department of the Academia Sinica held the first mechanics symposium.

February 11—15
The Academia Sinica held a Quarternary symposium in Beijing.

March 1
The Chinese Academy of Agricultural Sciences was officially established in Beijing.

March 11
The Academia Sinica and the Land Reclamation Ministry sponsored a symposium in Guangzhou on the utilization of tropical resources in South China.

April 30

The Academia Sinica invited its departmental committee members in Beijing for a discussion on the correct handling of contradictions among the people themselves.

May 4

Yu Dafu, Luo Zongluo and Jin Shanbao were chosen as corresponding academicians of the Lenin Agricultural Sciences Academy of the Soviet Union.

May 12

The 48th plenary session of the State Council approved the nominations of chairman and vice chairmen of its Science Planning Commission. The Commission had as its task the following seven items: "(1) supervising the implementation of the long-range plan, especially the priority projects, (2) compiling long-range as well as annual plans for scientific research which were to be incorporated into the state general plan, (3) co-ordinating major projects in scientific research between different administrative systems, (4) examining and solving major operational problems, such as books and instruments, in scientific research, (5) handling international co-operation, (6) controlling funds for scientific research, managing personnel problems in terms of training, job-assigning and making proper use of senior experts, and cultivating friendly relations with Chinese experts still working in the capitalist countries and arranging posts for them when they have returned to China."

The Commission's leading officers: Chairman, Nie Rongzhen; Vice Chairmen, Guo Moruo, Lin Feng, Li Siguang, Huang Jing; General Secretary, Fan Changjiang.

May 15

The tenth executive meeting of the Academia Sinica Standing Committee decided to divide the Biology-Geo-Science Department into the Biology Department and the Geo-science Department.

May 23–30

The Inter-departmental Committee of the Academia Sinica held its second plenary session in Beijing.

May 27

The All-China Association for the Dissemination of Scientific and Technological Knowledge held a national conference on publicity problems.

May 29

The twelfth executive meeting of the Academia Sinica Standing Committee approved the extension of departmental committee membership to 20 more Chinese scientists. (see the list elsewhere in this year book)

May 30
The Academia Sinica held an award-issuing ceremony for the 1956 winners (for natural sciences only).

June 13—15
The fourth enlarged meeting of the State Council Science Planning Commission ended with a decision to set up a national library center and compile a unified national book-catalogue.

July 1
The International Geo-physical Year began at zero hour GMT of the day.

August 15
The Four Country (China, the Soviet Union, Korea and Vietnam) Western Pacific Fishery Research Committee held its second plenary meeting in Moscow.

September 2
Academia Sinica President Guo Moruo was conferred in Beijing the title of honorary academician of the Rumanian Academy of Sciences.

September 12
The Academia Sinica, the Chinese People's Committee for World Peace, the Chinese People's Association for Cutural Relations with Foreign Countries and the All-China Federation of Literary and Artistic Circles jointly held a meeting in Biejing marking the 250th anniversary of the birth of the Swedish naturalist Carl de Linne.

September 29
China's first planetarium, Beijing Planetarium, was officially inaugurated.

October 15
Changjiang Bridge at Wuhan was completed and open to traffic.

November 1
The PRC scientific and technological delegations visited the Soviet Union, going to Moscow in several batches.

November 9
The Academia Sinica sponsored a meeting for the protection of paleovertebrate and paleoanthropological fossils.

November 10
The State Council Science Planning Commission worked out four plans to facilitate scientific research, concerning the national co-ordination of books, better archives and records, better manufacture, repair and supply of scientific instruments and better chemical reagents.

A chronicle of events in science and technology

December 14
China's national library center was established in Beijing. It comprised the Beijing Library and the libraries of the Academia Sinica, the Chinese Academy of Medical Sciences, the Chinese Academy of Agricultural Sciences, Beijing Agro-technical University, the Geological Ministry, the Chinese People's University, Qinghua University and Beijing University.

December 25
China's aircraft industry successfully manufactured a multi-purpose civilian-use aeroplane Model An-2.

December 31
Kexue Qingbao Gongzuo (Scientific Research Information) published its first issue.

1958

January 1
The Bao Cheng railway was opened to traffic.

January 8—15
The Chinese Academy of Agricultural Sciences held the first national field-crop strains conference in Beijing.

January 11—13
The Mathematics-Physics-Chemistry Department of the Academia Sinica held a symposium on progress in solid state physics and electronics theory.

January 29
The State Council held its 69th plenary session, which confirmed the procedures in appointing and removing the president and vice presidents of the Academia Sinica, and prepared a proposal to be submitted to the fifth session of the First National People's Congress.

February 11
The Chairman of the PRC issued a decree that Guo Muruo be appointed President of the Academia Sinica, and Chen Boda, Li Siguang, Zhang Jingfu, Tao Menghe, Zhu Kezhen, Wu Youxun appointed Vice Chairmen of the Academia Sinica in accordance with the decision made at the fifth session of the First National People's Congress.

February 13—15
The Academia Sinica called the directors of its research institutes to Beijing for a conference.

March 24—26
The Biology Department of the Academia Sinica and the Chinese Academy of Agricultural Sciences jointly sponsored a meeting in Beijing discussing co-ordination and co-operation in scientific research.

April 4—7

The Biology Department and the Geo-science Department sponsored a joint seminar on limnology.

April 19

A Sino-Soviet observation team conducted the observation of a total solar eclipse.

April 24

The Academia Sinica held a seminar on mineralization principles in the Qilian Shan area.

May 27

China successfully developed its first universal projective microscope.

June 20

Academia Sinica President Guo Moruo and Vice President Li Siguang were chosen academicians of the Soviet Academy of Sciences.

June 30

Xinhua News Agency announced that China's first experimental atomic reactor had gone into operation and a circular accelerator was completed, which marked China's stepping into the atomic age. Meanwhile a high-pressure electro-static accelerator was completed, too.

July 17—21

The Academia Sinica Geo-science Department and the State Bureau of Surveying and Cartography held a meeting marking the establishment of the National Atlas Compiling Committee.

August 1

China developed its first general digital electronic computer Model *August 1*.

August 18

The Sciences Department of the Academia Sinica held a symposium on the utilization of mountain snow and ice.

September 10—21

The Academia Sinica and other interested parties jointly sponsored the first national conference on minerals.

September 18—25

The All-China Federation of Scientific Societies and the All-China Association for the Dissemination of Scientific and Technological Knowledge convened a joint congress in Beijing and decided to amalgamate themselves into the Chinese Scientific and Technological Association.

A chronicle of events in science and technology

September 20

The Chinese Science and Technology University held its commencement ceremony.

Septemebr 22—24

An aquatic products conference was held in Wuhan.

September 24

Guo Moruo was appointed President of the Chinese Science and Technology University by the State Council.

September 27

A ceremony was conducted on the spot for China's first experimental atomic reactor and circular accelerator going into regular operation.

October 5 — November 12

The *Big Leap Forward* Exhibition of Natural Sciences Research Achievements was open in Beijing. Chairman Mao Zedong visited the exhibition and met the leaders of the Departments and research institutes of the Academia Sinica. Among the other visitors were Liu Shaoqi, Zhou Enlai, Zhu De, He Long, Chen Yi, Deng Xiaoping, Nie Rongzhen, Li Fuchun, Peng Zhen and Chen Yun.

October 5

Qinghua University in co-operation with the Beijing No. 1 Machine Tool Factory developed a computer-controlled programmable vertical miller.

October 12

The Beijing Industrial Technology College designed and manufactured China's first large planetarium.

October 17—19

The Academia Sinica and the Water Conservancy and Power Ministry jointly convened a national tidal power conference in Shanghai.

October 27

The Nei Menggu, Xinjiang, Ningxia Autonomous Regions and the Gansu, Shanxi, Qinghai provinces convened a joint conference on sand control in Hohhot.

October 29 — November 3

The Academia Sinica and the Constructional Engineering Ministry convened a joint meeting to promote new building materials and techniques in Harbin.

November 1—12

The State Council Science Planning Commission and the State Technological Commission jointly held a national meeting on scientific and technological information.

November 2—6
The Academia Sinica held a national spectroscopy conference.

November 23
The 102nd session of the Standing Committee of the National People's Congress decided on a proposal submitted by the State Council, to amalgamate the State Technological Commission and the State Council Science Planning Commission into the Science and Technology Commission. Chairman Mao appointed Vice Premier Nie Rongzhen to a concurrent post as the Chairman of the PRC Science and Technology Commission.

December 22 (1958) — January 4 (1959)
A national conference on local scientific and technological work was held in Shanghai and Comrade Nie Rongzhen delivered a speech on the occasion.

December 25
The Central Meteorological Bureau and the Chinese Academy of Agricultural Sciences jointly held a national conference on agricultural and meteorological work in Nanjing.

1959

January 20
The Chinese physicist Wang Ganchang, was chosen deputy director of the Institut Unifié de Recherches Nucleaires of the socialist countries.

January 21—28
Zhuang Xiaohui representing the Academia Sinica attended the 46th annual conference of the Indian Union of Science Conventions held in New Delhi.

January 23
The Nuclear Science Committee under the Academia Sinica and its Isotopic Application Committee were established.

February 5—11
Zhang Xi representing the Academia Sinica attended the 11th conference of the Pakistani Association for Science Promotion in Karachi.

February 14 — May 11
An Academia Sinica delegation headed by Wu Youxun toured the East European countries.

February 16—21
The Academia Sinica and the Water Conservancy and Power Ministry jointly held a working conference in Beijing on surveying and planning work for the grand project on South-to-North Water Transit in West China.

A chronicle of events in science and technology

February 23—27
The Academia Sinica held a conference on comprehensive surveying.

March 2—5
Zhong Shimo representing the Academia Sinica attended the second meeting of council members and advisors of the International Automation Association in Rome.

March 16—23
Yang Shixian and Liu Dagang representing the Academia Sinica attended the eighth Mendeleev conference on general and applied chemistry in Moscow.

March 17—22
The preparatory committee for the National Stratigraphical Conference held an ad hoc meeting in Lanzhou on the Strata in the Qilian Shan area.

March 20
The National Scientific and Technological Association held its first national working conference in Hanzhou.

April 1—5
The Academia Sinica and the State Surveying and Cartographical Bureau sponsored the second conference for the National Atlas Compiling Committee.

April 5—13
The Quarternary Geological Research Committee of the Academia Sinica sponsored discussions on the "Sanmen System" in the Sanmen Gorge reservoir area.

April 7—13
The Academia Sinica held a meeting in Beijing attended by the directors of its natural science research institutes. 43 priority projects were sorted out on a preliminary basis. Comrade Chen Yi and President Guo Moruo addressed the meeting.

April 18—28
The preparatory committee for the National Stratigraphical Conference held a meeting in Hangzhou, Changshan, Jiangshan and Shouchang on the stratum structure in West Zhejiang province.

May 6—14
Zhang Wenyou and other representatives from the Academia Sinica attended the centennial of the death of Alexander Humboldt in the DDR.

May 8—10
The State Metrological Bureau and the Academia Sinica Shanghai Branch jointly sponsored discussions on time frequency and the possibility for China to establish its own time frequency norm.

May 8 — June 12

The Academia Sinica Mathematics-Physics-Chemistry Department organized a natural gas research investigating team that toured 7 places including Lanzhou, Chongqing, Chengdu and Shanghai.

May 17

Xinhua News Agency release: The Academia Sinica Mountain Snow and Ice Utilization Research Team had started surveying and research in the major glacial areas of the Tian Shan Mountains in Xinjiang Autonomous Region.

May 21

The Water Conservancy Research Academy was renamed the Water Conservancy and Power Research Academy.

May 25—29

The Organic Chemistry Research Institute of the Academia Sinica and the Shanghai Pharmaceutical Industry Research Institute under the Ministry of Chemical Industry jointly convened the second national steroid hormone conference in Shanghai.

June 1

President Nesmiyanov of the Soviet Academy of Sciences, on behalf of his academy, conferred its overseas academician title to Guo Moruo, President of the Academia Sinica, and Li Siguang, Vice President of the Academia Sinica.

June 10—17

The Academia Sinica and the Ministries of Geology and Metallurgical Industry jointly held a national geological conference on rare and dispersed chemical elements.

June 11—20

The preparatory committee for the National Stratigraphical Conference held a meeting in Guiyang on the stratum structure in South Guizhou Province.

June 25 — July 4

Zhou Peiyuan representing the Academia Sinica attended the East-West Scientists Conference held in Baden, Austria.

June 25

The State Council promulgated a decree on unifying the metrological system.

July 1—12

The preparatory committee for the National Stratigraphical Conference held a meeting on the stratum structure in Shanxi Province.

A chronicle of events in science and technology

July 6—11

Zhang Wenyu, Peng Huanwu and Lu Min representing the Academia Sinica attended an international cosmic rays conference held in Moscow.

July 11

The eighth executive meeting of the Standing Committee of the Academia Sinica passed a directive that researchers should be given enough time for professional work.

July 18

Xinhua News Agency release: The Academia Sinica had recently held a metallurgical conference in Shanghai.

July 19

Xinhua News Agency release: China's first supersonics symposium was recently held in Wuchang.

August 2

Xinhua News Agency release: Soil Survey in Changjiang River basin had recently been completed after three years' work.

August 7

The preparatory committee for the National Stratigraphical Conference held discussions on the "Contrastive Summary Chart of China's Permian Period Strata."

August 8—20

The Academia Sinica Biology Department, the Chinese Academy of Agricultural Sciences and the National Soil Survey and Assessment Office jointly held a symposium in Harbin on soil survey, assessment and soil amelioration through deep ploughing.

August 23 — September 1

Liu Xuerong and Hu Zhining representing the Academia Sinica attended the international water conservancy and irrigational system conference held in Sofia.

September 10

The Academia Sinica Soil Research Institute, the Forestry Scientific Research Academy and the Central Meteorological Bureau jointly held a national symposium in Chengdu on forestry protection and meteorology.

September 14—19

Wang Ganchang, Li Yi and other Chinese scientists representing the Academia Sinica attended an international high-energy accelerator and instrumental control conference held in Geneva.

September 24—28

The Chinese Scientists Delegation headed by Academia Sinica Vice

President Zhu Kezhen attended the Sixth Congress of the World Scientists Union and its seminars held in Warsaw.

October 18
The Yuyuantan Experimental Hydropower Station, the first of its kind in China, was completed. Its construction was undertaken by the Water Conservancy and Power Scientific Research Academy.

November 11—15
The Mathematics-Physics-Chemistry Department and the Technical Sciences Department of the Academia Sinica jointly sponsored a catalysis research paper presentation in Dalian.

November 14—21
Under the auspices of the Academia Sinica, the Geological Ministry, the Ministry of Petroleum Industry, the Ministry of Coal Industry, the Ministry of Metallurgical Industry and the Chinese Geology Society, the National Stratigraphical Conference was convened. The National Stratigraphical Committee was established. at this Conference.

November 20 — December 4
The National Science and Technology Commission, the Ministry of Education and the Academia Sinica jointly convened a conference on natural science research work in the universities and colleges of China.

November 25 — December 1
The Academia Sinica and the Chinese Silicate Society jointly held a silicate research paper presentation in Shanghai.

December 2—8
The Academia Sinica Chemistry Research Institute and the Beijing Chemical Industry Research Institute jointly held a national plastic fibers research paper presentation.

December 2—11
The Water Conservancy and Power Research Academy and the Chinese Water Conservancy Society sponsored a conference discussing scientific and technological problems in water conservancy.

December 4—19
An Academia Sinica Chemistry delegation composed of Bian Boming and four other Chinese scientists visited the academies of Poland, Czechoslovakia, Hungary, the DDR and Rumania for academic investigation.

December 7
The Soviet ambassador to China, on behalf of the Praesidium of the Soviet Academy of Sciences, conferred a gold medal of the 1959 Karsky Award on Li Siguang, Vice President of the Academia Sinica.

A chronicle of events in science and technology

December 7—13

The Academia Sinica Biology Department held an embryology symposium in Shanghai.

December 8—12

The Academia Sinica and the Chinese Meteorology Society jointly sponsored an atmospheric circulation symposium.

December 19—28

The Academia Sinica Mathematics-Physics-Chemistry Department and the Chinese Physics Society jointly sponsored a national symposium on solid physics.

December 21—24

The Academia Sinica Geo-sciences Department held in Beijing a meeting marking the 30th anniversary of the discovery of the first cranium piece of Peking Man.

December 28 (1959) — January 3 (1960)

Ji Xianlin and Jian Zhuopo representing the Academia Sinica, attended the 50th anniversary of the Burmese Studies Association in Rangoon.

1960

The Academia Sinica sand control team surveyed the deserts in Northwest China and Nei Menggu, a total of six provinces and autonomous regions, and ascertained some natural conditions there conducive for building large production bases in agriculture, forestry and animal husbandry. Plans for the reclamation of some deserts such as the Gobi Desert were thus drawn up.

Fossils of *Homo sapiens* who lived 200,000 — 300,000 years ago were unearthed at Maba, Guangdong Province. The fossils were named after the place (Maba).

January 5—15

A national geographical symposium was held in Beijing.

January 7

It was reported that China's first national cardio-vascular disease research paper presentation was held in Xi'an.

February

Tha Academia Sinica held its first national biochemistry symposium in Shanghai.

February 28 — March 8

A national co-ordinating conference on geological science and technology was held in Beijing.

March 1—7

The first national remote control symposium was held in Beijing. It was decided that non-contact control equipment should receive major attention.

March

The Institut Unifie de Recherches Nucleaires in Moscow discovered, with the participation of Chinese scientists, a new fundamental particle, negative hyperon 3; the key device used was a propane gas bubble chamber designed under the instruction of the ex-deputy director Wang Ganchang, a famous Chinese pysicist.

April 15

Academia Sinica President Guo Muruo was chosen academician of the Czechoslovakian Academy of Sciences.

April 19

The third session of the Academia Sinica Inter-departmental Committee was held in Shanghai. The draft of a three-year program for China's natural science theoretical research was drawn up. Guo Moruo and Nie Rongzhen addressed the session.

May 25

The Chinese Mountaineering Team, for the first time known, climbed along the northern slope onto the highest peak of the world, Mount Qomolangma.

May

Assisted by the people of Zang nationality, a scientific surveying team, composed of 44 scientists and engineers, completed its preliminary survey of a 7,000 square-kilometer area east, west and north of Qomolangma.

July 22 — August 1

The Academia Sinica held a national operations research meeting in Jinan, Shangdong Province. Notes were compared on the application of operations and linear programming to traffic and transportation control, material distribution and the arrangement of rural labor.

Early October

The Taiyuan work team of the Academia Sinica Paleovertebrates and Paleoanthropological Research Institute found in Heiyukou valley, Xin County, an abundance of animal fossils, the majority of which were identified as fossils of odontae animals, belonging to the Ermayunian of the mesozoic era.

Middle October

The Academia Sinica Technological Sciences Department held a research paper presentation in Kunming on copper, nickel and tin ore dressing, refining and floatation dressing agents.

October

A microbial physiology group of the Academia Sinica Phytophysiology Research Institute discovered that RNA is a genetic transformation factor for microbes.

December 13

The State Council announced a directive that in scientific institutions, the eight-hour work-day system should be maintained and enough time should be appropriated for scientific research.

1961

January 6

Guangming Daily reported that the Academia Sinica sand control team had reached the center of the Taklamagan Desert and studied the water and soil resources there. This more-than-one-month survey cast some light on the mysteries of the desert.

January 20—28

A soil symposium sponsored by the Academia Sinica was held in Beijing.

January

The two-year general survey and positional observation, carried out by the Academia Sinica mountain snow and ice utilization research team, and

A scenery on the western part of the Taklamagan Desert

a few interested parties, provided tentative answers to what the bulk and distribution of the glaciers and snow and ice accumulation were in the Tian Shan area within the Xinjiang Autonomous Region, plus a basic understanding of the laws governing the accumulation and thawing of snow and ice there.

February 27 — March 4
The first national "karst" conference was held in Nanning.

March 10—17
The Forestry Ministry and the Chinese Forestry Scientific Research Academy jointly convened in Beijing, a national forest disease and pest insects symposium (North China section).

March 26—29
The Academica Sinica Forest and Soil Research Institute sponsored the forest hydrography and meteorology symposium, participated by all interested parties in China.

April 25
Xinhua News Agency release: Two autonomous regions, Ningxia and Nei Menggu, and two provinces, Gansu and Qinghai, had formed a research co-operative zone for studying ground water condition in terms of arid and semi-arid areas, grassland and desert. Its second conference was held in Yinchuan.

April 26
Scientific and technological circles in Beijing gathered to celebrate the centennial of the birth of Zhan Tianyou, a most outstanding patriotic engineer in China's democratic revolutionary period. Present at the meeting were Vice Premier Lu Dingyi and other leading comrades. Li Siguang, Chairman of the Chinese Scientific and Technological Association, delivered a speech on the occasion.

The Academia Sinica Taklamagan Desert research team started from Hetian to move along the Yulongkashi River and the Kalakashi River into the Desert while carrying out surveying and research work.

May 19
The Academia Sinica Biology Department held discussions on distant hybridization.

May 21
The Academia Sinica Biology Department held discussions on the concept of crop population.

May 21—25
The Academia Sinica held discussions on catalysis theory, focusing on the concept of surface bond.

May 22

Xinhua News Agency release: The Academia Sinica Shanghai Experimental Biology Research Institute Director, Zhu Xi, and its Associate Research Fellow, Wang Youlan, had started since 1951, the experiment of artificial parthenogenesis by blood-smear needling with toads. The world's first "fatherless" female toad born in March, 1959, had spawned in March, 1961. More than 800 tadpoles were generated and most of them had grown into small toads.

May

The Academia Sinica natural science history research group sponsored a symposium on the sectionalization of the natural science history of ancient China.

June

The Academia Sinica Applied Chemistry Research Institute, assisted by concerned units, developed China's first para-magnetic resonance spectroscope.

July 28

The Chinese Entomological Society and the preparatory committee for the Chinese Plants Protection Society, jointly sponsored an insect taxology seminar, exploring the significance insect taxology may have for actual production.

August 1

Wenhui Daily reported that Beijing's zoological circles had recently held discussions on the orientation of zoological taxology research and fauna studies. The Chinese Zoology Society had recently held a zoological regionalization symposium.

August 2–11

China's first differential equation symposium was held in Beijing.

August 21–29

The Academia Sinica held its third high molecules symposium in Changchun.

September 13

Xinhua News Agency release: Paleoanthropological cultural sites were discovered in Kehe Village, Ruicheng County, Shanxi Province. A preliminary study revealed that these newly discovered sites belonged to a period earlier than that of the Peking Man.

September

The leading Party groups of the State Science and Technology Commission and the Academia Sinica, promulgated the draft of the "14-Point Instruction on the Present Work of the Natural Science Research Institutions" and the "Provisional Regulations Concerning the Natural Science Research Institutes Under the Academia Sinica".

Zhu Xi experimenting on artificial parthenogenesis

The world's first shoal of small "grandfatherless" toads

The Academia Sinica and the Chinese Mathematics Society held separately in Beijing, Tianjin, Shanghai and other Chinese cities, seminars either on particular branches of mathematics or on specific mathematical problems, exploring the trends in basic mathematical and theoretical studies.

October 9—20
The Academia Sinica, the Chinese Metals Society and the Chinese Physics Society sponsored discussions on metal physics in Shenyang.

October 10—14
The Academia Sinica Technical Sciences Department and the Chinese Chemistry and Chemical Engineering Society held the first chromatography research paper presentation in Dalian.

November 10—18
A national antibiotics symposium was held in Shanghai.

November 23
Xinhua News Agency release: After many years' hard work by scientists, problems had been largely solved in the large-scale breeding of the castor-leaves-eating silkworms which originated in India.

December 1—8
The Chinese Metals Society and the Academia Sinica Technical Sciences Department jointly convened a national ore floatation dressing agent symposium in Changsha.

December
The Chinese Entomology Society held its 1961 symposium in Shanghai.

The Academia Sinica Experimental Biology Research Institute's Deputy Director Zhuang Xiaohui and his research assistant, Dai Rongxi discovered the embryonal epidermal conductive function in amphibious animals.

December 20—28
The Chinese Agricultural Crops Society held the first national crops strain breeding symposium in Changsha.

1962

January 16—28
A national forest science and technology conference was held in Guangzhou.

January 20—30
A national rice research seminar was held in Guangzhou.

January 27
Geologists in Beijing gathered to celebrate the 40th anniversary of the founding of the Chinese Geology Society which was the oldest natural science society in China.

A chronicle of events in science and technology

March 2
Guangming Daily reported that China's first oceanography conference had recently been held in Qingdao.

March 25
Xinhua News Agency release: After six years' of research and experiment. the Hubei Agricultural Scientific Research Institute found a way of preventing and controlling the pink boll-worms by pteromalidae, and many technical problems in raising the latter had been solved.

Guangming Daily reported that new progress had been made in the survey and research of China's glaciers and frozen earth. Ever since 1958, continuous surveying work had been done on the modern glaciers on Qilian Shan, Tian Shan and part of the Pamirs, and on the distribution of frozen earth on the Qinghai-Xizang (Tibet) Plateau, while research in relevant basic theories had been carried out. The special physico-geological phenomena and geomorphological types were also among the subjects of study.

March
A national scientific and technological work conference was held in Guangzhou. It was presided over by Vice Premier Nie Rongzhen and addressed by Premier Zhou Enlai and Vice Premier Chen Yi.

April 8
Xinhua News Agency release: Yu Zhenshan, a worker-mathematician, had recently succeeded in developing a new algorithm called "Gan-Zhu (Abacus) Algorithm" which was useful for popularizing mathematics.

May 6
Xinhua News Agency release: A more than 100-meter thick layer of perennially frozen earth was discovered underground in the Qinghai Xizang plateau by a scientific survey team, whose research demonstrated that the frozen earth layer explained the peculiar natural phenomena found on the surface.

June 20
A national marine-life symposium was held in Qingdao.

June 28
Xinhua News Agency release: A series of rather comprehensive traditional Chinese veterinary medicine books had been compiled by the country's famous veterinarians, and the series would go into publication on a book-by-book basis.

June
Having gone through the final tests, China's first 12,000-ton hydraulic press was put on trial-operation.

July 18
Xinhua News Agency release: A seven-year all-round survey of China's fruit-tree resources had been basically accomplished with satisfactory results. Many precious strains were identified, and an abundance of wild fruit trees was discovered.

August 22
Xinhua News Agency release: Thousands of geo-hydrological workers had over the past two years carried out on-the-spot surveys of the major agricultural and cattle-raising areas in northern and southern parts of China. The details of ground water distribution over an area of 100,000 square-kilometers were clarified and the geo-hydrological maps for farm irrigation of over 200 counties on the North China plain were drawn up.

End of August
The Chinese Geophysics Society and the Academia Geophysics Research Institute jointly held the second sun-earth symposium in Beijing.

September 4
Many famous scientists in Beijing gathered at the meeting hall of the Chinese People's Political Consultative Conference to pay tribute to the revered senior mathematician Xiong Qinglai for his distinguished contribution to teaching and research in mathematics in the past 40 years.

September 8
The Chinese Zoology Society and the Chinese Botany Society jointly sponsored the first cytology symposium in Beijing.

September 11
The Chinese Mechanics Society held a general mechanics symposium in Beijing.

Wenhui Daily reported that a marine biological resources survey team, whose operating range covered a vast area from the outlet of the Yalu River in the north to the Xisha Islands in the south, had carried out an 8-year on-the-spot investigation and obtained much information on China's marine animal and plant resources.

September 18
The sixth executive meeting of the Standing Committee of the Academia Sinica came to a decision for further implementation of the "14-point Instruction on the Present Work of Natural Science Research Institutions" and the "Provisional Regulations Concerning the Natural Science Research Institutes under the Academia Sinica".

September
The Chinese Entomology Society held a bee-keeping symposium in Hangzhou.

A chronicle of events in science and technology

October
Invited by the State Science and Technology Commission and the Ministry of Agriculture, more than 60 scientists representing different agricultural sciences gathered in Beijing and discussed the strengthening of agricultural science research. Premier Zhou delivered an important speech, maintaining that the concerned fronts and departments must be fully mobilized and properly arranged, and all should concentrate on the key task of agrotechnical innovations.

November 12
Xinhua News Agency release: A 12-year South China comprehensive biological resources survey had reached a successful completion.

November 23
The eighth executive meeting of the Standing Committee of the Academia Sinica passed a few regulations. Among them were the Detailed Administrative Regulations Concerning Research Students of the Academia Sinica and the Interim Regulations on the Establishment of Professional Assessment Archives for Natural Science Research and Technical Personnel of the Academia Sinica.

Late November
The Chinese Botany Society held a phytoecology and geophytology symposium in Beijing.

The Academia Sinica, the Chinese Metals Society and the Chinese Physics Society jointly sponsored its second national metal physics symposium.

November
The fauna, taxonomy and ecology sections of the Chinese Zoology Society held a conference in Guangzhou.

A second national semi-conductor symposium was held in Beijing.

The Academia Sinica Chemistry Research Institute and the Academia Sinica Southwest China Branch held a fourth high molecules symposium in Chengdu.

December 26
The Academia Sinica sponsored the "Technological Artifacts Exhibition".

December
In the light of the March Guangzhou Conference, a basic and technical sciences development program (1963-1971) had been worked out after five months' effort.

The Chinese Geography Society held a symposium on land hydrography in Changchun.

1963

Pterosaurier fossils were found for the first time in the Zhungaer Basin, Xinjiang Autonomous Region. The animals were believed to have existed 90—100 million years ago.

January 16
China produced an oral vaccine for infantile paralysis which showed good results.

January 29
At a scientific and technological conference held in Shanghai, Premier Zhou Enlai expounded on the great significance of modernization in science and technology and laid down major requirements.

February 8 — end of March
The Central Committee of the Chinese Communist Party and the State Council convened a national conference on agricultural science and technology in Beijing. A program was drawn up for the development of agricultural science and technology. During the conference, Chairman Mao and other state leaders received the conference participants and Premier Zhou Enlai, Vice Premier Tan Zhenlin, Vice Premier Nie Rongzhen delivered speeches.

February 26
Xinhua News Agency release: Chinese scientists had recently found and unearthed at Nanxiong County, Guangdong Province; paleovertebrates fossils in the stratum of "Red Petrological System". These fossils were very useful in identifying the geological age of the "Red Layer" and in the study of the systematic evolution of vertebrates".

Middle April
The Academia Sinica Mathematics-Physics-Chemistry Department, the Chinese Physics Society and the Chinese Electronics Society jointly held a magnetics and magnetic materials symposium at Wuxi.

April 22—30
The Academia Sinica's mountain snow and ice utilization research team and the Geography Research Institutes glacier and frozen earth research group held their first symposium in Lanzhou.

May 2
Xinhua News Agency release: The Ministry of Agriculture had convened in Beijing a national plants protection conference, at which Vice Premier Tan Zhenlin delivered a speech.

May 20
The 131st plenary session of the State Council confirmed the "Regulations on Forest Protection".

May 27

The State Council promulgated the Regulations on Forest Protection.

May

The second national on tropical forest bio-geographical community pedigree conference, presided over by Zhu Kezhen, Vice President of the Academia Sinica, was convened in Xishuangbanna, Yunnan Province.

June — August

The Academia Sinica Paleovertebrates Research Institute found in Lantian County, Shanxi Province, an intact ape-man's mandible fossil. It became the first ape-man fossil site of New China and the second ape-man site in China's history.

July 14

The Chinese Science and Technology University held a ceremony for its first 1,600 students, who were congratulated and addressed by Vice Premier Chen Yi, Vice Premier Nie Rongzhen and President Guo Moruo.

August 5—13

Xinhua News Agency release: Surgeons from the Shanghai Sixth People's Hospital successfully performed an operation rarely attempted: a com-

The first successful case of severed hand rejoined: Shanghai worker Wang Cunbo operating with his rejoined hand

pletely severed hand of a worker was rejoined at the wrist. On the 7th of August, Premier Zhou Enlai and Vice Premier Chen Yi received the performing surgeons Qian Yunqing, Chen Zhongwei, etc., and called their operation a creative job of great significance. On the 13th of the same month, the Ministry of Public Health held an award presentation for the successful operation.

August 26 — September 3
The first cotton symposium was held in Beijing.

August 28 — September 4
The Academia Sinica Mathematics-Physics-Chemistry Department held China's first symposium in Changchun on the structure of matter.

Early September
The Academia Sinica Technical Sciences Department and the Chinese Mechanical Engineering Society held the second national electrical processing conference in Beijing.

September 21—29
The Chinese Medical Society held its eighth national surgery symposium in Beijing.

September 22
Xinhua News Agency release: The Xin'anjiang Hydropower Station, which had been designed, constructed and installed by China itself, began to transmit power to Shanghai, Nanjing and Hangzhou.

September
The first national radiology symposium was held in Shanghai.

October 22—29
The second national catalysis research paper presentation was held in Lanzhou.

October 23
The 136th plenary session of the State Council confirmed the "Regulations on Awards for Inventions" and the "Regulations on Awards for Technical Innovations".

October 24—31
The State Science and Technology Commission computational technology group, the Chinese Electronics Society and the Chinese Mathematics Society jointly convened the third computer technology notes-comparing meeting in Xi'an.

November 3
The State Council promulgated the "Regulations on Awards for Inventions" and the "Regulations on Awards for Technical Innovations".

A chronicle of events in science and technology

Early November

The first national electro-chemistry research paper presentation was held in Changchun.

The Chinese Zoology Society held a national experimental zoology symposium in Shanghai.

November 13

Xinhua News Agency release: The Forestry Ministry recently called a special meeting on the protection and hunting of wild animals.

Xinhua News Agency release: The Chinese Mechanics Society recently held the first fluid-dynamics symposium in Shanghai.

December

The Chinese Medical Society held the first national neurology and psychiatry symposium in Guangzhou.

The Academia Sinica Central-South China Physics Research Institute built China's first cosmic rays observation station.

The Academia Sinica Xizang comprehensive survey team held an academic summary meeting.

The Chinese Silicate Society held the first glass fibre symposium in Shanghai.

1964

The Water Conservancy and Power Science Academy completed compiling the "Hydrological Maps of China".

"The Natural Geographical Maps of the People's Republic of China" was completed.

January 1

The Beijing Scientists' Club was officially opened.

January

The Chinese Academy of Agricultural Sciences held the second national crops strain breeding conference.

Early March — Early May

The Chinese Mountaineering Team-Cum-Scientific Survey Group climbed Mount Xixiabangma and collected precious data and samples.

Middle March

The Chinese Plants Protection Society and the Chinese Entomology Society jointly held the first national biological prevention and elimination of farm and forest pests Symposium in Wuhan.

March 19—26
Two academic specialty sections, soil science and phytophysiology, from the Academia Sinica, held their assist-agriculture working conference.

April 4
The Standing Committee executive meeting of the Academia Sinica came to a decision that the Regulations on Academia Sinica Work (natural sciences) be promulgated and trial-implemented.

A national conference on schistosomiasis research ended in Shanghai.

April 6
China's first electronic blood-corpuscle counter was developed by a Shanghai worker named Zhang Genfu.

April 7
The first national haematology conference was held in Tianjin.

April 9
Xinhua News Agency release: China had newly developed an oral non-antimonial medicine for acute schistosomiasis.

May 1
China's first general digital electronic computer, on trial and produced by the Academia Sinica Computer Research Institute and other concerned units, went into regular operation. It had a speed of 50,000 calculations per minute and a capacity of 16,348 words.

May 18
The fifth executive meeting of the Standing Committee of the Academia Sinica approved in principle that the draft of the Provisional Rules on Natural Science Research Achievements in the Academia Sinica could be implemented within the scope of the Academia Sinica.

May 22
The Paleovertebrates and Paleoanthropology Research Institute found at Lantian; fossils of an ape-man's skull, three ape-man's teeth and many other mammals.

May 23
The Academia Sinica held an academy-wide notes-comparing meeting and issued the title of "advanced work-unit" to the Shanghai Astronomial Observatory time-latitude research group and 11 other units, and the title of "advanced worker" was conferred on 33 comrades. Moreover, 112 scientific research projects and technical jobs obtained citations.

June 23—30
The Chinese Medical Society held the sixth national pediatrics symposium in Beijing.

June 29

The State Planning Commission and the State Economic Commission jointly issued awards on new industrial products. 82 of them went to the Academia Sinica.

June

The Academia Sinica Assist-Agriculture Office held a farm pesticides research conference in Shanghai.

July

The State Science and Technology Commission held a conference on standardization in agriculture in Beijing.

The Chinese Meteorology Society held a weather and dynamic meteorology symposium in Lanzhou.

July 21

On the north slope of Mount Xixiabangma at an altitude of 5,700 meters above sea level, the Chinese Mountaineering Team scientific survey group found tree-leaf fossils of "high mountain oak" and "long-petioled dogwood" that were a million years old.

August 14

Xinhua News Agency release: The State Council had recently approved and circulated the "Regulations on Protection and Reproduction of Aquatic Resources" (draft), drawn up by the Ministry of Aquatic Production, and asked the central government departments and local governments concerned to examine and trial-implement it.

Middle August

An ultrasonic diagnosis symposium was held in Shanghai.

Late August

The Chinese Botany Society and the Chinese Oceanography and Limnology Society jointly convened the first national algology symposium.

August 21–31

The 1964 Beijing Science Symposium was convened in Beijing with an overseas participation of 367 scientists from 44 countries and regions in Asia, Africa, Latin America and Oceania. President Guo Moruo addressed the opening session and Vice Premier Nie Rongzhen delivered a congratulatory speech.

August 29

Xinhua News Agency release: After its initial success, the Shanghai Sixth People's Hospital succeeded in rejoining another severed arm of a worker.

September 23

The Academia Sinica Geophysics Research Institute Tibetan geomagnetic survey team obtained geomagnetic data at the highest location ever in the world's geomagnetic history. The height of Mount Qomolangma was 6,300 meters above sea level.

Early October
The Chinese Forest Society and the Chinese Pedology Society jointly held the first national forest soil symposium in Shenyang.

October 16
By exploding its first atom bomb, China successfully carried out its first nuclear test.

Late October
The Chinese Geography Society held an arid area geography symposium in Lanzhou.

November 16
Guangming Daily reported that the second national Quarternary symposium and the Lantian Cenozoic meeting had recently been held in Xi'an.

Middle November
China's first soil microbiology symposium was held in Wuhan.

December
The Academia Sinica Electronics Research Institute developed a high-voltage long-arc xenon lamp, which filled a gap in China's light source research.

The Academia Sinica Mathematics-Physics-Chemistry Department called an assessment meeting, examining China's first solar telescope, trial-produced by the Shanghai Scientific Instruments Factory and the Nanjing Scientific Instruments Factory.

The Academia Sinica held the first national refractory metals symposium in Shenyang.

China's first 1,500 kilo-watt turbine generator was trial-produced by the Nanjing Turbine Power Generator Factory.

The first national ship dynamics symposium was held in Shanghai.

A capillary chromatograph and an electronic orbit meter were trial-produced by scientific research institutions and manufacturing units in cooperation.

1965

The Academia Sinica Applied Chemistry Research Institute developed new plastics materials: polytetrafluorothylene and polysexafluoropropane.

The successful explosion of China's first atomic bomb

The Academia Sinica Applied Chemistry Research Institute developed polyimine.

The Shanghai Scientific Instruments Factory trial-produced China's first high-resolution nuclear-magnetic resonance spectroscope.

The preparatory section for the Beijing Astronomical Observatory under the Academia Sinica developed a 3.2-centimeter band solar radiotelescope.

The Academia Sinica Automation Research Institute and the Shanghai Scientific Instruments Factory co-developed China's first large analogue electronic computer.

The Academia Sinica Geology Research Institute prospecting team ascertained a large rare-element ore deposit, consisting of more than 20 rare elements.

January
The Academia Sinica decided to give its 1964 awards to 140 projects.

The Academia Sinica Zijinshan Astronomical Observatory discovered two comets in succession.

February 15
The first agricultural sciences academy in the Xinjiang Autonomous Region was established.

Middle February
The first national symposium on pharmacology was held in Shanghai.

February
The Academia Sinica Paleovertebrates and Paleoanthropology Research Institute succeeded in repairing, modelling and framing the *Mamenchisaurus hechuanensis,* first discovered by a petroleum-geological team in Sichuan Province.

The Academia Sinica Geophysics Research Institute developed a large size vacuum test equipment.

The Chinese Chemical Engineering Society held a farm pesticides symposium in Shanghai. Papers presented at the symposium indicated that China had 18 new farm insecticides trial-produced in 1964, and 5 of the emulsifiers that China used to import were then produced through its own efforts.

March
The second national oncology symposium was held in Shanghai.

The State Council held a national agricultural scientific experiment in Beijing at which Premier Zhou Enlai and Vice Premier Li Xiannian delivered speeches.

The Academia Sinica Scientific Instruments Factory developed an electron-emitting microscope.

April 26
Xinhua News Agency release: A new supersonic thickness gauge that China first developed had gone into mass production in Shanghai.

April 28
Xinhua News Agency release: The Shanghai Electric Generator Plant developed a double-circulation, internal water-cooling steam turbine generator.

May 14
By exploding another atomic bomb, China successfully conducted its second nuclear test.

May 31
The Academia Sinica Geo-sciences Department held a symposium on the first research achievements concerning the Lantian ape-man, announcing that *Homo etectus lantianensis* belonged to the earliest type of ape-man.

Late May
The State Assessment Committee examined with satisfaction, China's first large fast-speed transistorized general digital computer.

May
The Shanghai Scientific Instruments Factory, co-operating with some interested parties from the Academia Sinica, trial-produced China's first horizontal electronic diffractometer.

China's first self-designed 500,000-volt standard capacitor was trial-produced in Xi'an.

China's first high-precision frequency-changing electric source was developed in Shanghai.

Polytetrafluoroethylene (Teflon) was trial-produced in Shanghai.

June
Xinhua News Agency release: After 8 years of repeated field investigation and research, the students and staff of the Geology Department of Nanjing University came to the conclusion that the granite in South China belonged to four different geological periods challenging the traditional assumption, which had been held for more than 40 years, that granite in this area belonged to one and the same geological period.

The molecular sieve, a highly effective separatory material, was developed and went into mass production in China.

June — July

China's synthetic insulin (Chain A and Chain B), after primary depuration, proved through animal tests to have a steady 10% activity of natural insulin. Among the participants were the Academia Sinica Biochemistry Research Institute, Organic Chemistry Research Institute and Beijing University.

July 9

Xinhua News Agency release: China's first 24-order medium-sized analogue electronic computer was manufactured by the Tianjin Electronic Instruments Factory.

August 2

Xinhua News Agency release: China's first large electron microscope was trial-produced in Shanghai.

Niu Jinyi from the Academia Sinica Biochemistry Research Institute reading his paper on the synthetic insulin

August 18

The "Warm-Cloud Rain-Fall Fluctuation Theory," put forth by junior Chinese meteorologists, broke away from the traditional view held by international meteorological circles for more than ten years about how clouds turn into rain.

Middle August

The first national bast-fiber plants symposium was held in Xiaoshan County, Zhejiang Province.

September

The Peking Steel Wire Factory manufactured a precision product, the FeCrAl resistance element, which only few countries in the world could manufacture then.

The Changhai Hospital in Shanghai successfully planted an artificial heart valve in one of its patients.

The precision of time-measurement in China's astronomical observatories had reached advanced world levels, with a difference of less than two thousandths of a second.

China's first large urea-equipment, designed and manufactured by herself, went into production in Shanghai.

October 8

Xinhua News Agency release: after 15 years' effort, the western section of the Northeast shelter belt for farmland, the largest of its kind in China, had come to cover an area of more than 200,000 hectares, protecting 1,300,000 hectares of farmland and tens of thousands hectares of grassland.

October
A symposium on national rice disease, pest insects and borers was held in Changsha.

The third national ENT symposium was held in Nanjing.

The Chinese Forest Society held the first national afforestation and sand control symposium in Minqin County, Gansu Province.

China's first high-precision semi-automatic universal circular grinder was trial-produced by the Shanghai Machine Tools factory.

China's first arc-plasmatic jet spraying gun was developed in Shanghai.

November
The first national obstetrics and gynaecology symposium was held in Beijing. Premier Zhou Enlai gave important instructions at the meeting.

The Academia Sinica Organic Chemistry Research Institute developed a non-ferrous metal extractant N-235.

The State Assessment Committee examined the world's first synthetic crystalline protein of bovine insulin made in China and concluded that it had attained approximation to natural bovine insulin both in crystalline shape and biological activity. This testified to China's leading position in this aspect of research.

China's first 8,820-horsepower heavy, low-speed diesel engine was manufactured in Shanghai.

December
The Academia Sinica convened the first national piezo-electric ceramics symposium.

The Chinese Meteorology Society held a complementary weather forecast symposium in Guilin.

The Chinese Medical Society held the first national opthalmology symposium in Wuhan.

The preparatory committee for the Chinese Corrosion and Protection Society held, in Shanghai, a symposium on the corrosion and correct use of stainless steel, and intergranular corrosion.

The Academia Sinica Shanghai Silicate Research Institute produced a large synthetic mica crystal.

The Academia Sinica Zijinshan Astronomical Observatory and the Nanjing Astronomical Instruments Factory trial-produced China's first 43/60/80 refractive-reflective-telescope.

A chronicle of events in science and technology

1966

January
The Shanghai Astronomical Observatory developed an ammonia molecular clock with a short-term stability of 5×10^{-11} and a long-term stability of 5×10^{-10}

February 24–26
The assessment committee, founded by the Ministry of Metallurgical Industry and the Academia Sinica, examined the process of stainless steel tube making with a molybdenum-tipped ram. The new process was a result of the co-operation between the Academia Sinica Metals Research Institute, the Anshan Iron and Steel Complex Seamless Tube Factory and the Anshan Iron and Steel Complex Research Institute.

Early March
A national forestry conference was held in Beijing.

March 13
Xinhua News Agency release: The Chinese Geology Society has recently held a national "karst" symposium in Guilin.

Middle April
The third national farm-crop strain breeding conference was held in Beijing.

May 9
China conducted a thermo-nuclear explosion over its western area.

May 20
Xinhua News Agency release: The Beijing Union Hospital achieved gratifying results in treating chorioepithelium cancer and chorioadenoma.

July 23–31
The 1966 summer physics seminar of the Beijing Scientific Symposium was held in Beijing with an overseas participation of over 140 scientists from more than 30 countries, in Asia, Africa, Latin America and Oceania. Premier Zhou Enlai cabled a special message to the seminar Vice Premier Nie Rongzhen, on behalf of the Chinese Government, people and scientists, who delivered a congratulatory speech at its opening ceremony. On the afternoon and evening of the 31st, Chairman Mao and Liu Shaoqi, Zhu De, Dong Biwu, Deng Xiaoping, Chen Yi, Li Fuchun, He Long and other state leaders separately met the participants from various countries.

October 27
China conducted a nuclear-warheaded guided missile test.

December 28
China conducted a nuclear test.

Vice Premier Nie Rongzhen addressing the 1966 summer physics seminar of the Beijing Science Symposium

1967

June 17
China successfully exploded a H-bomb over its western area.

1968

December 27
China exploded another H-Bomb.

1969

September 23
China successfully carried out its first underground nuclear test,

September 29
China exploded a H-bomb.

1970

April 24
China launched its first artificial earth satellite, with a mass of 173 kilograms. It had an orbit with an apogee of 2,384 kilometers, a perigee of 439 kilometers, and an orbital inclination of $68.5°$. It took 114 minutes for the satellite to complete one orbit.

1971

Hu Han, an associate research fellow, and his research team from the Academia Sinica Genetics Research Institute succeeded for the first time in China, and in the world, in wheat reproduction through pollen culture.

March 3
China successfully launched a scientific experimental satellite, with a mass of 221 kilograms, that sent back all kinds of scientific data during flight. It had an orbit with a perigee of 266 kilometers, an apogee of 1826 kilometers, and an orbital inclination of 69.9. It took 106 minutes for the satellite to complete an orbit.

April 29
Academia Sinica Vice President Li Siguang passed away.

September
Chinese scientists, using an X-ray diffractive method, carried out measurements of pig's insulin crystalline structure with a resolution of 2.5 A. This was another major achievement in protein research.

November 18
China conducted a nuclear test.

The successful explosion of China's first H-bomb.

1972

January 7
China conducted a nuclear test.

January
A large dinosaur fossil was found in Zhucheng County, Shandong Province, which offered reliable evidence for the identification of the late Cretaceous strata in east Shandong.

The fossils of a pair of 3-meter long tusks, and other parts of the *Nama Paleoelephas* were discovered in Zhengzhou, Henan Province, which provided material evidence for the study of the geological features, climate and vertebrates in the Henan area 400,000 years ago.

1973

June 27
China conducted a H-bomb test.

July 31
The Academia Sinica Paleovertebrates and Paleoanthropology Research Institute discovered many paleocene mammal fossils in the "Red Layer" in Nanxiong of Guangdong, Qianshan and Toncheng of Anhui, Dayu of Jiangxi and Chaling of Hunan.

September 3
It was reported that China's first photoelectric astrolabe for astronomical

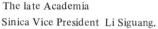
The late Academia Sinica Vice President Li Siguang.

time and altitude measuring had been developed, and had proved to have functioned well for two years.

1974

Chinese scientists, who had continued their large-scale multidisciplinary comprehensive survey on the Qinghai Xizang Plateau, and carried out complementary investigations around Mount Qomolangma, were rewarded with a considerable amount of very precious data.

February 7
Academia Sinica Vice President Zhu Kezhen passed away.

June 17
China conducted a nuclear test.

July 8
It was reported that scientists from the Academia Sinica Lanzhou Glacier-Frozen earth-Desert Research Institute, together with the maintenance workers of the Yumen Railway Administration Bureau Branch, had succeeded in setting up sand-break tree belts after much experimentation, and in keeping sand storms from the railway.

October 9
China succeeded in applying traditional medicine for total anaesthesia. It was a complex anaesthetic (a combination of Chinese and Western medicine), with *Flos daturae* as the primary ingredient, which was good for large and medium-sized operations.

1975

China developed an analogous satellite-ground communication station.

January
China developed a new type of radio telescope, a 450 megahertz radio complex interferometer for high resolution solar observation.

February
Tha Academia Sinica Zijinshan Astronomical Observatory discovered a planetoid with a slightly abnormal orbit. Its solar orbit had a radius slightly smaller than that of an ordinary planetoid while its eccentricity was bigger.

The Academia Sinica Physics Research Institute developed a small experimental device for controlled thermo-nuclear reactions with a quasi-stable state, ring-shaped strong magnetic field. It had gone into operation after debugging.

March
China's first phase high-precision laser distance gauge was developed.

The late Academia Sinica Vice President Zhu Kezhen

April

China's desert scientists had accomplished fundamental investigations on the area, distribution, natural conditions and resources of the deserts in the country and worked out a set of measures for protecting the farmland from sands and protecting the railway sections from the Gobi winddrifts. Meanwhile, a prototype reverse osmosis desalinator with a daily output of ten-ton fresh water was developed for desert use.

June

Chinese scientists, assisted by the Chinese Mountaineering Team, accomplished a multidisciplinary comprehensive scientific survey in the Qomolangma area.

July 27

China launched an artificial earth satellite that completed each circle around the earth in 91 minutes. It had an orbit with a perigee of 186 kilometers and an apogee of 464 kilometers, and an orbital inclination of 69°.

August

Hua Guofeng, Vice Premier of the State Council, invited a few scientists for a discussion on the policy of "Let a Hundred Flowers Bloom and a Hundred Schools Contend".

August 30

The Academia Sinica Beijing Astronomical Observatory discovered a nova in the Swan Constellation.

September

The Academia Sinica Beijing Scientific Instruments Factory trial-produced a high resolution scanning electron microscope.

The Academia Sinica Nanjing Soil Research Institute developed a set of ion-selective electrodes.

October

A new cotton strain was developed by the Academia Sinica Genetics Research Institute. It achieved a production increase of 20—30% in comparison with the best cotton strains that China was then popularizing.

Chinese scientists succeeded in producing new rice strains by haploid breeding with pollen culture.

October 27

China conducted an underground nuclear test.

October 31

Li Tianci, a junior researcher of the Academia Sinica Zijinshan Astronomical Observatory, invented the "quick finder for Western and Chinese calender date matching". It constituted a new auxiliary tool for studies of Chinese science and history.

November 8
Chinese scientists found many of paleozen mammal fossils in Naomugen People's Commune, Siziwang Banner, Nei Menggu Autonomous Region.

November 26
China launched an artificial earth satellite that had an orbit with a perigee of 173 kilometers and an apogee of 483 kilometers, and an orbital inclination of 63°. It took 91 minutes for the satellite to complete an orbit.

December 3
It was reported that the scientific survey, the largest China had ever done on the Qinghai-Xizang Plateau brought back an abundance of first-hand data and samples. More than 240 scientists took part in this project.

December 5
The Chemistry Department of Xiamen (Amoy) University developed China's first complex electro-chemical test instrument.

December 16
China launched an artificial earth satellite.

1976

January 23
China conducted a nuclear test.

February 18
The Academia Sinica Microbiology Research Institute succeeded in producing fumaric acid by liquid paraffin fermentation.

March 6
Arising from the co-operation between the Academia Sinica Chemistry Research Institute and the Beijing Analytical Instruments Factory, a new specific surface area and pority diameter distribution meter, with distinct Chinese characteristics, was successfully produced. Its mass production and popularization were to follow.

March 8
A big meteoric shower, rare in the world's history, fell on a location in China's Jilin Province. Of the 100 odd meteors collected, the largest weighed 1,770 kilograms, and the shower covered an area of more than 500 square kilometers, the investigation team from the Academia Sinica reported.

A chronicle of events in science and technology

June
The Qinghai Xizang Plateau comprehensive scientific survey team from the Academia Sinica, assisted by the PLA and civil aviation departments, flew over the world's highest peak, Mount Qomolangma (8,848.13 meters above sea level) five times while taking geographical and geomorphological photos of the top, the south and north slopes, and the east and west sides of Mount Qomolangma. The three big glaciers, East Rongbu, Middle Rongbu and West Rongbu, together with the glacial lakes and other features between the glaciers, were photographed too.

July 28
Strong earthquakes struck the Tangshan area, with 7.8 magnitude and 11 degrees epicentral intensity.

August 30
China launched an aritifical earth satellite.

September 26
China conducted a nuclear test.

October 17
China successfully conducted an underground nuclear test.

November 17
China successfully exploded another H-bomb.

December 7
China launched an artificial earth satellite that returned to the earth surface and was retrieved, all according to a precisely scheduled plan.

1977

Aeroplanes for agriculture and forest use were designed and manufactured by China.

A scientific survey ship "Experiment", from the Academia Sinica South China Sea Oceanography Research Institute, sailed across the South China Sea and accomplished a scientific survey of the area south of China's Zhongsha Islands, Xisha Islands and north of China's Nansha Islands.

January 3
Xinhua News Agency release: A large-scale survey on geothermal resources in Xizang conducted by Chinese scientists indicated that Xizang was rich in geo-thermal resources.

January 14
China's first single channel, telefacsimile communication network was built for public use within the bounds of Yidu County, Hubei Province.

January 20
In the Changdu area, Xizang, 4,200 meters above sea level, Chinese scientists found for the first time, fossils of dinosaurs which existed 140–160 million years ago.

January
Liu An Radio Factory in Anhui Province developed a high-precision fully-automatic, transistorized, continuous-operating nuclear gyromagnetometer, most suitable for China's seismic stations.

February 5
The Chuanguang Instruments Factory in Henan Province developed a China-designed large, multi-use optical transitive function meter.

February 13
China's post and telecommunication units developed the second model of a 960-channel microwave telecommunication system.

February 25
Yang Le and Zhang Guanghou, researchers from the Academia Sinica Mathematics Research Institute, co-operated in the study of function theory and discovered for the first time in the world, the structural relationship between "deficient value" and "singular directions", two major concepts in the theory of functional value distribution.

February
China's archaelogists found two teeth fossils of apeman in Bailong Cave, Yunxi County, Hubei Province.

The historical site of a large shipyard that existed during the Qin and Han dynasties was found for the first time in Guangzhou. This greatly benefited the research in China's ancient ship-building and marine navigation.

Team 909 of the Airborne Geophysical Prospecting Detachment under the State Geological Administration Bureau accomplished prospecting work over vast prescribed sea areas in China's South Yellow Sea, East Sea and South Sea from June 1974 to February 1977.

March 11
The Academia Sinica Zijinshan Astronomical Observatory and Beijing Observatory discovered that Uranus has a ring too.

Late March
The Academia Sinica held a high-energy physics symposium in Beijing.

March
The Ministry of Agriculture and Forestry held a national conference on forestry and aquatic production in Beijing.

April 9

In Lufeng County, Yunnan Province, the fossil of an intact mandible of a paleopithecus was found for the first time in China, by the combined investigation team from the Academia Sinica Paleovertebrates and Paleoanthropology Research Institute and the provincial museum.

April

The Academia Sinica theoretical study group held its first seminar, discussing the relations between natural science and philosophy. The absurd substitution of natural science by philosophy advocated by the "Gang of Four" was repudiated.

May 28

A new drilling technique with artificial diamonds was developed as a national priority project of major significance.

May

The Henan Provincial Geological Bureau found a large good-quality natural soda mine and named it Wucheng Mine, in Tongbo County, Nanyang Prefecture.

May — June

China's forest scientists found two rare tree species, the Guangxi crytomeria and "Wangtian" tree in the Yunnan Province and Guangxi Autonomous Region.

The National Science and Technology Association held six symposiums. Its Vice Chairman Zhou Peiyuan delivered a speech entitled, "On the Two-Line Struggle Over the Basic Theory of Natural Science".

June 14

It was reported that a Shixia culture of the neolithic period was found in the Shixia district near the "Maba-man" cave-site in Qujiang County, Guangdong Province.

June 20 — July 7

The Academia Sinica held a working conference in Beijing to discuss the draft of the academy's academic subjects development program. Party Chairman Hua Guoteng, Vice Chairman Ye Jianying and other Party and state leaders received the conference participants on July 3.

June 22

After four years' hard work, Chinese scientists accomplished a comprehensive field survey of the Qinghai Xizang Plateau. The Academia Sinica convened a special conference, discussing the papers from this scientific survey.

June

Professor Tang Aoqing from the Jilin University Chemistry Department, succeeded in enhancing a qualitative stage to reach a semi-quantitative stage in the theory of molecular orbits symmetry conservation. Together with his fellow colleagues, Professor Tang developed his molecular orbit pattern theory. The two achievements enriched and developed molecular theory in quantum chemisty.

June — July
The Yellow River harnessing task group, the Water Conservancy and Power Ministry and the Agriculture and Forestry Ministry, jointly held in Yenan, Shanxi Province, a working conference on soil and water conservation in areas around the middle reaches of the Yellow River.

The fossils of a fairly complete carnivorous dinosaur were found in a reservoir construction site in the Yongchuan County, Sichuan Province.

July 13
It was reported that China had developed a new antibiotic, globo-erythromycin.

July 18
China's first electron microscope that could magnify an object 800,000 times was trial-produced in Shanghai. Its design and materials were all of Chinese origin.

July 19
It was reported that two Chinese ocean-going scientific research ships, Xiangyanghong No. 5 and Xiangyanghong No. 12, had successfully conducted their second survey mission over a vast Pacific area.

July 20
The State Council promulgated the Metrological Regulations of the People's Republic of China for trial implementation.

July 25
The Chinese Mountaineering Team climbed the highest peak of the Tianshan Mountains, Mount Tomul.

July
The Public Health Ministry held a national tumour prevention and cure symposium in Beijing.

August 3—8
Vice Premier Deng Xiaoping called a meeting in Beijing to discuss China's scientific and educational work.

August 8
The Chinese Writing System Reform Committee and the State Standardization and Metrology Administrative Bureau, jointly promulgated a list of Chinese characters to be used for the names of some measures and weights in order to better implement the Metrological Regulations of the People's Republic of China.

A chronicle of events in science and technology

August 18
It was reported that Chinese scientists, combining mountaineering with scientific surveys, had recently accomplished a multidisciplinary comprehensive survey of Mount Tomul, the highest peak of the Tianshan Mountains, and its vicinity, and obtained a considerable amount of first-hand scientific data and samples.

August 25—28
For three days on end, a few Chinese scientists and model workers held talks with young science enthusiasts from the middle schools in Beijing, counselling and encouraging them to have a good command of the basic sciences such as mathematics, physics and chemistry, so as to make great achievements in science and technology for the four modernizations of their motherland. The number of young participants exceeded 7,000.

August 28
The Chinese biologist, Tong Dizhou, and his colleagues, in co-operation with the American biologists Niu Manjiang and his wife, discovered that cell-plasma's informational RNA has an apparent effect on the growth and heredity of animals, while the DNA in the cell nucleus has an inducing effect on the growth and heredity of animals with respect to distant relations.

August
The Academia Sinica convened a symposium on the Jilin meteoric shower. The participating scientists learnt approximately as to where and how the world's biggest meteoric shower came and fell, and the scientists produced their models for its formation and evolution.

The Academia Sinica held a fundamental particles seminar in the Huangshan Mountains, Anhai Province, at which the American physicist Prof. Yang Zhenning gave a lecture.

An astrophysics symposium was held in the Huangshan Mountains.

The Ministry of Agriculture and Forestry held a coordinating meeting on wheat and rye scientific research in Wudu County, Gansu Province.

The Academia Sinica held a working conference for the Chinese Science and Technology University.

September 17
China conducted another nuclear test.

September 18
The Central Committee of the Communist Party of China released an announcement on the convening of the National Science Conference in the spring of 1978 in Beijing, and on the restoration of the State Science and Technology Commission.

September 25
Chairman Hua Guofeng, Vice Chairmen Ye Jianying, Deng Xiaoping, Li Xiannian and Wang Dongxing, and other Party and State leaders received the members of the preparatory committee for the National Science Conference.

September 27 — October 31
A national conference on natural science planning was held in Beijing, with the participation of 1,200 scientists, professors and experts. They drew up a development plan for the basic sciences such as mathematics, physics, chemistry, astronomy, earth-sciences, biology, and then drafted a program.

September 28
The National Science and Technology Association of the People's Republic of China held discussions on how the Association and its academic societies should go into active operation in order to meet the needs of the rapid development of China's scientific and technological base.

September
The Academia Sinica Beijing Astronomical Observatory recorded three big, continuous solar explosions, the biggest of which lasted for four hours on September 20 and covered one thousandth of the sun's visible surface.

October 3
According to the Party Central Committee's instructions for the restoration of titles of professional posts, the Academia Sinica conducted the following promotions: Chen Jingrun from Assistant Research Fellow to Research Fellow, Yang Le and Zhang Guanghou from Research Assistant to Associate Research Fellow.

October 19
Entrusted by the Academia Sinica, the Chinese Science and Technology University established a graduate school in Beijing.

October 22
Ziranbianzhengfa Tongxun (The Journal of the Dialectics of Nature), affiliated to the Academia Sinica, resumed operation.

October
The Petroleum and Chemical Industry Ministry, the Agriculture and Forestry Ministry, the Academia Sinica and the National Supply and Marketing Co-operative, co-sponsored a national conference on humid acid manure research achievements in Changchun.

The Chinese Academy of Agricultural and Forestry Sciences held a Northwest-China afforestation and sand control notes-comparing session in Wuwei Prefecture, Gansu Province.

A chronicle of events in science and technology

The operation of transplanting a free fibule with blood vessels attached was accomplished in Shanghai, which constituted a significant progress in microscopic surgery in the osteological field for China.

November 3—4
The Academia Sinica Zijinshan Astronomical Observatory discovered an obscure fast-moving celestial body.

November 8
China designed and developed its first digital satellite-ground telecommunication station.

November 20 — early December
Directors of China's meteorological bureaus gathered in Beijing for a conference.

November 30
Academia Sinica Vice President and National Science and Technological Association Vice Chairman Wu Youxun passed away in Beijing.

November
The Education Ministry convened a planning conference in Beijing for academic subjects in the applied sciences and new technology in China's universities and colleges. A draft program for 14 subjects in the applied sciences and new technology was worked out.

December 10 (1977) — January 16 (1978)
A national science and technology planning conference was held in Beijing.

December 11—15
A conference for the prevention and cure of schistosomiasis in twelve provinces and one autonomous region in South China, was convened in Shanghai.

December 21
An extraordinarily large rough diamond weighing 156.7860 carats was found by the Changlin Production Brigade, Jishan People's Commune, Linshu County, Shandong Province. It was later named "Changlin Diamond" by Chairman Hua Guofeng.

December 27
At the seventh enlarged Standing Committee session of the Fourth National Committee of the Political Consultative Conference, Comrade Fang Yi made a speech entitled, "On the Present Situation in Scientific and Educational Fields."

December
The Chinese Science and Technology Association of the People's Republic of China sponsored symposiums for its five academic societies: the Chinese

The late Academia Sinica Vice President Wu Youxun

Zoology Society, the Chinese Geography Society, the Chinese Metals Society, the Chinese Aviation Society and the Chinese Forest Society.

The State Seismological Bureau convened a conference in Beijing on the Learn-From-Daching Campaign concerning the seismological front.

With approval from the State Council, the Academia Sinica abrogated the administrative system of revolutionary-committee-in-charge and restored the system of director-in-charge, in all its scientific research institutes, and then appointed a few directors and deputy directors.

1978

January 2
A ceramic reflective mirror surface developed by Chinese scientists, was for the first time installed on the astronomical telescopes in the Academia Sinica Zijinshan Astronomical Observatory.

January 5
The third national agricultural mechanization conference was convened in Beijing.

The Chinese Mechanics Society held the first national pyrodynamics symposium on Huangshan Mountain, Anhui Province.

January 10-14
The Academia Sinica held a discussion on the nation's natural science journals.

January 13
It was reported that according to the results of a survey organized by the Changjiang River Basin Planning Office, the Changjiang River has its source not at the southern foot of the Bayankala Mountain, but at the Tuotuo River southwest of the Geladandong Snow Mountains, that is the major range of the Tanggula Mountains. The total river length is not 5,800 kilometers but 6,300 kilometers.

January 20
It was reported that hundreds of Chinese oceanographical scientists had accomplished a comprehensive oceanographical survey of the East China Sea and obtained first-hand scientific data and samples. They were quite useful for the study of the geological structure and geomorphological genesis and development history of the continental shelf in the East China Sea, and also for the exploitation of oil and natural gas in this region.

January 26
China launched another artificial earth satellite, which having fulfilled its planned mission of scientific experiments, had returned to the earth surface.

A chronicle of events in science and technology

February 4—5
The enlarged Praesidium meeting of the National Science and Technology Association reviewed the 1977 work of all its national academic societies and discussed their policies and tasks for the year 1978.

Early February
The sixth national rice hybridization research co-operation meeting was held in Nanchang.

February 16
China's first "Biggest Possible Rainstorm Isogram" was finalized. This isogram meant for national use had been examined and approved by the Water Conservancy and Power Ministry and the Central Meteorological Bureau.

February 21
It was reported that the junior Chinese mathematician Zhang Guanghou had found in the course of his function theory research, the structural relationship between the "deficient values", "asympototic values" and the "Julia direction" (a singular direction) for integral functions and meromorphic functions.

February 22
The Academia Sinica held discussions on photosynthesis research work in Guangzhou.

March 2
The State Council approved and circulated a report from the Education Ministry that the national priority universities and colleges were to be restored, expanded and well run. Out of 86 "priorities", 60 of them, including Beijing University, Fudan University and Jilin University, were merely to be restored while 28 were newly promoted.

March 7
The Standing Committee of the National People's Congress appointed Fang Yi, Li Chang, Zhou Peiyuan, Tong Dizhou, Hu Keshi, Yan Jici, Hua Luogeng and Qian Sanqiang as Vice Presidents of the Academia Sinica.

March 10
The National Geology Society resumed activities and held a discussion on its present tasks.

March 14
Huang Tonghian and his colleagues engaged in machine gear measurements, developed China's new and unique wheel measuring technique. They formed their own concept of "integral error of gear dynamics".

The Academia Sinica Scientific Instruments Factory designed and manufactured China's first scanning ion-probe mass-spectrum microanalyser Model LT-1.

30 years' review of China's science and technology

"Machines, Natural Forces and Scientific Application", by Karl Marx was published by the People's Publishing House with a nation-wide circulation.

March 15
China conducted a new nuclear test.

At the 20th anniversary of the founding of the PLA Academy of Military Science, Chairman Hua Guofeng and other Party and state leaders received all comrades from the Academy.

March 18—31
Presided over by Chairman Hua Guofeng, the National Science Conference was convened in Beijing. Chairman Hua addressed the conference with an important speech entitled "Enhance the Science and Cultural Level of the Whole Nation." Vice Chairman Deng Xiaoping delivered an important speech too. In addition, there were reports by Vice Premier Fang Yi explaining the national science and technological development (1978—1985) program (draft), a written speech by President Guo Moruo, a work report by Vice President Li Chang and a closing speech by Vice Premier Ji Dengkui. Cited at the conference were 826 advanced collectives, 1192 advanced scientists and technicians, and also units and individuals from the 7657 excellent project achievements. Honour certificates were handed out by Vice Premier Fang Yi to the cited individuals and units.

March
Lou Shibo, a junior teacher from the Shanghai Railway College, made achievements in basic mathematical theory such as multi-valued logic. The Chairman of the American Mathematical Society offered him membership.

April 2
The Party Chairman Hua Guofeng and Vice Chairmen Ye Jianying, Deng

The late Academia Sinica President Guo Moruo

The Commencement of the National Science Conference

Xiaoping, Li Xiannian, Wang Donxing and other Party and state leaders received the 6,000 participants of the National Science Conference.

April 3
Historical data about drought and water-logging over the past 500 years were sorted out, and graded drought and water logging distribution maps were drawn up in Nanjing. These maps were then the world'd first drought and water-logging climatic maps that covered such a large area and over such a long historical period.

April 10—14
Yang Le and Zhang Guanghou from the Academia Sinica Mathematics Research Institute, attended the international analysis symposium in Zurich, Switzerland and presented papers.

April 13—14
The Beijing members of the Standing Committee of the National People's Congress gathered for a briefing by Vice Premier Fang Yi on the state of science and technology.

April 17
The Academia Sinica Scientific Instruments Factory at Beijing trial-produced a double-channel X-ray spectroscope Model X-3F.

(From left to right) Huo Luogeng, Cheng Jingrun, Yang Le and Zhang Guanghou at the National Science Conference

April 24
Yang Le and Zhang Guanghou read academic papers at the Imperial College of Science and Technology in London.

April 30
The State Science and Technology Commission officially announced that the National Science and Technology Association and its academic societies were to resume operation.

May 9—12
A delegation from the Chinese Mechanical Engineering Society attended the 17th International Heat Treatment Conference in Spain.

May 15
Ziran Zazhi (Nature Journal) published its first issue in Shanghai.

May 25—30
A Sino-Australian plant tissue culture symposium was held in Beijing with the participation of 50 scientists from 10 countries.

May 29
China developed its medium and high vibration frequency bases, both of which reached international standards in terms of precision.

May
Ground was broken in Beijing in the construction of a prototype research model of China's first 30—50 billion electron-volt ring-shaped proton synchrotron.

Surgeons from the Beijing Jishuitan Hospital performed a blood vessel transplant operation, restoring blood circulation and hand function, and thus saved an arm from being amputated. The patient had had deep burns on his forearm and wrist as a result of an electric shock. This case indicated a new way of treating electric burns on an arm.

June 2—12
A national health and medical science conference was held in Beijing. At the conference Chairman Hua gave important instructions.

June 9 — July 11
Upon invitation, an Academia Sinica delegation headed by its Vice President Li Chang, visited Holland and West Germany.

June 12
The Vice Chairman of the Standing Committee of the National People's Congress and President of the Academia Sinica, Guo Moruo, died. He was considered a staunch revolutionary who fought for the communist cause all his life, and was an outstanding cultural fighter for the proletariat.

A chronicle of events in science and technology

June 17 – July 12
An Academia Sinica delegation, headed by Qian Sanqiang, visited Belgium and France.

June 19
An award-giving ceremony was held for the winners of a mathematics contest sponsored by the Ministry of Education and the National Science and Technology Association. The winners were middle school students from a few provinces and municipalities.

June
An Academia Sinica delegation attended the 11th International Soil Science Conference held in Canada.

An Academia Sinica delegation attended the third international symposium on nitrogen-fixation held in the United States.

The State Planning Commission, the State Science and Technology Commission, the Ministry of Civil Administration and the State Statistics Bureau jointly announced a nation-wide general survey of scientific and technological personnel.

July 5–21
The preparatory committee for the Chinese Dialectics of Nature Society sponsored a national summer symposium on the dialectics of nature.

July 10
The second theoretical seminar on the relationship between theory and practice was held in Beijing, under the auspices of the Academia Sinica theoretical study group and the Chinese Dialectics of Nature Society.

August 6
The Academia Sinica held the fourth laser symposium in Guangzhou.

August 7
Rocks as old as 3.6 billion years were found in Taipingzhai, Qianxi County, Hebei Province.

September 12–29
An Academia Sinica delegation headed by Vice President Zhou Peiyuan toured Japan.

September
China's "Man and Biosphere" national committee was established.

October 5 – November 6

Upon invitation, an Academia Sinica delegation, headed by Vice President Hu Keshi, visited England and Sweden.

October 6

It was reported that the Academia Sinica had its nine branches established in succession in Shanghai, Xinjiang, Chengdu, Hefei, Lanzhou, Guangzhou, Shenyang, Changchun and Wuhan.

October 6–13

Under the auspices of the preparatory group of the Academia Sinica Academic Committee and the editorial board of the Journal of Dialectics of Nature, a seminar on the ideological history of microcosmic physics was held in Guilin, Guangxi.

October 12

Professor Hou Zhenting, a young mathematician from China's Changsha Railway College, won the 1978 Davidson Prize for having solved the criteria for uniqueness in the Q-Process, which had puzzled mathematicians for the past 40 years.

October 21–27

The Chinese Medical Society held the ninth national surgical symposium with the participation of surgical experts from countries like Rumania and Yugoslavia, and experts from countries like the United States, France and Switzerland who belonged to the International Surgery Association.

Late October

An assessment committee was formed by the Academia Sinica and the Fourth Machine Building Ministry etc., and it examined the ECL 256X1 bit, full-decoding quick random storage and FAMOS 2408 bit full-decoding programmable non-erasable storage, developed by the Academia Sinica Shanghai Metallurgical Research Institute.

October

With approval from the Party's Central Committee, the Academia Sinica began to grant subsidies for scientific research.

November 6–10

China's "Man and Biosphere" national committee held its first meeting in Beijing and after discussions, drew up a preliminary plan for the "Man and Biosphere" research projects in China.

November 9–16

The National Science and Technology Association held the second enlarged session of its first conference in Beijing.

November 21

Qinghua University conferred an honorary professorship on Prof. Ren Zhigong, deputy director of the applied physics research center at the John Hopkins University.

November 27 — December 11
Upon invitation, an Academia Sinica delegation, headed by Qin Lisheng, visited the Philippines.

November
The "Regulations on Awards for Technical Innovations" formerly promulgated by the State Council in 1963, was now reprinted and circulated by the State Economics Commission, the State Science and Technology Commission, the Agriculture and Forestry Ministry and the Public Health Ministry.

November — December
A Chinese science and technology delegation, headed by Yu Wen, conducted a tour of the United States for a month, focusing on American science and technology management.

December 28
The State Council announced the promulgation of the revised "Regulations on Awards for Inventions". It said that "in order to encourage inventions and promote science and technology in our drive for an earlier realization of the four modernizations, the Regulations on Awards for Inventions, originally promulgated in 1963, has now been revised and repromulgated, and at the same time the original document is abrogated".

December 30
The State Science and Technology Commission hosted a New Year tea party in honour of the Chinese scientists who had returned from abroad.

Two famous Chinese biologists Tong Dizhou and Bei Shizhang were honourably admitted into the Communist Party of China.

December
The State Council approved the great project of building the North-China "Great Green Wall", a shelter belt system in Northwest China, the northen part of North China and the western part of Northeast China.

1979

January 5
Xinhua News Agency release: The Academia Sinica co-operated with three provinces in establishing three agricultural modernization research institutes: one in Taoyuan County, Hunan Province, one in Luancheng County, Hebei Province, and the third in Hailun County, Heilongjiang Province. These three counties served as bases for conducting comprehensive scientific experiments in agricultural modernization, and their initial achievements were reported.

January 6

Two Chinese mathematicians Chen Jiangrun and Wu Wenjun, invited by Princeton University, started for the United States for a three-month research stint there.

January 23

A Spring Festival tea party was given for scientific and technological circles in Beijing by the National Science and Technology Association and the science and technology section of the Chinese People's Political Consultative Conference.

January 26

The Academia Sinica Departmental Committee members in Beijing attended a tea party in the Great Hall of the People. Leading comrades Wang Zhen, Fang Yi and Deng Yingchao were present and delivered speeches.

January 31

It was reported that Chinese metallurgical scientists, after assiduous and co-ordinated research since 1958, had eventually developed a new method for separating titanium from iron.

February 3

It was reported that a Xinjiang Autonomous Region survey team for uncultivated land had completed a three-year investigation in the northern part of the Tarim Basin, the largest basin in China.

February 4

The Jinniu District Hospital in Chengdu, Sichuan Province, developed an artificial lenticular body made of a transparent organosilicon rubber.

February 5 — March 1

The State Aquatic Production General Bureau held a national conference on aquatic products.

February 7

The State Science and Technology Commission, the Education Ministry and the Agriculture and Forestry Ministry jointly sponsored a national conference on scientific research in institutions of higher learning.

February 10

The State Council of the People's Republic of China promulgated the "Regulations on the Protection and Reproduction of Aquatic Resources".

February 14

Young pioneers from Changzhou, Jiangsu Province, came up with a suggestion that children of the whole nation go in for a We-Love-Science campaign. This suggestion drew warm responses from the Youth League's Central Committee, the National Science and Technology Association and the Education Ministry.

A chronicle of events in science and technology

February 19

Yang Le, a junior mathematician, had solved a very difficult problem in the study of functional value distribution. By relating in his research, the singular directions (Borel's directions) with derivative and multiple values, he made significant progress.

February 19–21

The People's Republic of China's UNESCO national committee was established and held its first meeting. The committee consisted of members from the Ministry of Education, the State Science and Technology Commission, the Ministry of Culture, the Academia Sinica, the Chinese Academy of Social Sciences, the Ministry of Foreign Affairs, the Ministry of Economic Relations with Foreign Countries, the Ministry of Finance, the Ministry of Water Conservancy and Power, the State Cultural and Historical Relics Administrative Bureau, the State Publishing Administrative Bureau, Xinhua News Agency, the Central Broadcasting Administrative Bureau, the State Geological General Bureau, the State Seismological Bureau, the State Ocean Bureau, the National Science and Technology Association, the All-China Federation of Literary and Artistic Circles, the National Pedagogical Association, etc.

February 20–22

The National Science and Technology Association, the Chinese Physics Society and the Chinese Astronomy Society, held a joint conference to celebrate the centennial of the birth of Albert Einstein. Zhou Peiyuan and Yu Guangyuan delivered speeches (see elsewhere in this year-book) and the Ministry of Post and Telecommunications issued commemoration stamps.

February 23

The sixth session of the Standing Committee of the Fifth National People's Congress approved the "PRC Forest Law" (for trial implementation) and decided, on a suggestion from the State Council, that the 12th of March be made the National Tree Planting Day in order to mobilize the Chinese people into planting more trees and quicken the process of afforesting the whole country.

February

The Chinese Farm Machinery Society held a national conference in Beijing on the dissemination of science and the popularization of farm machinery.

The Chinese Academy of Social Sciences, the Chinese Academy of Forest Sciences and the Propaganda Department of the Provincial Party Committee of Heilongjiang, jointly sponsored a national conference on the planning of forest economic science.

March 2–5

The Youth League's Central Committee and the Ministry of Forestry jointly convened a national tree-planting conference for young people in Yenan.

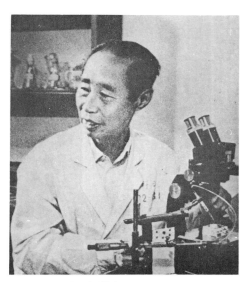

The late Academia Sinica Vice President Tong Dizhou

March 12
It was China's first Tree Planting Day. Hua Guofeng, Deng Xiaoping, Li Xiannian and other leading comrades went to Beijing suburbs and planted trees with the cadres and masses.

The Chinese Futurology Society was established.

March 24
Early in the morning, a partial lunar eclipse took place.

March 20 — April 20
A Scientific and Educational Film Festival was held for a month throughout China.

March 22 — April 5
An Academia Sinica Delegation, headed by Qian Sanqiang, went to Australia and attended the commemoration of the 25th anniversary of the founding of the Australian Academy of Sciences.

March 24
The Chinese Academy of Agricultural Sciences held a conference in Hefei on scientific research in farm-crop strain resources.

March 29 — April 11
The Chinese Water Conservancy Society held a symposium in Tianjin on the proposed grand project of South-to-North Water Transit.

March 30
Academia Sinica Vice President Tong Dizhou passed away.

Late March
The Chinese Geographical Place Names Committee held its first conference in Beijing since the founding of New China.

March
The Ministry of Culture, the Ministry of Education and the National Science and Technology Association, jointly held a conference on the planning of scientific and educational film production.

China's first wild animal and birds survey team found four kinds of tropical birds for the first time in the southeast border area of Yunnan Province.

China's first environmental acoustics symposium was held.

The State Science and Technology Commission, the Ministry of Agriculture and Forestry, the State Forestry General Bureau and the Ministry of Water Conservancy and Power, jointly held in Xi'an, a scientific research conference on water and soil conservation and comprehensive development in farming, forests and animal husbandry on the loess highland in Northwest China.

A chronicle of events in science and technology

April 3—7
A national conference on the agricultural natural resources survey and agricultural regionalization was held in Beijing under the auspices of the State Agricultural Commission, the State Science and Technology Commission, the Ministry of Agriculture and the Academia Sinica. It was announced at the conference that with the approval from the State Council, a national committee for agricultural natural resources survey and agricultural regionalization would be officially established.

April 15 — May 10
China's first high energy astrophysics symposium was held in Nanjing.

April 16
Professor Ge Dingbang, an overseas Chinese in the United States donated to his motherland the fossil of "Yuan's Santai Dinosaur".

April 17
Two major scientific achievements: the synthesis of large area Fluorophlogopite and a new deoxidizing catalyst won the second and third state invention awards respectively.

April
A national standardization conference was held in Beijing.

May 6
Invited by the University of Birmingham, England; Hua Luogeng, Academia Sinica Vice President and director of the Chinese Mathematical Society, and his party arrived in London to deliver lectures.

May 18—24
The Ministry of Public Health held a national symposium in Shanghai on Chinese traditional medicine.

May 28
The Science and Technology Committee of the Ministry of Agriculture was established.

June 1
With approval from the Party's Central Committee, the Academia Sinica decided to dissolve its political department in accordance with the national policy shift.

June 1—5
The Chinese Medical Society and the Chinese Traditional Medicine Society jointly convened a national symposium on acupuncture, moxibustion and acupuncture anaesthesia, with the participation of 150 invited medical experts and friends from more than 30 countries.

Early June

The State Science and Technology Commission and the Academia Sinica jointly held a high energy experimental physics symposium in Beijing.

June 8

Fudan University President Su Buqing and Professor Xia Daoxing were invited to read academic papers at a seminar sponsored by the Mathematics Department of Tokyo University, Japan.

SECTION II

BRIEF INTRODUCTION TO PERIODICALS AND NEWSPAPERS OF THE NATURAL SCIENCES

I. Natural Sciences Periodicals of General Interest

KEXUE HUABAO/*Scientific Pictorial*
Popular science magazine.
First published in August 1933.
Semi-monthly from August 1933 — December 1937; Monthly from 1938.
Publication suspended from August 1937 — October 1937; June 1966 — February 1972; resumed in March 1972.
Published by Shanghai Scientific and Technical Publishers, 450 Ruijin Second Road, Shanghai, China.
Formerly: Kexue Pujiziliao (March 1972 — December 1974).
        Kexue Puji   (January 1975 — December 1977).

KEXUE SHIJIE/*Science World*
Popular science publication.
First published in 1979. Monthly.
Published by the Association of Creative Popular Science of Guangdong Province, the Association for the Popularization of Science and Technology of Guangdong Province, Dongfeng Second Road, Guangdong Province, China.

KEXUE SHIYAN/*Scientific Experiments*
Popular science magazine.
First published in April 1971. Monthly.
Published by Science Press, 137 Chaonei Street, Beijing, China.

KEXUE TONGBAO/*Science Bulletin*
Comprehensive academic journal of the natural sciences.
First published in May 1950. Semi-monthly.
Publication suspended from second half of 1966 — June 1973; resumed in July 1973.
Published by Science Press, 137 Chaonei Street, Beijing, China.

KEXUE WENYI/*Science Literature*
Popular science publication.
First published in May 1975.
No fixed period of publication.
Published by People's Publishing House of Sichuan, 39 Xuedao Street, Chengdu, Sichuan Province, China.

KEXUEZHICHUN/*The Spring of Science*
Popular science publication.
First published in March 1979. Monthly.
Published by the Association for Creative Popular Science of Guangdong Province, the Association for the Popularization of Science and Technology of Guangdong Province, Dongfeng Second Road, Guangzhou, Guangdong Province, China.

SHIJIE KEXUE YIKAN/*World Science Translation Journal*
Comprehensive translation publication of the natural sciences.
First published in January 1979. Monthly.
This journal was published on the basis of two former publications, Shijie Kexue and Kexue Yikan.
Published by Shanghai Scientific and Technical Publishers, 450 Ruijin Second Road, Shanghai, China.

XIANDAIHUA/*Modernization*
Popular science publication.
First published in March 1979. Monthly.
Published by Popular Science Press, Friendship Hotel, Beijing, China.

ZHISHI JIUSHI LILIANG/*Knowledge is Power*
Popular science magazine.
First published in March 1956. Monthly.
In May 1979, Kezue Dazhong (January 1963 — April 1979) merged with Zhishi Jinshi Liliang and retained the latter name.
Published by Popular Science Press, Friendship Hotel, Beijing, China.

ZHONGGUO KEXUE/*Scientia Sinica*
Comprehensive academic journal of the natural sciences.
First published in 1950. Monthly.
Chinese edition suspended from 1952–1971; publication resumed in 1972.
English edition began in 1952.
Published by Science Press, 137 Chaonei Street, Beijing, China.

ZHONGXUE KEJI/*Science and Technology for Middle Schools*
Popular science reading.
First published in July 1973.
Bimonthly (no fixed period of publication from July 1973 – August 1976).
Published by Shanghai Educational Press, 123 Yongfu Road, Shanghai, China.
Formerly: Zhongxiaoxue Kejihuodong Ziliao (July 1973 – September 1974).
Zhongxue Kejihuodong Ziliao (October 1974 – August 1976).

ZIRAN BIANZHENGFA TONGXIN/*Dialectics of Nature*
First published in January 1979. Quarterly.
Published by Shanghai Scientific and Technical Publishers, Unit 20, Northwestern District, Friendship Hotel, Beijing, China.

ZIRAN ZAZHI/*Nature Journal*
Comprehensive magazine of the natural sciences.
First published in May 1978. Monthly.
Published by Shanghai Scientific and Technical Publishers, 450 Ruijin Second Road, Shanghai, China.

II. Acta and Journals of Universities and Colleges

QINGHUA DAXUE XUEBAO/*Journal of Qinghua University*
University journal.
First published in December 1955.
Quarterly (No fixed date of publication from December 1955 – December 1977).
Publication suspended from 1966 – 1972; resumed in 1972.
Published by Beijing Press, Qinghua University, Beijing, China.
Formerly: Qinghua Beida Ligong Xuebao (October 1974 – May 1979).

BEIJING DAXUE XUEBAO (ZIRAN KEXUE BAN)/*Acta Scientiarum Naturalium Universitatis Pekinensis*
University journal.
First published in September 1955. Quarterly.
Publication suspended from 1966 – 1972; resumed in 1973.
Published by Beijing Press, Beijing University, Haidian District, Beijing, China.
Formerly: Qinghua Beida Ligong Xuebao (October 1974 – May 1979).

SHANGHAI JIAOTONG DAXUE XUEBAO/*Journal of Shanghai Jiaotong University*
University journal.
First published in January 1957. Quarterly.
Publication suspended from 1961 – 1962; April 1966 – June 1978; resumed in July 1978.

Published by Shanghai Jiaotong University, 1954 Huashan Road, Shanghai, China.
Formerly: Jiaotong Daxue Xuebao (January 1957 – 1960).

XI'AN JIAOTONG DAXUE XUEBAO/*Journal of Xi'an Jiaotong University*
University Journal.
First published in March 1960. Quarterly.
Before 1960, Jiaotong Daxue Xuebao was jointly published by Shanghai Jiaotong University and Xi'an Jiaotong University.
Published by Xi'an Jiaotong University, Xi'an, Shanxi Province, China.

FUDAN XUEBAO (ZIRAN KEXUE BAN)/*Fudan Journal (Natural Science)*
University journal.
First published in April 1955. Quarterly.
Publication suspended in 1961; 1967 – 1972; resumed in 1973.
Published by Fudan University, 220 Handan Road, Shanghai, China.
Formerly: Fudan Daxue Xuebao (Ziran Kexue Ban) (1962 – 1966).

TONGJI DAXUE XUEBAO/*Journal of Tongji University*
University journal.
First published in February 1956. Quarterly.
Publication suspended from 1966 – April 1978; resumed in May 1978.
Published by Tongji University, 1239 Siping Road, Shanghai, China.

NANJING DAXUE XUEBAO (ZIRAN KEXUE BAN)/*Journal of Nanjing University (Natural Science Edition)*
University journal.
First published in January 1955.
Quarterly (Semi-annually: 1974 – 1977).
The inaugural issue was the general edition of liberal arts and science. From then on, the journal was published in several separate editions of mathematics, physics, chemistry, astronomy, geography and biology annually, semi-annually and quarterly.
Publication suspended 1960 – 1961; 1966 – 1973; resumed 1974.
Published by Nanjing University, Hankou Road, Nanjing, Jiangsu Province, China.

SHANXI DAXUE XUEBAO (ZIRAN KEXUE BAN)/*Journal of Shanxi University (Natural Science Edition)*
University journal.
First published in July 1978. Quarterly.
Published by Shanxi University, Taiyuan, Shanxi Province, China.

HANGZHOU DAXUE XUEBAO (ZIRAN KEXUE BAN)/*Journal of Hangzhou University (Natural Science Edition)*
University journal.
First published in January 1963. Quarterly.

Publication suspended from July 1966 — first half of 1977; resumed in second half of 1977.

Published by Hangzhou University, Tianmushan Road, Hangzhou, Zhejiang Province, China.

XIAMEN DAXUE XUEBAO (ZIRAN KEXUE BAN)/*Universitatis Amoiensis Acta Scientiarum Naturalium*

University journal.

First published in 1952. Quarterly.

From 1953, the journal was published in several separate editions of mathematics, biology, finance and economics, marine biology, literature and history. It was published as a natural science edition starting from No. 4 of 1954.

Publication suspended from 1966 — 1973; resumed in 1974.

Published by Xiamen University, Xiamen, Fujian Province, China.

WUHAN DAXUE XUEBAO (ZIRAN KEXUE BAN)/*Wuhan University Journal (Natural Science Edition)*

University journal.

Resumed publication in 1973.

Quarterly (No fixed period of publication before 1966).

Published by Wuhan University, Wuchang Luojiashan, Wuhan, Hubei Province, China.

ZHONGSHAN DAXUE XUEBAO (ZIRAN KEXUE BAN)/*Acta Scientiarum Universitatis Sunyatseni*

University journal.

First published in November 1955.

Quarterly (No fixed period of publication from 1955 — 1966).

Publication suspended from 1967 — 1972; resumed in 1973.

Published by Zhongshan University, Guangzhou, Guangdong Province, China.

GUANGXI DAXUE XUEBAO (ZIRAN KEXUE BAN)/*Journal of Guangxi University (Natural Science Edition)*

University journal.

First published in April 1976.

No fixed period of publication.

Published by Guangxi University, Xixiangtang Road, Nanning, Guangxi Zhuang Autonomous Region, China.

NEI MENGGU DAXUE XUEBAO (ZIRAN KEXUE BAN)/*Journal of University of Inner Mongolia (Natural Science Edition)*

University journal.

First published in September 1959. Semi-annually.

Published by University of Inner Mongolia, Huhehaote, Inner Mongolia Autonomous Region, China.

HEILONGJIANG DAXUE XUEBAO (ZIRAN KEXUE BAN)/*Journal of Heilongjiang University (Natural Science Edition)*

University journal.
First published in June 1978. Semi-annually.
Published by Heilongjiang University, Xuefu Road, Nangang District, Harbin, Heilongjiang Province, China.

ZHENGZHOU DAXUE XUEBAO (ZIRAN KEXUE BAN)/*Journal of Zhengzhou University (Natural Science Edition)*
University journal.
First published in 1960.
Semi-annually (No fixed period of publication from 1960 — 1965).
Publication suspended from 1966 — 1973; resumed in 1974.
Published by Zhengzhou University, Room 321, Jiaoyi Building, Daxue Road, Zhengzhou, Henan Province, China.

HEBEI DAXUE XUEBAO (ZIRAN KEXUE BAN)/*Journal of Hebei University (Natural Science Edition)*
University journal.
First published in 1960.
Semi-annually (Quarterly: 1960 — 1965).
Published by Hebei University, 1 Hezuo Road, Baoding, Hebei Province, China.

XIBEI DAXUE XUEBAO (ZIRAN KEXUE BAN)/*Journal of Northwestern University (Natural Science Edition)*
University journal.
First published in September 1957.
Three issues yearly (Quarterly from September 1957 — 1959).
Published by Xibei University, Xiaonanmenwai, Daxue East Road, Xi'an, Shanxi Province, China.

ZHONGGUO KEXUEJISHU DAXUE XUEBAO/*Journal of Chinese University of Science and Technology*
University journal.
First published in February 1965.
No fixed period of publication.
Publication suspended from second half of 1966 — June 1975; resumed in July 1975.
Published by Chinese University of Science and Technology, Hefei, Anhui Province, China.

SHANGHAI KEXUEJISHU DAXUE XUEBAO/*Journal of Shanghai University of Science and Technology*
University journal.
First published in June 1978.
No fixed period of publication.
Published by Shanghai University of Science and Technology, Information Department of Science and Technology, Nanmen, Jiading County, Shanghai, China.

Acta and journals of universities and colleges

JILIN DAXUE XUEBAO (ZIRAN KEXUE BAN)/*Acta Scientiarum Naturalium Universitatis Jilinensis*
University journal.
First published in 1955. Quarterly.
Publication suspended from summer of 1966 — 1972; resumed in 1973.
Published by Jilin University, 75 Jiefang Da Road, Changchun, Jilin Province, China.
Formerly: Jilindaxue Ziran Kexue Xuebao (1955 — Spring 1966).

BEIJING SHIFANDAXUE XUEBAO (ZIRAN KEXUE BAN)/*Journal of Beijing Normal University (Natural Science Edition)*
University journal.
First published in 1956. Quarterly.
Publication suspended from 1966 — 1974; resumed in 1975.
Published by Beijing Normal University, Baitaipingzhuang, Waidajie, Xin Jiekou, Beijing, China.

FUJIAN SHIDA XUEBAO (ZIRAN KEXUE BAN)/*Journal of Fujian Normal University (Natural Science Edition)*
University journal.
First published in 1975. Semi-annually.
Published by Fujian Normal University, Department of Science Research and Production, Fuzhou, Fujian Province, China.

GANSU SHIDA XUEBAO (ZIRAN KEXUE BAN)/*Journal of Gansu Normal University (Natural Science Edition)*
University journal.
First published in June 1957.
No fixed period of publication.
Publication suspended from 1966 — 1973; resumed in 1974.
Published by Gansu Normal University, Lanzhou, Gansu Province, China.
Formerly: Xibei Shifanxueyuan Xuebao (June 1957 — 1958).

WUHAN SHIFANXUEYUAN XUEBAO (ZIRAN KEXUE BAN)/*Journal of Wuhan Normal College (Natural Science Edition)*
College journal.
First published in June 1978. Semi-annually.
Published by Wuhan Normal College, Baojian, Wuchang, Hubei Province, China.
Formerly: Wushi Keji (1974 — May 1978).

XINAN SHIFAN XUEBAO (ZIRAN KEXUE BAN)/*Journal of Southwestern Normal College (Natural Science Edition)*
College journal.
First published in May 1957.
No fixed period of publication.
Publication suspended in 1966; resumed in 1967.
Published by Xinan Normal College, Beibei, Chongqing, Sichuan Province, China.

HUANAN SHIYUAN XUEBAO (ZIRAN KEXUE BAN)/*Journal of South China Normal College (Natural Science Edition)*
College journal.
First published in 1954. Semi-annually.
Publication suspended from second half of 1966 – 1973; resumed in 1974.
Published by South China Normal College, Shipai, Guangzhou, Guangdong Province, China.
Formerly: Guangdong Shiyuan Xuebao (Ziran Kexue Ban).

KAIFENG SHIYUAN XUEBAO (ZIRAN KEXUE BAN)/*Journal of Kaifeng Normal College (Natural Science Edition)*
College journal.
First published in September 1975.
No fixed period of publication.
Published by Kaifeng Normal College, Kaifeng, Henan Province, China.

JIANGSU SHIYUAN XUEBAO (ZIRAN KEXUE BAN)/*Journal of Jiangsu Normal College (Natural Science Edition)*
College journal.
First published in 1960.
No fixed period of publication.
Publication suspended from June 1964 – 1973; 2 issues were published since 1974: one in 1975 and the other in 1978.
Published by Jiangsu Normal College, 1 Hongqi East Road, Suzhou, Jiangsu Province, China.

NANJING SHIYUAN XUEBAO (ZIRAN KEXUE BAN)/*Journal of Nanjing Normal College (Natural Science Edition)*
College journal.
First published in March 1975. Semi-annually.
Published by Nanjing Normal College, 122 Ninghai Road, Nanjing, Jiangsu Province, China.

ZHONGGUO YIKEDAXUE XUEBAO/*Journal of Chinese Medical College*
College journal.
First published in August 1972. Quarterly.
Published by Chinese Medical College, 3 Wuduan, Nanjing Street (section 5), Heping District, Shenyang, Liaoning Province, China.
Formerly: Yixue Yanjiu (Shengyang Medical College)(August 1972 – 1978, 24 issues).

HAYIDA XUEBAO/*Journal of Harbin Medical College*
College journal.
First published in 1952. Quarterly.
Publication suspended from second half of 1966 – 1973; resumed in 1973.
Published by Harbin Medical College, Department of Information on Medical Research, Harbin, Heilongjiang Province, China.

Acta and journals of universities and colleges

BAIQIUEN YIKEDAXUE XUEBAO/*Journal of Bethune Medical College*
College journal.
First published in February 1975. Quarterly.
Published by Bethune Medical College, 6 Xinmin Street, Changchun Jilin Province, China.
Formerly: Jilin Yikedaxue Xuebao (1959 — January 1975).

FUJIAN YIDA XUEBAO/*Journal of Fujian Medical College*
College journal.
First published in October 1959. Quarterly.
Publication suspended from 1966 — 1972; resumed in 1973.
Published by Fujian Medical College, Quanzhou, Fujian Province, China.
Formerly: Fujian Yixueyuan Xuebao (October 1959 — 1965).

SHANGHAIYDIYI YIXUEYUAN XUEBAO/*Acta Academiae Mediciane Primae, Shanghai*
College journal.
First published in June 1956.
Bimonthly (Quarterly: June 1956 — March 1958, with 8 issues in all bound up into one volume; 1964 — March 1966).
Publication suspended from April 1966 — July 1978; resumed in August 1978.
Published by Shanghai Scientific and Technical Publishers, Shanghai First Medical College, Fenglinqiao, Shanghai, China.
Formerly: Shangyi Xuebao (jointly published with Shanghai Second Medical College) (the second quarter of 1958 — 1963).

GUANGXI YIXUEYUAN XUEBAO/*Journal of Guangxi Medical College*
College journal.
First published in March 1977. Quarterly.
Published by Guangxi Medical College, Yan'an Road, Nanning, Guangxi Zhuang Autonomous Region, China.

XINJIANG YIXUEYUAN XUEBAO/*Journal of Xinjiang Medical College*
College journal.
First published in 1959.
No fixed period of publication.
Publication suspended from 1966 — February 1978; resumed in March 1978.
Published by Xinjiang Medical College, Beijing Road, Vrumai, Xinjiang Vighur Autonomous Region, China.
Formerly: Xinyi Xuebao when inaugurated.

ANYI XUEBAO/*Journal of Anhui Medical College*
College journal.
First published in 1958. Bimonthly.
Publication suspended from 1966 — 1971; resumed in 1972.
Published by Anhui Medical College, Hefei, Anhui Province, China.

XI'AN YIXUEYUAN XUEBAO/*Journal of Xi'an Medical College*
College journal.
First published in August 1955. Quarterly.
Publication suspended from 1966 – 1974 and in 1978; resumed in 1979.
Published by Xi'an Medical College, Nanjiaoxiaozhai, Xi'an, Shanxi Province, China.

SHANDONG YIXUEYUAN XUEBAO/*Journal of Shandong Medical College*
College journal.
First published in February 1956. Quarterly.
Publication suspended from 1966 – 1970; resumed in 1971.
Published by Shandong Medical College, Department of Scientific Research, Jinan, Shandong Province, China.
Formerly: Yiyao Xuebao (1971 – 1977).

WUHAN YIXUEYUAN XUEBAO/*Journal of Wuhan Medical College*
College journal.
First published in 1957. Quarterly.
Publication suspended from 1963 – 1974; resumed in 1975.
Published by Wuhan Medical College, Hangkong Road, Wuhan, Hubei Province, China.

BEIJING YIXUEYUAN XUEBAO/*Journal of Beijing Medical College*
College journal.
First published in 1959. Quarterly.
Publication suspended from second half of 1966 – 1973; resumed in 1973.
Published by Beijing Medical College, Yueyuan Road, Beijiao, Beijing, China.

QINGHAI YIXUEYUAN TONGXUN/*Communications of Qinghai Medical College*
College journal.
First published in October 1978.
No fixed date of publication.
Published by Qinghai Medical College, South First Floor, 65 Kunlun Road, Xining, Qinghai Province, China.

NINGXIA YIXUEYUAN XUEBAO/*Journal of Ningxia Medical College*
College journal.
First published in January 1979. Semi-annually.
Published by Ningxia Medical College, Nanmenwai, Yinchuan, Ningxia Hui Autonomous Region, China.

YUNNAN ZHONGYI XUEYUAN XUEBAO/*Journal of Yunnan College of Traditional Chinese Medicine*
College journal.
First published in January 1978. Quarterly.

Published by Yunnan College of Traditional Chinese Medicine, Baita Road, Kunming, Yunnan Province, China.

SHANDONG ZHONGYI XUEYUAN XUEBAO/*The Academic Journal of Shandong College of Traditional Chinese Medicine*
College journal.
First published in February 1977. Quarterly.
Published by People's Publishing House of Shandong, Shandong College of Traditional Chinese Medicine, Jinan, Shandong Province, China.

SHANXI ZHONGYI XUEYUAN/*Journal of Shanxi College of Traditional Chinese Medicine*
College journal.
First published in March 1978. Quarterly.
Published by Shanxi College of Traditional Chinese Medicine, Weiyang Road, Xianyang, Shanxi Province, China.

HENAN ZHONGYI XUEYUAN XUEBAO/*Journal of Henan College of Traditional Chinese Medicine*
College journal.
First published in January 1976. Quarterly.
Published by Henan College of Traditional Chinese Medicine, Zhengzhou, Henan Province, China.

GANSU NONGDA XUEBAO/*Journal of Gansu Agricultural University*
University journal.
First published in 1959.
No fixed date of publication (Semi-annually: 1959 – 1965).
Published by Gansu Agricultural University, Huangyang Town, Wuwei County, Gansu Province, China.

DONGBEI LINXUEYUAN XUEBAO/*Journal of Northeastern Forestry Institute*
College journal.
First published in July 1957.
Quarterly (from 1957 – first half of 1979).
Publication suspended from 1958 – 1962, 1966 – 1976; resumed in 1977.
Published by Northeastern Forestry Institute, Hexing Road, Harbin, Heilongjiang Province, China.

SHANDONG HAIYANGXUEYUAN XUEBAO/*Journal of Shandong Oceanographical Institute*
College journal.
First published in October 1959. Semi-annually.
Publication suspended from second half of 1965 – 1977; resumed in 1978.
Published by Shandong Oceanographical Institute, 5 Yushan Road, Qingdao, Shandong Province, China.

HEBEI DIZHIXUEYUAN XUEBAO/*Journal of Hebei Geological Institute*
College journal.
First published in 1978.
No fixed date of publication.
Published by Hebei Geological Institute, Xuanhua, Hebei Province, China.

CHANGCHUN DIZHIXUEYUAN XUEBAO/*Journal of Changchun Geological Institute*
College journal.
First published in 1959. Quarterly.
It resumed publication in 1975.
Published by Changchun Geological Institute, Dizhigong, Changchun, Jilin Province, China.

ZHONGNAN KUANGYE KUEYUAN XUEBAO/*Journal of Central-South Institute of Mining and Metallurgy*
Institute journal.
First published in 1956. Quarterly.
Publication suspended from 1964 – 1972; resumed under current title in 1977.
Published by Central–South Institute of Mining and Metallurgy, Yuelushan, Changsha, Hunan Province, China.
Formerly: Kuangye Keji (1973 – 1976, 10 issues in all).

III. Periodicals in Mathematics, Physics and Chemistry

SHUXUE XUEBAO/*Acta Mathematica Sinica*
Specialized journal.
First published in March 1951.
Bimonthly (Quarterly: March 1951 – December 1963).
Published by Science Press, Chinese Society of Mathematics, Institute of Mathematics, Academia Sinica, Zhongguancun, Beijing, China.

SHUXUE TONGHAO/*Bulletin of Mathematics*
Specialized academic publication.
First published in November 1951. Monthly.
Publication suspended from July 1960 – March 1961; resumed in April 1961.
Published by Geological Press, Beijing Normal University, Baitaipingzhuang, Xingjiekouwai Street, Beijing, China.
Formerly: Zhongguo Shuxue Zazhi (November 1951 – December 1952).

YINGYONG SHUXUE XUEBAO/*Acta Mathematicae Applacatae Sinica*
Specialized journal.
First published in July 1976. Quarterly.
Published by Science Press, Institute of Mathematics, Academia Sinica, Zhongguancun, Beijing, China.

SHUXUE DE SHIJIAN YU RENSHI/*Mathematics in Practice and Cognition*
Specialized academic publication.
First published in 1971. Quarterly.
Published by Science Press, Institute of Mathematics, Academia Sinica, Zhongguancun, Beijing, China.

JISUAN SHUXUE/*Mathematica Numerica Sinica*
Specialized journal.
First published in 1978. Quarterly.
Published by Science Press, Computing Center, Academia Sinica, Zhongguancun, Beijing, China.

JISUANJI XUEBAO/*Chinese Journal of Computers*
Specialized academic publication.
First published in July 1978. Quarterly.
The journal is published as the transactions of the Electronic Computer Society attached to the Chinese Society of Electronics.
Published by Science Press, Institute of Computing Technology, Academia Sinica, Zhongguancun, Beijing, China.
Formerly: Chinese Journal of Physics (1933 — 1952).

WULI XUEBAO/*Acta Physica Sinica*
Specialized journal.
First published in 1933 under the title *Chinese Journal of Physics* in English. Changed to present title in 1953 and published in Chinese.
Quarterly (1933 — 1954); six issues yearly 1955 — 1958; Monthly (1959 — September 1966).
Publication suspended from October 1966 — 1973; resumed (six issues yearly) in 1974.
Published by Science Press, Institute of Physics, Academia Sinica, Zhongguancun, Beijing, China.

GAONENGWULI YU HEWULI/*Physica Energiae Fortis et Physica Nuclearis*
Specialized journal.
First published in November 1977. Bimonthly.
Published by Science Press, Institute of High Energy Physics, Academia Sinica, 19 Yuquan Road, Beijing, China.

GAONENG WULI/*High Energy Physics*
Popular science publication.
First published in September 1976. Quarterly.
Published by Science Press, Institute of High Energy Physics, Academia Sinica, 19 Yuquan Road, Beijing, China.

WULI/*Journal of Physics*
Specialized academic publication.
First published in June 1972.
Bimonthly (Quarterly: June 1972 — 1973).
Published by Science Press, P.O. Box 603, Beijing, China.

WULIYANJIU TONGXUN/Communications in Physics Research
Specialized academic publication.
First published in June 1978.
No fixed date of publication.
Published by Chongqing College of Architectural Engineering, Chongqing, Sichuan Province, China.

LIXUE XUEBAO/*Acta Mechanica Sinica*
Specialized journal.
First published in February 1957. Quarterly.
Publication suspended from fourth quarter of 1960 — 1961 and from 1966 → 1977; resumed in 1978.
Published by Science Press, Institute of Mechanics, Academia Sinica, Zhongguancun, Beijing, China.

LIXUE YU SHIJIAN/*Mechanics in Practice*
Specialized journal.
First published in February 1979. Quarterly.
Published by Science Press, Institute of Mechanics, Academia Sinica, Zhongguancun, Beijing, China.

LIXUE QINGBAO/*Mechanics*
Specialized academic publication.
First published in 1961.
No fixed date of publication.
Published by Institute of Mechanics, Academia Sinica, Zhongguancun, Beijing, China.
Formerly: Keji Qingbao (No. 1, 1961 — No. 12).

SHENGXUE XUEBAO/*Acta Acustica*
Specialized journal.
First published in September 1964. Quarterly.
Publication suspended October 1966 — January 1979; resumed in February 1979.
Published by Science Press, P.O. Box 2712, Beijing, China.

YUANZINENG KEXUE JISHU/*Science and Technology of Nuclear Energy*
Specialized academic publication.
First published in 1959.
Six issues yearly (Quarterly: 1975 — first half of 1979).
Publication suspended from 1967 — 1974; resumed in 1975.
Published by Nuclear Energy Publishing House, P.O. Box 275–65, Beijing, China.

HUAXUE XUEBAO/*Acta Chimica Sinica*
Specialized journal.
First published in 1932.
Quarterly (six issues yearly: 1932 — May 1966).

Publication suspended from June 1966 — 1974; resumed in 1975.

Published by Shanghai Scientific and Technical Publishers, Institute of Organic Chemistry, Academia Sinica, 245 Lingling Road, Shanghai, China.

HUAXUE TONGBAO/*Chemistry*

Specialized academic publication.

First published in January 1943.

Six issues yearly (Monthly: July 1952 — June 1966).

Publication suspended from July 1966 — July 1973; resumed in August 1973.

Published by Science Press, P.O. Box 2709, Beijing, China.

Formerly: Huaxue (January 1934 — June 1952).

HUAXUE TONGXUN/*Communications in Chemistry*

Specialized academic publication.

First published in 1976. Semi-annually.

Published by Guangzhou Institute of Chemistry, Academia Sinica, Wushan Fei'eling, Guangzhou, Guangdong Province, China.

GAOFENZI TONGXUN/*Communications in Polymer Science*

Specialized journal.

First published in March 1957.

Six issues yearly (Quarterly: August 1978 — December 1978).

Publication suspended from June 1966 — July 1978; resumed August 1978.

Published by Science Press, P.O. Box 2709, Beijing, China.

FENXI HUAXUE/*Analytical Chemistry*

Specialized academic publication.

First published in 1973.

Six issues yearly (Quarterly: 1973).

Published by People's Press of Jilin Province, Changchun Institute of Applied Chemistry, Academia Sinica, Changchun, Jilin Province, China.

HUAXUE SHIJI/*Chemical Reagents*

Specialized academic publication.

First published in April 1979.

Six issues yearly.

Published by Scientific and Technical Information Center for Chemical Reagents, Ministry of Chemical Engineering, Beijing Chemical Factory, Huagong Road, Dongjiao, Beijing, China.

Formerly: Fenxihuaxue Yu Shiji Yicong (before April 1979). The title of the journal was changed to the present one as the contents of the journal are mainly concerned with domestic information.

HEJISHU/*Nuclear Technology*

Specialized academic publication.

First published in November 1978. Quarterly.

Published by Shanghai Scientific and Technical Publishers, P.O. Box 8204, Shanghai, China.

IV. Periodicals in Astronomy and Earth Science

DIQIU WULI XUEBAO/*Acta Geophysica Sinica*
Specialized journal.
First published in June 1948. Quarterly.
The journal was published in English from June 1948 — 1966.
Publication suspended from 1967 — 1972; resumed in 1973.
Published by Science Press, P.O. Box 928, Beijing, China.

DIQIU HUAXUE/*Geochimica*
Specialized journal.
First published in March 1973. Quarterly.
Published by Science Press, Institute of Geochemistry, Academia Sinica, 67 Guanshui Road, Guiyang, Guizhou Province, China.

TIANWEN XUEBAO/*Acta Astronomica Sinica*
Specialized journal.
First published in August 1953.
Quarterly (Semi-annually: 1953 — 1977).
Publication suspended from 1960 — 1961, second half of 1966 — 1974; resumed in 1975.
Published by Science Press, Zijinshan Observatory, Nanjing, Jiangsu Province, China.

TIANWEN TONGXUN/*Communications in Astronomy*
Specialized academic publication.
First published in April 1978.
No fixed period of publication.
Published by Zijinshan Observatory, Nanjing, Jiangsu Province, China.

TIANWEN ALHAOZHE/*The Amateur Astronomer*
Popular science publication.
First published in April 1958.
Monthly (Bimonthly: April 1958 — May 1960; August 1978 — June 1979).
Publication suspended from June 1960 — June 1963, June 1966 — July 1978; resumed in August 1978.
Published by Amateur Astronomer Press, 138 Xiwai Street, Beijing, China.

DAQI KEXUE/*Scientia Atmospherica Sinica*
Specialized journal.
First published in September 1976. Quarterly.
Published by Science Press, Institute of Atmospheric Physics, Academia Sinica, Qijiahuozi, Deshengmenwai, Beijing, China.

QIXIANG/*Meteorological Monthly*
Specialized academic publication.

First published in October 1974. Monthly.
Published by Central Bureau of Meteorology, 46 Baishiqiao Road, Beijing, China.

DILI XUEBAO/*Acta Geographica Sinica*
Specialized journal.
First published in September 1934. Quarterly.
Publication suspended from second half of 1966 — August 1978; resumed in September 1978.
Published by Science Press, Institute of Geography, Academia Sinica, 917 Building, Beishatan, Deshengmenwai, Beijing, China.

DILI ZHISHI/*Geographical Knowledge*
Popular science publication.
First published in January 1950. Monthly.
Publication suspended from July 1960 — December 1960, June 1966 — September 1972; resumed in October 1972.
Published by Science Press, Institute of Geography, Academia Sinica, 917 Building, Beishatan, Deshengmenwai, Beijing, China.
Formerly: Dili (Specialized academic publication) (1961 — March 1966).

DIZHI XUEBAO/*Acta Geologica Sinica*
Specialized journal.
First published in 1922. Quarterly.
Published by Science Press, Chinese Academy of Geology, Baiwanzhuang, Fuwai, Beijing, China.
Formerly: Zhongguo Dizhi Xuehuizhi (1922 — 1951).

DIZHI PINGLUN/*Geological Review*
Specialized academic publication.
First published in 1936. Bimonthly.
Publication suspended from 1966 — August 1979; resumed in September 1979.
Published by Geology Press, Chinese Academy of Geology, Baiwanzhuang, Fuwai, Beijing, China.

DIZHI KEXUE/*Scientia Geologica Sinica*
Specialized academic publication.
First published in 1958.
Quarterly (Bimonthly: 1958; Monthly: 1959 — February 1960).
Publication suspended from March 1960 — 1962, October 1966 — 1972; resumed in 1973.
Published by Science Press, Institute of Geology, Academia Sinica, Qijiahuozi, Deshengmenwai, Beijing, China.

SHUIWEN DIZHI GONGCHEN DIZHI/*Hydrogeology and Engineering Geology*
Specialized academic publication.
First published in 1957. Bimonthly.

Publication suspended from June 1960 — 1965, June 1966 — 1978; resumed in 1979.

Published by Geology Press, National General Bureau of Geology, Xisi, Beijing, China.

DIZHI YU KANTAN/Geology and Exploration

Specialized academic publication.

First published in January 1957.

Twice per quarter (Semi-monthly at inaugural time; Bimonthly: 1964 — August 1966).

Publication suspended from June 1960 — 1963, September 1966 — April 1972; resumed in May 1972.

Published by P.O. Box 103, Guilin, Guangxi Zhuang Autonomous Region, China.

DICENGXUE ZAZHI/Acta Stratigraphica Sinica

Specialized academic publication.

First published in March 1966. Quarterly.

Publication suspended from October 1966 — August 1978; resumed in September 1978.

Published by Science Press, Institute of Geological Palaeontology, Academia Sinica, 39 Beijing East Road, Nanjing, Jiangsu Province, China.

XIBEI DIZHEN XUEBAO/Northwestern Seismological Journal

Specialized journal.

First published in March 1979. Quarterly.

Published by Lanzhou Institute of Seismology, National Bureau of Seismology, Panxuan Road, Lanzhou, Gansu Province, China.

DIZHEN YANJIU/Journal of Seismological Research

Popular science publication.

First published in May 1967.

Quarterly (No fixed date of publication from May 1967 — 1973).

Published by Science Press, Sanlihe, Beijing, China.

DIZHEN ZHANXIAN/Journal of Seismology

Popular science publication.

First published in May 1967.

Bimonthly (No fixed date of publication from May 1967 — 1973).

Published by Science Press, Sanlihe, Beijing, China.

BINGCHUAN DONGTU/Glaciers and Frozen Soil

Specialized academic publication.

First published in 1978.

Semi-annually. The issue of 1978 was a trial issue.

Published by Lanzhou Institute of Glacier and Frozen Soil, Academia Sinica, 14 Dongganf West Road, Lanzhou, Gansu Province, China.

HAIYANG/*Oceanography*
Popular science publication.
First published in December 1975.
Monthly (Bimonthly: December 1975 – 1978).
Published by Oceanographical Press, National Bureau of Oceanography, 31 Dongchangan Street, Beijing, China.
Formerly: Haiyang Zhanxian (December 1975 -- 1978).

HAIYANG SHIJIAN/*Oceanography in Practice*
Specialized academic publication.
First published in March 1978. Quarterly.
Published by Second Institute of Oceanography, National Bureau of Oceanography, P.O. Box 75, Hangzhou, Zhejiang Province, China.

HAIYANG YU HUZHAO/*Oceanologia et Limnologia Sinica*
Specialized academic publication.
First published in November 1957. Semi-annually.
Published by Science Press, Institute of Oceanography, Academia Sinica, 7 Nanhai Road, Qingdao, Shandong Province, China.

KAOGU XUEBAO/*Acta Archaeologia Sinica*
Specialized journal.
First published in 1950. Quarterly.
Publication suspended from 1966 – 1971; resumed in 1972.
Published by Science Press, Institute of Archaeology, Chinese Academy of Social Sciences, 27 Wangfujing Street, Beijing, China.

KAOGU/*Archaeology*
Specialized academic publication.
First published in 1955. Bimonthly.
Publication suspended from second half of 1966 – 1971; resumed in 1972.
Published by Science Press, Institute of Archaeology, Chinese Academy of Social Sciences, 27 Wangfujing Street, Beijing, China.

HUANJING BAOHU/*Environmental Protection*
Popular science publication.
First published in February 1974. Bimonthly.
Published by Beijing Press, Bei'erxiang, Fuwai, Beijing, China.

HUANJING KEXUE/*Environmental Science*
Specialized academic publication.
First published in September 1976.
Six issues yearly (Quarterly: September 1976 – 1977).
Published by Institute of Environmental Chemistry, Academia Sinica, P.O. Box 934, Beijing, China.

HUANJING BAOHU ZHISHI/*Environment Protection*
Popular science publication.

First published in January 1978.
Bimonthly (Quarterly: January 1978 — December 1978).
Published by Environment Protection Magazine Publishers, 1 Yiheng Road, Nonglinshang Road, Guangzhou, Guangdong Province, China.

V. Periodicals in the Biological Sciences

DONGWU XUEBAO/*Acta Zoologica Sinica*
Specialized journal.
First published in May 1935. Quarterly.
Publication suspended from July 1966 — 1972; resumed in 1973.
Published by Science Press, Institute of Zoology, Academia Sinica, Zhongguancun, Xijiao, Beijing, China.

DONGWU FENLEI XUEBAO/*Acta Zootaxonomica Sinica*
Specialized academic publication.
First published in July 1964. Quarterly.
Publication suspended from July 1966 — 1978; resumed in 1979.
Published by Science Press, Institute of Zoology, Academia Sinica, Zhongguancun, Xijiao, Beijing, China.

DONGWUXUE ZAZHI/*Chinese Journal of Zoology*
Specialized academic publication.
First published in 1957. Quarterly.
Publication suspended in 1961, 1967 — 1973; resumed in 1974.
Published by Science Press, Institute of Zoology, Academia Sinica, Zhongguancun, Xijiao, Beijing, China.

KUNCHONG XUEBAO/*Acta Entomologica Sinica*
Specialized journal.
First published in September 1950. Quarterly.
Published by Science Press, Institute of Zoology, Academia Sinica, Zhongguancun, Xijiao, Beijing, China.

KUNCHONG ZHISHI/*Entomological Knowledge*
Specialized academic publication.
First published in 1955.
Bimonthly (Quarterly when inaugurated).
Published by Science Press, Institute of Zoology, Academia Sinica, Zhongguancun, Xijiao, Beijing, China.

ZHIWU XUEBAO/*Acta Botanica Sinica*
Specialized journal.
First published in 1952. Quarterly.
Publication suspended from 1967 — 1972; resumed in 1973.
Published by Science Press, Institute of Botany, Academia Sinica, 141 Xiwai Street, Beijing, China.

ZHIWU FENLEI XUEBAO/*Acta Phytotaxonomica Sinica*
Specialized academic publication.

First published in March 1951.

Quarterly (Semi-annually: 1976 — 1977).

Publication suspended from October 1966 — September 1973; resumed in 1974.

Published by Science Press, Institute of Botany, Academia Sinica, 141 Xiwai Street, Beijing, China.

ZHIWU ZAZHI/*Journal of Botany*

Popular science publication.

First published in 1974.

Bimonthly (Quarterly: 1974 — 1975).

Published by Science Press, Institute of Botany, Academia Sinica, 141 Xiwai Street, Beijing, China.

Formerly: Zhiwuxue Zazhi (Specialized academic publication) (1974 — 1976).

ZHIWU SHENGI XUEBAO/*Acta Phytophysiologia Sinica*

Specialized journal.

First published in July 1964. Quarterly.

Published by Shanghai Scientific and Technical Publishers, Institute of Phytophysiology, Academia Sinica, 300 Fenglin Road, Shanghai, China.

WEISHENGWU XUEBAO/*Acta Microbiologica Sinica*

Specialized journal.

First published in 1953.

Quarterly (Semi-annually: 1953 — 1956, 1973 — 1974).

Publication suspended from June 1966 — 1972; resumed in 1973.

Published by Science Press, Institute of Microbiology, Academia Sinica, Zhongguancun, Beijing, China.

WEISHENGWUXUE TONGBAO/*Bulletin of Microbiology*

Specialized academic publication.

First published in March 1974.

Bimonthly (Quarterly: March 1974 — 1977).

Published by Science Press, Institute of Microbiology, Academia Sinica, Zhongguancun, Beijing, China.

GUSHENGWU XUEBAO/*Acta Palaeontologica Sinica*

Specialized academic publication.

First published in 1953.

Bimonthly (Quarterly: 1953 — 1978).

Publication suspended from second half of 1966 — 1975; resumed in 1976.

Published by Science Press, Institute of Geological Palaeontology, 39 Beijing East Road, Nanjing, Jiangsu Province, China.

GUJIZHUI DONGWU YU GURENLEI/*Vertebrata Palasiatica*

Specialized journal.

First published in 1957. Quarterly.

Publication suspended from second half of 1966 — 1972; resumed in 1973.

Published by Science Press, Institute of Vertebrata Palasiatica, Academia Sinica, 142 Xiwai Street, Beijing, China.

Formerly: Gujizhui Dongwu Yu Gurenlei (Chinese Edition) (1959 — 1960): In 1961, the above two journals were combined under the present title.

HUASHI/*Fossils*
Popular science publication.
First published in 1972.
Quarterly (Semi-annually: 1973 — 1975).
Published by Science Press, Institute of Vertebrata Palasiatica, Academia Sinica, 142 Xiwai Street, Beijing, China.

YICHUAN XUEBAO/*Acta Genetica Sinica*
Specialized journal.
First published in June 1974. Quarterly.
Published by Science Press, Institute of Genetics, Academia Sinica, 917 Building, Beishatan, Deshengmenwai, Beijing, China.

YICHUAN/*Hereditas*
Specialized academic publication.
First published in January 1979. Bimonthly.
Published by Science Press, Institute of Genetics, Academia Sinica, 917 Building, Beishatan, Deshengmenwai, Beijing, China.

SHENGWU HUAXUE YU SHENGWU WULI XUEBAO/*Acta Biochimica et Biophysica Sinica*
Specialized journal.
First published in 1958. Quarterly.
Publication suspended from fourth quarter of 1966 — 1974; resumed in 1975.
Published by Shanghai Scientific and Technical Publishers, Shanghai Institute of Biochemistry, Academia Sinica, 320 Yueyang Road, Shanghai, China.
Formerly: Shenhua Xuebao (1958 — 1960).

SHENGWU HUAXUE YU SHENGWU WULI JINZHAN/*Journal of Biochemistry and Biophysics*
Specialized academic publication.
First published in 1974.
Bimonthly (Quarterly: 1974 — 1976).
Published by Science Press, Institute of Biophysics, Academia Sinica, Zhongguancun, Beijing, China.

SHIYAN SHENGWU XUEBAO/*Acta Biologiae Experimentais Sinica*
Specialized academic journal.
First published in May 1936.

Quarterly (Semi-annually: December 1954 — 1963).
Publication suspended from 1945 — September 1950, June 1951 — November 1954, 1966 — 1977; resumed in 1978.
Published by Shanghai Scientific and Technical Publishers, Shanghai Institute of Cytobiology, Academia Sinica, 320 Yueyang Road, Shanghai, China.
Formerly: Zhongguo Shiyan Shengwuxue Zazhi, mainly in foreign language, (May 1936 — May 1951).

SHENGLI XUEBAO/*Physiologica Sinica*
Specialized academic journal.
First published in January 1927. Quarterly.
Publication suspended from 1941 — November 1948, June 1966 — 1977; resumed in 1977.
Published by Shanghai Scientific and Technical Publishers, Shanghai Institute of Physiology, 320 Yueyang Road, Shanghai, China.
Formerly: Zhongguo Shenlixue Zazhi (January 1927 — 1952).

VI. Periodicals in Applied Sciences

HANGKONG ZHISHI/*Applied Science Aerospace*
Popular science publication.
First published in January 1964. Monthly.
Publication suspended from September 1966 — 1973; resumed in 1974.
Published by Aerospace Journal Press, Beijing Aeronautics College, Xueyuan Road, Haidian District, Beijing, China.
(1958 — July 1960, published by Beijing Aeronautics College).

JIANCHUAN ZHISHI/*Naval and Merchant Ships*
Popular science publication.
First published in 1979. Bimonthly.
Published by Naval and Merchant Ships Journal Press, P.O. Box 818, Beijing, China.

ZIDONGHUA XUEBAO/*Acta Automatica Sinica*
Specialized academic publication.
First published in 1963. Quarterly.
Publication suspended from July 1966 — 1978; resumed in 1979.
Published by Science Press, Institute of Automation, Academia Sinica, Zhongguancun, Beijing, China.

ZIDONGHUA/*Automation*
Specialized academic publication.
First published in 1957.
No fixed date of publication.
Publication suspended from 1960 — 1976; resumed in 1977.
Published by Science Press, Institute of Automation, Academia Sinica, Zhongguancun, Beijing, China.

DIANZI XUEBAO/*Acta Electronica Sinica*
Specialized academic journal.
First published in September 1962. Quarterly.
Publication suspended from second half of 1965 — August 1978; resumed in September 1978.
Published by National Defence Industry Publishers, P.O. Box 750, Beijing, China.

DIANZI KEXUE JISHU/*Electronics Science and Technology*
Specialized scientific and technical publication.
First published in July 1964. Monthly.
Publication suspended from July 1966 — 1973; resumed in 1974.
Published by Electronic Science and Technology Editorial Committee, Chinese Society of Chinese Electronics, No. 74 Lugucun, Xijiao, P.O. Box 750, Beijing, China.
Formerly: Wuxian Dian Ji Shu (July 1964 — 1977).

DIANZI JISHU/*Electronics Technology*
Specialized scientific and technical publication.
First published in November 1964. Monthly.
Publication suspended from February 1967 — 1978; resumed in 1979.
Published by Electronics Journal Press, Shanghai Society of Electronics, Room 434, Post and Telecommunications Building, 276 Beisuzhou Road, P.O. Box 253, Shanghai, China.

DIANZIXUE TONGXUN/*Communications in Electronics*
Specialized academic journal.
First published in May 1970.
Quarterly (Before 1972: No fixed date of publication).
Published by Science Press, P.O. Box 2702 — 55, Beijing, China.
Formerly: Dianzi Qijian Yiwen Yu Wenzhai (May — December 1970), Weibo Qijian (1971 — 1974).

JIGUANG/*Laser*
Specialized academic publication.
First published in September 1974. Monthly.
Published by Shanghai Scientific and Technical Publishers, P.O. Box 8211, Shanghai, China.

WUXIANDIAN/*Radio*
Popular science publication.
First published in July 1955. Monthly.
Publication suspended from July 1960 — June 1961, April 1967 — September 1973; resumed in October 1973.
Published by People's Telecommunications Publishers, Beijing, China.

WUXIANDIAN YU DIANZHI/*Radio and Television*
Popular science publication.
First published in July 1958.

Monthly (July 1958 — June 1960); no fixed date of publication from 1978.
Publication suspended from July 1960 — 1977; resumed in 1978.
Published by Shanghai Scientific and Technical Publishers, 450 Ruijin Road, Shanghai, China.

ZHENKONG/*Vacuum*
Specialized academic publication.
First published in 1964. Bimonthly.
Publication suspended from 1966 — 1970; resumed in 1971.
Published by Shenyang Institute of Vacuum Technology, No. 2, Section 3, Zhongshan Road, Heping District, Shenyang, Liaoning Province, China.
Formerly: Zhenkong Jishu Huoye Wenxuan (1964 — 1965); Zhenkong Jishu Baodao (1971 — 1977).

JILIANG JISHU/*Metrology*
Specialized academic publication.
First published in 1962. Bimonthly.
Published by Technological Standard Publishers, Information Department of Chinese Institute of Metering Science, Xiaohuangzhuang, Andingmenwai, Beijing, China.
Formerly: Jiliang Gongzuo (1962 — 1975).

JIANZHU XUEBAO/*Architectural Journal*
Specialized academic journal.
First published in June 1954.
Bimonthly (Quarterly: June — September 1954; Monthly: July 1956 — June 1960; Quarterly: 1975 — 1978).
Publication suspended from October 1954 — July 1955; July — September 1960, 1961 — 1965, July 1966 — September 1973; resumed in October 1974.
Published by Chinese Architectural Industry Publishers, Baiwanzhuang, Beijing, China.

VII. Periodicals in the Medical Sciences

ZHONGHUA WAIKE ZAZHI/*Chinese Journal of Surgery*
Specialized medical publication.
First published in January 1951. Bimonthly.
Publication suspended from September 1966 — July 1977; resumed in August 1977.
Published by Chinese Medical Association, 42 Dongsi-xi Street, Beijing, China.
Formerly: Waike Xuebao (January 1951 — December 1952).

ZHONGHUA ERKE ZAZHI/*Chinese Journal of Paediatrics*
Specialized medical publication.
First published in July 1950.
Quarterly (Bimonthly: 1955 — July 1960, September 1962 — May 1970).

Publication suspended from August 1960 — August 1962, June 1966 — July 1978; resumed in August 1978.

Published by Chinese Medical Association, 42 Dongsi-xi Street, Beijing, China.

ZHONGHUA ERBIYANHOUKE ZAZHI/*Chinese Journal of Otorhinolaryngology*

Specialized medical publication.

First published in March 1958.

Quarterly (Bimonthly: January 1958 — June 1960, January 1964 — August 1966).

Publication suspended from July 1960 — April 1963, September 1966 — August 1978; resumed in September 1978.

Published by Chinese Medical Association, 42 Dongsi-xi Street, Beijing, China.

ZHONGHUA YANKE ZAZHI/*Chinese Journal of Ophthalmology*

Specialized medical publication.

First published in October 1950.

Quarterly (Bimonthly: May 1964 — June 1966).

Publication suspended from July 1960 — April 1964, July 1966 — July 1978; resumed in August 1978.

Published by Chinese Medical Association, 42 Dongsi-xi Street, Beijing, China.

ZHONGHUA FUNCHANKE ZAZHI/*Chinese Journal of Obstetrics and Gynaecology*

Specialized medical publication.

First published in 1953. Quarterly.

Publication suspended from second half of 1960 — 1962, the second half of 1966 — first half of 1978; resumed in second half of 1978.

Published by Chinese Medical Association, 42 Dongsi-xi Street, Beijing, China.

ZHONGHUA SHENJING JINSHENKE ZAZHI/*Chinese Journal of Neurology and Psychiatry*

Specialized medical publication.

First published in March 1955.

Quarterly (Bimonthly: 1958 — 1966).

Publication suspended from second half of 1960 — first half of 1963, second half of 1966 — the first half of 1978; resumed in second half of 1978.

Published by Chinese Medical Association, 42 Dongsi-xi Street, Beijing, China.

ZHONGHUA ZHONGLIU ZAZHI/*Chinese Journal of Oncology*

Specialized medical publication.

First published in January 1979. Quarterly.

Published by the Chinese Medical Association, 2 Yabao Road, Beijing, China.

ZHONGHUA FANGSHEXUE ZAZHI/*Chinese Journal of Radiology*
Specialized medical publication.
First published in September 1953.
Quarterly (Bimonthly: April 1964 – May 1966).
Publication suspended from June 1960 – March 1964, June 1966 – the second quarter of 1978; resumed in second half of 1978.
Published by Chinese Medical Association, 42 Dongsi-xi Street, Beijing, China.

ZHONGHUA XINXIEGUANBING ZAZHI/*Chinese Journal of Cardiovascular Diseases*
Specialized medical publication.
First published in 1973. Quarterly.
Published by Chinese Medical Association, Fuwai Hospital, Chinese Academy of Medical Science, North Lishi Road, Beijing, China.
Formerly: Xinzangxieguan Jibing (1973 – 1978).

ZHONGHUA ILLIAO ZAZHI/*Chinese Journal of Physiotherapy*
Specialized medical publication.
First published in August 1978. Quarterly.
Published by Physiotherapeutic Hospital, Tanggangzi, Anshan, Liaoning Province, China.

ZHONGHUA JIEHE HE HUXIXI JIBING ZAZHI/*Chinese Journal of Tuberculosis and Respiratory Diseases*
Specialized medical publication.
First published in September 1978. Quarterly.
Published by Chinese Medical Association, 42 Dongsi-xi Street, Beijing, China.
Formerly: Zhonghua Jiehebingke Zazhi (1953). This publication was combined with Zhongguo Fanglao in December 1959, taking the title of Zhongguo Fanglao Zazhi and was suspended in August 1966.

HULI ZAZHI/*Journal of Nursing*
Specialized medical publication.
First published in 1954.
Bimonthly (Quarterly: 1954 – 1955).
Publication suspended from August 1960 – April 1963, August 1966 – 1977; resumed in third quarter of 1977.
Published by Chinese Nurses Association, 42 Dongsi-xi Street, Beijing, China.

DAZHONG YIXUE/*Popular Medicine*
Popular science publication.
First published in 1948. Monthly.
Publication suspended from 1960 – June 1978; resumed in July 1978.
Published by Shanghai Scientific and Technical Publishers, 450 Ruijin Road, Shanghai, China.

JIEFANGJUN YIXUE ZAZHI/*Chinese Medical Journal of the People's Liberation Army*
Comprehensive medical publication.
First published in May 1964. Bimonthly.
Publication suspended from July 1966 — July 1979; resumed in August 1979.
Published by People's Military Surgeons Publishers, A3, 22, Fuxing Road, Beijing, China.

SHANGHAI YIXUE/*Shanghai Medicine*
Comprehensive medical publication.
First published in January 1978. Monthly.
Published by Chinese Medical Association, Shanghai Branch, 1623 Beijing West Road, Shanghai, China.
Formerly: Yixue Qingkuang Jiaoliu

XINYYIXUE/*New Medicine*
Comprehensive medical publication.
First published in August 1969. Monthly.
Published by Zhongshan Medical College, Zhongshan Second Road, Guangzhou, Guangdong Province, China.

TIANJIN YIYAO/*Tianjin Medicine*
Comprehensive medical publication.
First published in January 1974.
Monthly (Bimonthly in 1973).
Published by Tianjin Medical Science and Technology Information Station, 167 Chengdou Dao, Heping District, Tianjin, China.

TIANJIN YIYAO GUKE FUKAN/*Tianjin Medicine and Bone Department Sub-Journal*
Specialized medical publication.
First published in February 1958. Quarterly.
Publication suspended from 1967 — the second quarter of 1978; resumed in the third quarter of 1978.
Published by Tianjin Medical Science and Technological Information Station, Tianjin Hospital, Tianjin, China.

TIANJIN YIYAO ZHONGLIUXUE FUKAN/*Tianjin Medicine Oncology Supplement*
Specialized medical publication.
First published in March 1963. Quarterly.
Published by Tianjin Medical Science and Technological Information Station, 167 Chengdu Dao, Heping District, Tianjin, China.

JIANGSU YIYAO/*Jiangsu Medicine*
Comprehensive medical publication.
First published in January 1975.
Monthly (Bimonthly: 1975 — 1976).

Published by Jiangsu People's Publishers, 300 Hanzhong Road, Nanjing, Jiangsu Province, China.

JIANGSU YIYAO (ZHONGYI FENCE)/*Jiangsu Medicine (Traditional Chinese Medical Science Sub-Volume)*
Comprehensive medicine publication.
First published in 1976.
Quarterly (Semi-annually: 1976 and 1978).
Publication suspended in 1977; resumed in 1978.
Published by Jiangsu Medical Journal Publishers, 300 Hanzhong Road, Nanjing, Jiangsu Province, China.

SHANDONG YIYAO/*Shandong Medicine*
Comprehensive medical publication.
First published in 1971.
Monthly (No fixed date of publication: 1971 — 1977).
Published by Bureau of Hygiene, Jinan, Shandong Province, China.
Formerly: Shandong Yikan (Suspended in the second semester of 1966).

GUIZHOU YIYAO/*Guizhou Medicine*
Comprehensive medical publication.
First published in February 1979. Bimonthly.
Published by Bureau of Hygiene of Guizhou Province and Chinese Medical Association, Guizhou Branch, Shibeicun, Guiyang, Guizhou Province, China.

SHANXI YIYAO ZAZHI/*Shanxi Medical Journal*
Comprehensive medical publication.
First published in 1956. Bimonthly.
Published by Chinese Medical Association, Shanxi Branch, Bureau of Hygiene of Shanxi Province, 8 Jianshe North Road, Taiyuan, Shanxi Province, China.
Formerly: Shanxi Yixue Zazhi (Suspended in 1966).

LIAONING YIYAO/*Liaoning Medicine*
Comprehensive medical publication.
First published in 1970. Bimonthly.
Publication suspended from 1972 — 1973; resumed in 1974.
Published by Chinese Medical Association, Liaoning Branch Section, 4 Heping Street, Shenyang, Liaoning Province, China.

FUJIAN YIYAO ZAZHI/*Fujian Journal of Medicine*
Comprehensive medical publication.
First published in January 1979. Bimonthly.
Published by Fujian Medical Journal Editorial Department, 11 Mishusiang, Wusi Road, Fuzhou, Fujian Province, China.
Formerly: Fujian Yiyao Weisheng and Fujian Chijiaoyisheng.

SHANXI XINYIYAO/*Shanxi New Medicine*
Comprehensive medical publication.

First published in January 1972.
Monthly (Bimonthly: 1972 — 1978).
Published by Shanxi People's Publishers, 20 Xihuamen, Xi'an, Shanxi Province, China.

XINYI YAOXUE ZAZHI/*New Medical Journal*
Comprehensive medical publication.
First published in January 1955. Monthly.
Published by New Medicine Journal Publishers, Institute of Traditional Chinese Medicine, 3 Haiyuncang, Dongzhimennei, Beijing, China.

SHANGHAI ZHONGYIYAO ZAZHI/*Shanghai Traditional Chinese Medical Science*
Comprehensive medical publication.
First published in June 1955.
Bimonthly (Monthly: June 1955 — July 1966).
Publication suspended from August 1966 — October 1978; resumed in November 1978.
Published by Shanghai Society of Traditional Chinese Medicine and Shanghai Traditional Chinese Medical College, 530 Lingling Road, Shanghai, China.

XINZHONGYI/*New Traditional Chinese Medicine*
Comprehensive medical publication.
First published in December 1969.
Bimonthly (No fixed date of publication: December 1969 — 1972).
Published by Guangzhou Traditional Chinese Medical College, Sanyuanli, Guangzhou, Guangdong Province, China.

ZHEJIANG ZHONGYIYAO/*Zhejiang Traditional Chinese Medicine*
Comprehensive medical publication.
First published in December 1956. Monthly.
Publication suspended from July 1960 — June 1963, August 1966 — 1974; resumed in 1975.
Published by Zhejiang People's Publishers, Qingchun Road, Hangzhou, Zhejiang Province, China.
Formerly: Zhejiang Zhongyi Zazhi (December 1956 — June 1960, July 1963 — July 1966).

LIAONING ZHONGJI YIKAN/*Liaoning Middle Grade Medical Journal*
Comprehensive medical publication.
First published in January 1978.
Monthly (Bimonthly: January — December 1978).
Published by Liaoning Middle Grade Medical Journal Editorial Office, 251 Tianjin Street, Zhongshan District, Luda, Liaoning Province, China.
Formerly: Yiyao Weisheng (published by Luda Bureau of Hygiene).

CHIJIAOYISHENG ZAZHI/*Barefoot Doctor Journal*
Popular science publication.

First published in January 1973. Monthly.

Published by People's Hygiene Publishers, 100 Yingxin Street, Xuanwu District, Beijing, China.

GUANGXI WEISHENG/*Guangxi Hygiene*

Comprehensive medical publication.

First published in January 1979. Bimonthly.

Published by Chinese Medical Association, Guangxi Branch and Guangxi Medical Science Information Institute, 19 Liaoyuan Road, Nanning, Guangxi Zhuang Autonomous Region, China.

Formerly: Guangxi Weisheng (1972 — June 1976) and Guangxi Chijiao-yishen (July 1976), published by the Bureau of Hygiene, Guangxi Zhuang Autonomous Region.

YAOXUE XUEBAO/*Acta Pharmaceutica Sinica*

Specialized pharmaceutical publication.

First published in January 1936.

Monthly (Semi-annually: 1953 — 1961).

Publication suspended from 1948 — 1952, 1966 — 1978; resumed in 1979.

Published by Chinese Pharmaceutical Association, 1 Xiannongtan Street, Beijing, China.

Formerly: Zhonghua Yaoxue Zazhi and Zhongguo Yaoxue Zazhi (1936 — 1952).

YAOXUE TONGBAO/*Pharmaceutical Bulletin*

Comprehensive pharmaceutical publication.

First published in January 1953. Monthly.

Publication suspended from July 1960 — March 1963, September 1966 — June 1978; resumed in July 1978.

Published by Chinese Pharmaceutical Association, 42 Dongsi-xi Street, Beijing, China.

ZHONGZHENGYAO YANJIU/*Chinese Prepared Medicine Research*

Comprehensive medical publication.

First published in January 1978.

Bimonthly (Quarterly: January 1978 — June 1979).

Published by Traditional Chinese Prepared Medicine Information Center, National General Medicine Administration, 324 Renmin Road, Shanghai, China.

ZHONGOAOYAO TONGXUN/*Communications on Traditional Chinese Herbal Medicine*

Specialized medical publication.

First published in February 1970.

Monthly (Bimonthly: February — December 1970).

Publication suspended in first half of 1971; resumed in second half of 1971.

Published by Institute of Hunan Medicine Industrial Research, Shaoyang, Hunan Province, China.

YIXUE YANJIU TONGXUN/*Communications on Medical Research*
Specialized medical publication.
First published in March 1972.
Monthly (No fixed date of publication: March 1972 — December 1977).
Published by Institute of Medical Information, Chinese Academy of Medical Science, 9 Dongdan-santiao, Beijing, China.

PIFUBING FANGZHI YANJIU TONGXUN/*Communications of Skin Diseases Prevention and Cure Research*
Specialized academic journal.
First published in 1972.
Quarterly (No fixed date of publication: 1972 — second quarter of 1976).
Published by Institute of Skin Diseases, Prevention and Cure, Chinese Academy of Medical Science, 2 Jiankang Street, Taizhou, Jiangsu Province, China.

SHENJING JINGSHEN JIBING ZAZHI/*Nervous and Mental Disease Journal*
Specialized medical publication.
First published in May 1975.
Bimonthly (Two issues published in 1975; quarterly in 1976).
Published by Zhongshan Medical College, Guangzhou, Guangdong Province, China.
Formerly: Xinyixue Shenjin Xitong Jibing Fukan (total number 2).

SHUXIE JI XIEYEXUE/*Blood Transfusion and Blood Subject*
Specialized academic publication.
First published in November 1977. Quarterly.
Published by Chinese Academy of Medical Science, Sichuan Branch, P.O. Box 6, Jianyang, Sichuan Province, China.

KANGSHENGSU/*Antibiotics*
Specialized medical publication.
First published in 1976.
Bimonthly (Quarterly: 1976 — 1977; No fixed date of publication in 1978).
Published by Sichuan Institute of Antibiotics Industry, Sanbanqiao, Chengdou, Sichuan Province, China.
Formerly: Kangjunsu (1976 — 1978).

VIII. Periodicals in Agricultural Science

ZHONGGUO NONGYE KEXUE/*Scientia Agricultura Sinica*
Specialized academic publication.
First published in 1960.
Quarterly (Monthly: 1960 — September 1966).
Published by Agriculture Publishers, 30 Baishiqiao Road, Xijiao, Beijing, China.

ZUOWU XUEBAO/*Acta Agronomica Sinica*
Specialized academic journal.
First published in February 1962. Quarterly.
Publication suspended from August 1966 — July 1979; resumed in August 1979.
Published by Agriculture Publishers, Chinese Society of Agriculture, Hepingli, Beijing, China.

YUANYI XUEBAO/*Acta Horticulturae Sinica*
Specialized academic journal.
First published in May 1962. Quarterly.
Publication suspended from August 1966 — July 1979; resumed in August 1979.
Published by Agriculture Publishers, Chinese Society of Agriculture, Hepingli, Beijing, China.

ZHIWU BAOHU XUEBAO/*Acta Phytophyiacica*
Specialized academic journal.
First published in February 1962. Quarterly.
Publication suspended from June 1966 — 1978; resumed in 1979.
Published by Agriculture Publishers, Chinese Society of Agriculture, Hepingli, Beijing, China.

ZHIWU BINGLI XUEBAO/*Plant Pathology Academic Journal*
Specialized academic journal.
First published in June 1959. Semi-annually.
Publication suspended from 1966 — May 1979; resumed in June 1979.
Published by Agriculture Publishers, Chinese Society of Agriculture, Hepingli, Beijing, China.

XUMU SHOUYI XUEBAO/*Acta Veterinaria et Zootechnica Sinica*
Specialized academic journal.
First published in July 1956. Quarterly.
Publication suspended from 1960 — 1961, in 1966; resumed in 1967.
Published by Agriculture Publishers, Chinese Society of Agriculture, Hepingli, Beijing, China.

ZHONGGUO XUMU ZAZHI/*Chinese Journal of Animal Husbandry*
Specialized popular science publication.
First published in July 1963. Bimonthly.
Publication suspended from May 1966 — March 1979; resumed in April 1979.
Published by Agriculture Publishers, Chinese Society of Agriculture, Hepingli, Beijing, China.

ZHONGGUO SHOUYI ZAZHI/*Chinese Journal of Veterinary Medicine*
Specialized popular science publication.
First published in July 1962. Monthly.
Publication suspended from May 1966 — September 1978; resumed in October 1978.

Published by Agriculture Publishers, Chinese Society of Agriculture, Hepingli, Beijing, China.

ZHIWU BAOHU/*Plant Protection*
Specialized publication.
First published in July 1963. Bimonthly.
Publication suspended from May 1966 – September 1978; resumed in October 1978.
Published by Agriculture Publishers, Chinese Society of Agriculture, Hepingli, Beijing, China.

NONGYE KEJI TONGXUN/*Communications in Agricultural Science and Technology*
Popular science publication.
First published in January 1972. Monthly.
Published by Agriculture Publishers, 30 Baishiqiao Road, Beijing, China.

NONGCUNKEXUE SHIYAN/*Agricultural Science Experiment*
Popular science publication.
First published in January 1978. Monthly.
Published by Science Press, 137 Chaonei Street, Beijing, China.

SHANGHAI NONGYE KEJI/*Shanghai Agricultural Science and Technology*
Comprehensive agricultural science and technology publication.
First published in May 1971. Monthly.
Published by Shanghai Society of Agriculture and Shanghai Academy of Agricultural Science, 2901 Beidi Road, Shanghai, China.
Formerly: Nongye Keji Jianbao (May 1971 – 1972); Nongye Keji Tongxun (1973 – 1976).

KEXUE ZHONGTIAN/*Scientific Farming*
Popular science publication.
First published in March 1972. Monthly.
Published by Shanghai Scientific and Technical Publishers, Ruijin Second Road, Shanghai, China.

JIANSU NONGYE KEXUE/*Jiansu Agricultural Science*
Comprehensive agricultural science and technology publication.
First published in 1973. Bimonthly.
Published by Academy of Agricultural Science of Jiangsu Province, Xiaolingwei, Nangjing, Jiangsu Province, China.
Formerly: Jiangsu Nongye Keji (1973 – 1978).

ZHEJIANG NONGYE KEXUE/*Zhejiang Agricultural Science*
Comprehensive agricultural science and technology publication.
First published in June 1960.
Bimonthly (Monthly: June 1960 – July 1966).
Publication suspended from August 1966 – January 1972; resumed in February 1972.

Published by Zhejiang People's Publishers, Scientific and Technological Research Department, Academy of Agricultural Science of Zhejiang Province, Gongnong Road, Hangzhou, Zhejiang Province, China.

ANHUI NONGLIN KEXUE SHIYAN/*Anhui Scientific Experiments of Farming and Forestry*
Popular science publication.
First published in 1972. Monthly.
Published by Academy of Farming and Forestry Science of Anhui Province, Silihe, Hefei, Anhui Province, China.

LIAONING NONGYE KEXUE/*Liaoning Agricultural Science*
Comprehensive agricultural science and technology publication.
First published in June 1963. Bimonthly.
Publication suspended from October 1966 — August 1970; resumed in September 1970.
Published by the Society of Agriculture of Liaoning Province and Academy of Agricultural Science of Liaoning Province, Maguanqiao, Dongling, Shenyang, Liaoning Province, China.

XIN NONGYE/*New Agriculture*
Popular science publication.
First published in February 1971. Semi-annually.
Published by Shenyang Agricultural College, Dongling, Shenyang, Liaoning Province, China.

HUNAN NONGYE KEJI/*Hunan Agricultural Science and Technology*
Comprehensive agricultural science and technology publication.
First published in 1970. Bimonthly.
Published by Hunan Society of Agriculture and Academy of Agricultural Science of Hunan Province, East Suburbs, Mapoling, Changsha, Hunan Province, China.

HUBEI NONGYE KEXUE/*Hubei Agricultural Science*
Comprehensive agricultural science and technology publication.
First published in August 1962. Monthly.
Publication suspended from November 1966 — 1971; resumed as Hubei Nongye Keji in 1972.
Published by Hubei Agricultural Science Editorial Committee, Academy of Agricultural Science of Hubei Province, Wuchang, Hanhu, Wuhan, Hubei Province, China.
Formerly: Hubei Nongye Kexue was combined with Hubei Nongye Kexue Jishu Tongxun of the Agricultural Department of Hubei Province under the name Hubei Nongye Kexue Jishu from July to October 1966.

FUJIAN NONGYE KEJI/*Fujian Agricultural Science and Technology*
Comprehensive agricultural science and technology publication.
First published in October 1970.
Bimonthly (No fixed date of publication from October 1970 — 1974).

Published by Academy of Agricultural Science of Fujian Province, Butang, Fuzhou Suburbs, Fujian Province, China.
Formerly: Nongye Keji Jianbao (October 1970 — 1972).

GUANGDONG NONGYE KEXUE/*Guangdong Agricultural Science*
Comprehensive agricultural science and technology publication.
First published in September 1956. Bimonthly.
Publication suspended from June 1966 — July 1974; resumed in August 1974.
Published by Academy of Agricultural Science of Guangdong Province and South China Agricultural College, Shipai, Guangzhou, Guangdong Province, China.

GUANGXI NONGYE KEXUE/*Guangxi Agricultural Science*
Comprehensive agricultural science and technology publication.
First published in October 1974. Monthly.
Published by Academy of Agricultural Science of Guangzizhuang, Guangxi, Xixiangtang, Nanning, Guangxi Province, China.
Formerly: Guangxi Nongye (January 1964 — July 1966).

YUNNAN NONGYE KEJI/*Yunnan Agricultural Science*
Comprehensive agricultural science and technology publication.
First published in February 1972.
Bimonthly (No fixed date of publication from February 1972 — 1975).
Published by Academy of Agricultural Science of Yunnan Province, Longtou Street, North Suburbs of Kunming, Yunnan Province, China.

GUIZHOU NONGYE KEJI/*Guizhou Agricultural Science*
Comprehensive agricultural science and technology publication.
First published in 1973. Monthly.
Published by Academy of Agricultural Science of Guizhou Province, West Suburbs of Guiyang, Guizhou Province, China.

XINJIANG NONGYE KEXUE/*Xinjiang Agricultural Science*
Comprehensive agricultural science and technology publication.
First published in 1959.
Bimonthly (Monthly: 1959 — August 1966). Published in Han and Wei languages.
Publication suspended in first quarter of 1961, September 1966 — 1976; resumed in 1977.
Published by Xinjiang People's Publishers, Xinjiang Academy of Agricultural Science, Laomanzheng, Urumqi, Uighur Autonomous Region, Xinjiang, China.

NONGYE KEXUE SHIYAN/*Scientific Experiments in Agriculture*
Popular science publication.
First published in May 1975. Semi-monthly.
Published by Inner Mongolia People's Publishers, Inner Mongolia Bureau of Agriculture, Huhehaote Inner Mongolia Autonomous Region, China.

MIANHUA/*Cotton*
Specialized scientific and technological publication.
First published in February 1974. Bimonthly.
Published by Institute of Cotton, Chinese Academy of Agricultural Science, Baibi, Anyang, Henan Province, China.

ZHONGGUO YOULIAO/*Chinese Oil Crops*
Specialized scientific and technological publication.
First published in February 1971. Quarterly.
Published by Institute of Oil Crops, Chinese Academy of Agricultural Science, Baoji'an, Wuchange, Wuhan, Hubei Province, China.
Formerly: Youliao Zuowu Keji (February 1971 — February 1979).

CANSANG TONGBAO/*Silkworm and Mulberry Bulletin*
Popular science publication.
First published in 1956.
Quarterly (Bimonthly: 1956 — 1959).
Publication suspended from 1960 — 1975, resumed in 1976.
Published by Scientific and Technological Association of Zhejiang Province, Silkworm and Mulberry Department of Zhejiang University, Huajiachi, Zhejiang Province, China.
Formerly: Cansi Tongbao (1956 — 1959).

CANYE KEXUE/*Silkworm Science*
Specialized academic publication.
First published in 1963. Quarterly.
Publication suspended from July 1966 — 1978; resumed in 1979.
Published by Jiangsu People's Publishers, Institute of Sericulture, Chinese Academy of Agricultural Science, Sibaidu, Zhenjiang, Jiangsu Province, China.

LINYE KEXUE/*Scientia Silvae Sinicae*
Specialized academic publication.
First published in 1955. Quarterly.
Published by Science Press, Chinese Institute of Forestry, behind Wanshoushan, West Suburbs, Beijing, China.
Formerly: Zhongguo Linye Kexue (October 1975 — 1977).

LINYE KEJI TONGXUN/*Communications in Forestry Science and Technology*
Specialized scientific and technological publication.
First published in 1958. Monthly.
Publication suspended 1968 — 1971; resumed in 1972.
Published by Chinese Forestry Science Research Institute, Information Institute of Chinese Forestry Science Research Institute, 30 Baishiqiao Road, West Suburbs, Beijing, China.
Formerly: Linye Kuaibao (1958 — 1967).

ZHONGGUO GANJU/*Chinese Orange*
Specialized scientific and technological publication.

First published in 1972. Quarterly.
Published by Institute of Orange, Chinese Academy of Agricultural Science, Xiemachang, Beibei, Chongqing, Sichuan Province, China.
Formerly: Ganju Keji Tongxun (1972 – 1978).

REZUO KEJI TONGXUN/*Tropical Crops Science and Technology*
Specialized scientific and technological publication.
First published in July 1973. Bimonthly.
Published by Chinese Tropical Crops Association and South China Tropical Crops Research Institute, Information Institute of Science and Technology, Baodaoxincun, Dan County, Hainan Island, Guangdong Province, China.

DANSHUI YUYE/*Fresh Water Fishing*
Specialized scientific and technological publication.
First published in 1971.
Monthly (Bimonthly: 1978).
Published by Danshui Yuye Editorial Department, Chinese Aquatic Products Association, Changjiang Institute of Aquatic Products, National Aquatic Products Main Bureau, Shashi, Hubei Province, China.
Formerly: Danshui Yuye Keji Dongtai, Danshui Yuye Keji Zazhi (1971 – 1978).

TURANG XUEBAO/*Acta Pedologica Sinica*
Specialized academic journal.
First published in 1948. Quarterly.
Publication suspended from July 1966 – June 1978; resumed in July 1978.
Published by Science Press, Nanjing Institute of Soil, Academia Sinica, 71 Beijing East Road, Nanjing, Jiangsu Province, China.
Formerly: Zhongguo Turang Xuehui Huizhi (1948 – 1951).

TURANG TONGBAO/*Soil Bulletin*
Specialized scientific and technological publication.
First published in October 1957. Bimonthly.
Publication suspended from June 1966 – January 1979; resumed in February 1979.
Published by Liaoning People's Publishers, Henyang Agricultural College, Shenyang, Liaoning Province, China.

TURANG/*Soil*
Specialized scientific and technological publication.
First published in February 1974. Bimonthly.
Published by Jiangsu People's Publishers, Nanjing Institute of Soil, Academia Sinica, 71 Beijing East Road, Nanjing, Jiangsu Province, China.

TURANGXUE JINZHAN/*Progress of Pedology*
Specialized scientific and technological publication.

Periodicals in agricultural science

First published in June 1973. Bimonthly.
Published by Nanjing Institute of Soil, Academia Sinica, 71 Beijing East Road, Nanjing, China.
Formerly: Turang Nonghua (June 1973 — 1978).

TURANG FEILIAO/*Soil and Fertilizer*
Specialized scientific and technological publication.
First published in August 1964. Bimonthly.
Published by Institute of Soil and Fertilizer, Chinese Academy of Agricultural Science, 30 Baishiqiao Road, West Suburbs, Beijing, China.
Formerly: Gengzuo Yu Feiliao (August 1964 — the first half of 1966), Tufie Yu Kexue Zhongtian (1972).

NONGCUN KEXUE SHIYAN (JILIN PROVINCE)/*Scientific Experiments in the Farm*
Popular science publication.
First published in 1974. Monthly.
Published by People's Publishers of Jilin Province, Scientific and Technological Association of Jilin Province, 8 Minkang Road, Changchun, Jilin Province, China.

IX. Local Technical Newspapers

BEIJING KEJIBAO/*Beijing Newspaper of Science and Technology*
Weekly (published every Friday).
Published at 140 Xizhimenwai Street, Beijing, China.

SICHUAN KEJIBAO/*Sichuan Newspaper of Science and Technology*
First published in June 1979.
Weekly (published every Thursday; Trial issues in 1977; Semi-monthly during 1978).
Published by Scientific and Technological Association of Sichuan Province, Chengdu, Sichuan Province, China.

GUANGDONG KEJIBAO/*Guangdong Newspaper of Science and Technology*
First published in March 1973.
Weekly (published every Friday).
Published by Scientific — Technical Commission of Guangdong Province, 125 Yuexiu Central Road, Guangzhou, Guangdong Province, China.

HUBEI KEJIBAO/*Hubei Newspaper of Science and Technology*
First published in July 1976.
Three issues monthly.
Published by Editorial Department of Hubei Kejibao, Xiaohongshan, Wuchang, Wuhan, Hubei Province, China.

YUNNAN KEJIBAO/*Yunnan Newspaper of Science and Technology*
First published in September 1972.

Weekly (published every Friday).
Published by Information Institute of Science and Technology of Yunnan Province, 20 Huguo Road, Kunming, Yunnan Province, China.
Formerly: Keji Xiaobao.

LIAONING KEJIBAO/*Liaoning Newspaper of Science and Technology*
First published in July 1978.
Weekly (published every Monday).
Published by Scientific and Technological Association of Liaoning Province, 33 Section 4, Zhongshan Road, Heping District, Shenyang, Liaoning Province, China.

GUANGXI KEJIBAO/*Guangxi Newspaper of Science and Technology*
First published in July 1974.
Three issues monthly.
Published by Scientific and Technical Commission of Guangxi Zhuang Autonomous Region, Xinmin Road, Nanning, Guangxi, China.
Formerly: Guangxi Kepu (1974 — 1978).

QINGHAI KEJIBAO/*Qinghai Newspaper of Science and Technology*
First published in May 1979.
Weekly (published every Friday).
Published by Scientific and Technological Association of Qinghai Province, 18 Xisanlou, People's Government of Qinghai Province, Xi Street, Xining, Qinghai Province, China.

HEBEI KEJIBAO/*Hebei Newspaper of Science and Technology*
First published in January 1959.
Weekly (published every Friday).
Publication suspended from July 1967 — 1975; resumed in 1976.
Published by Scientific and Technological Association of Hebei Province, Fuqiang Street, Shijiazhuang, Hebei Province, China.
Formerly: Kexue Yu Jishu and Hebei Keji Xiaobao (January 1959 — July 1967).

HENAN KEJIBAO/*Henan Newspaper of Science and Technology*
First published in September 1974.
Weekly (published every Thursday).
Published by Scientific and Technological Committee of Henan Province, Building No. 73, Weier Road, Xingzheng-qu, Zhengzhou, Henan Province, China.

ANHUI KEJIBAO/*Anhui Newspaper of Science and Technology*
First published in December 1978.
Weekly (published every Friday).
Published by Anhui Newspaper of Science and Technology Publishers, 47 Changjiang Road, Hefei, Anhui Province, China.

FUJIAN KEJIBAO/*Fujian Newspaper of Science and Technology*
First published in September 1978.

Weekly (published every Friday).
Published by Fujian Newspaper of Science and Technology Publishers, Scientific and Technical Commission of Fujian, Fuzhou, Fujian Province, China.

GANSU KEJIBAO/*Gansu Newspaper of Science and Technology*
First published in January 1979.
Three issues monthly.
Published by Scientific and Technological Association of Gansu Province, 65 Dingxi South Road, Lanzhou, Gansu Province, China.

GUIZHOU KEJIBAO/*Guizhou Newspaper of Science and Technology*
First published in January 1976.
Three issues monthly (Semi-monthly: January 1976 – 1978).
Published by Scientific and Technological Association of Guizhou Province, 68 Kexue Road, Guiyang, Guizhou Province, China.

SHANXI KEJIBAO/*Shanxi Newspaper of Science and Technology*
First published in September 1978.
Weekly (published every Wednesday).
Published by Editorial Department of Shanxi Newspaper of Science and Technology, 55 Maoerxiang, Taiyuan, Shanxi Province, China.

JIANGSU NONGYE KEJIBAO/*Jiangsu Newspaper of Agricultural Science and Technology*
First published in May 1975.
Three issues monthly.
Published by Academy of Agricultural Science of Jiangsu Province, Xiaolingwei, Nanjing, Jiangsu Province, China.

JIANGXI KEJIBAO/*Jiangxi Newspaper of Science and Technology*
First published in December 1975.
Three issues monthly.
Published by Editorial Department of Jiangxi Newspaper of Science and Technology, Building of People's Government Newspaper of Science and Technology, Jiangxi Province, Nanchang, Jiangxi Province, China.

SHANDONG KEJIBAO/*Shandong Newspaper of Science and Technology*
Weekly (published every Thursday).
Published by Shandong Scientific and Technological Association, 120 Ganshiqiao South Street, Jinan, Shandong Province, China.

ZHEJIANG KEJIBAO/*Zhejiang Newspaper of Science and Technology*
Six issues monthly.
Published by Editorial Department of Zhejiang Newspaper of Science and Technology, 18 Hangzhou, Zhejiang Province, China.

SHANGHAI KEJIBAO/*Shanghai Newspaper of Science and Technology*
First published in May 1971.

Weekly.
Published by Shanghai Scientific and Technological Association, 47 Nanchang Road, Shanghai, China.
Formerly: Shanghai Keji (May 1971 — 4th June 1976).

KEXUE YUANDI/*Scientific Garden*
Three issues monthly.
Published by Tianjin Scientific and Technological Association, 37 Heping Road, Tianjin, China.

TIANJIN KEJI XIAOXI/*Tianjin Scientific and Technological News*
First published in 1974.
Monthly (No fixed date of publication from 1974 — 1977).
Published by Tianjin Institute of Science and Technology, 379 Heping Road, Tianjin, China.

ZHONGGUO ZIRAN BIANZHENGFA YANJIUHUI TONGXUN/*Communications of the Chinese Natural Dialects Research Society*
First published in February 1978.
Semi-monthly (published on the 10th and 25th of the month).
Published by Chinese Preparatory Committee of Natural Dialecticals (trans. Bao Yujin, Shao Benqiu), Research Society, Sanlihe, Beijing, China.

SECTION III

NAME LIST OF MEMBERS OF ACADEMIA SINICA DEPARTMENTS

(in the fields of the natural and technical sciences)

I

Order Issued by the State Council of the People's Republic of China:

The name list of members of Academia Sinica Departments (altogether 233 people) was approved by the Tenth Plenary Session of the State Council held on May 31, 1955, and it is published now.

Premier Zhou Enlai
June 3, 1955.

(1) Department of Mathematics, Physics and Chemistry

Wang Zhuxi	Wang Ganchang	Wang Xianghao	Jiang Zehan
Yu Ruihuang	Wu Youxun	Wu Xuezhou	Li Fangxun
Li Guoping	Wang You	Zhou Tongqing	Zhou Peiyuan
Shi Ruwei	Ke Zhao	Liu Dagang	Duan Xuefu
Ji Yupan	Hu Ning	Tang Aoqing	Yuan Hanqing
Zhang Qinglian	Zhang Yuzhe	Liang Shuquan	Zhuang Changgong
Xu Baolu	Chen Jiangong	Lu Xueshan	Fu Ying
Peng Huanwu	Yun Ziqiang	Zeng Zhaolun	Hua Luogeng
Huang Ziqing	Huang Kun	Huang Minglong	Yang Shixian
Ye Qisun	Ge Tingsui	Yu Hongzheng	Zhao Zhongyao
Zhao Chengu	Lu Jiaxi	Qian Sanqiang	Qian Weichang
Qian Linzhao	Yan Jici	Su Buqing	Rao Yutai

(2) Department of Biology and Geoscience

Ding Ying	Yin Zanxun	Wang Jiaji	Wang Yinglai
Tian Qijun	Wu Xianwen	Zhu Xi	He Zuolin

Wu Yingkai	Wu Zhengyi	Li Siguang	Li Lianjie
Li Qingkui	Li Jitong	Shen Qizhen	Bei Shizhang
Zhou Zezhao	Meng Xianmin	Cheng Dan'an	Lin Qiaozhi
Lin Rong	Wu Heng	Bing Zhi	Zhu Kezhen
Jin Shanbao	Hou Guangjiong	Hou Defeng	Yu Dafu
Yu Jianzhang	Hu Jingfu	Xia Jianbai	Sun Yunzhu
Yin Hongzhang	Tu Zhi	Tu Changwang	Qin Renchang
Ma Wenzhao	Zhang Wenyou	Zhang Xiaoqian	Zhang Jingyue
Zhang Zhaoqian	Zhang Xijun	Liang Xi	Liang Boqiang
Xu Jie	Chen Wengui	Chen Shixiang	Chen Zhen
Chen Huanyong	Chen Fengtong	Si Xingjian	Tang Peisong
Sheng Tongsheng	Cheng Yuqi	Tong Dizhou	Feng Depei
Feng Zefang	Huang Jiqing	Huang Bingwei	Huang Jiasi
Yang Weiyi	Yang Zhongjian	Ye Juquan	Pei Wenzhong
Zhao Jiuzhang	Zhao Hongzhang	Liu Chengzhao	Liu Chongle
Le Senxun	Pan Shu	Cai Banghua	Cai Qiao
Deng Shuqun	Zheng Wanjun	Xiao Longyou	Zhu Futang
Qian Chongshu	Xie Jiarong	Zhong Huilan	Dai Fanglan
Dai Songen	Wei Xi	Luo Zongluo	Gu Gongxu

II

The following is the name list of more scientists invited by the Chinese Academy of Sciences (in the fields of the natural sciences) to be members of Academia Sinica Departments; the name list being recommended at the meetings of the Academia Sinica Departments Committee and approved at the 12th meeting of the Standing Committee of the Administrative Affairs of Academia Sinica held in May 29, 1957:

(1) Department of Mathematics, Physics and Chemistry

Wu Wenjun	Guo Yonghuai	Qian Xuesen	Wang Dezhao
Zhang Wenyu	Zhang Zongsui	Cai Liusheng	

(2) Department of Biology

Wang Shanyuan	Zhang Xiangtong	Feng Lanzhou	Tang Feifan
Liu Sichi			

(3) Department of Technical Sciences

Wang Daheng	Wang Zhixi	Shi Zhiren	Zhu Wuhua
Wu Xuelin	Li Wencai	Li Qiang	Li Guohao
Li Xun	Wang Huzhen	Zhou Ren	Zhou Zhihong
Meng Zhaoying	Shao Xianghua	Hou Xianglin	Hou Debang
Mao Yisheng	Sun Dehe	Ma Dayou	Zhang Dayu
Zhang Guangdou	Zhang Wei	Zhang Deqing	Liang Sicheng
Zhang Mingtao	Tao Hengxian	Cheng Xiaogang	Huang Wenxi
Yang Tingbao	Ye Zhupei	Lei Tianjue	Jin Shuliang
Zhao Feike	Liu Xian	Liu Dunzhen	Cai Fangyin
Chu Yinghuang	Qian Lingxi	Qian Zhidao	Yan Kai

(3) Department of Geosciences

Wang Zhuquan   Feng Jinglan   Fu Chengyi

(4) Department of Technical Sciences

Wu Zhonghua   Zhao Zongyu   Wang Juqian

III

The defunct members of Academia Sinica Departments before December of 1979:

Wu Youxun	Li Fangxun	Zhang Zongsui	Zhuang Changgong
Xu Baolu	Chen Jiangong	Fu Ying	Yun Ziqiang
Zeng Zhaolun	Huang Minglong	Ye Qisun	Guo Yonghai
Yu Hong Zheng	Zhao Chenggu	Rao Yutai	Ding Ying
Wang Zhuquan	Wang Jiaji	Tian Qijun	Zhu Xi
He Zuolin	Li Sunguang	Li Jitong	Meng Xianmin
Cheng Dan'an	Bing Zhi	Zhu Kezhen	Hu Jingfu
Xia Jianbai	Sun Yunzhu	Tu Zhi	Tu Changwang
Ma Wenzhao	Zhang Jingyue	Zhang Zhaoqian	Liang Xi
Liang Bo Qiang	Chen Wengui	Chen Zhen	Chen Huanyong
Si Xingjian	Tong Dizhou	Feng Lanzhou	Feng Zefang
Feng Jinglan	Tang Feifan	Yang Weiyi	Yang Zhongjian
Zhao Jiuzhang	Liu Chengzhao	Liu Chongle	Deng Shuqun
Xiao Longyou	Qian Chongshu	Xie Jiarong	Dai Fanglan
Luo Zongluo	Shi Zhiren	Zhou Ren	Wang Juqian
Hou Debang	Zhang Deqing	Liang Sicheng	Cheng Xiaogang
Ye Zhupei	Jin Shu Liang	Zhao Feike	Liu Xianzhou
Liu Dunzhen	Cai Fangyin		

SECTION IV

LIST OF PAST SCIENTISTS

The following persons, all of whom died recently, were widely recognised for their scientific contributions to the Republic of China.

Bing Zhi (1886–1965): Alias Nongshan. Native of Kaifeng, Henan Province. Zoologist. Member of Academia Sinica Department of Biology. Before liberation, Professor at Nanjing Higher Normal School, Southeastern University, National Central University, Xiamen (Amoy) University and Fudan University. He, together with Hu Xiansu, founded the Institute of Biology, Science Society of China and Jingsheng Biological Survey, both contributing to the development of Chinese botany and Zoology. After liberation, Senior Researcher of Institute of Aquatic Biology, and Institute of Zoology, Academia Sinica. He was a specialist in systematic zoology, zoomorphology, zoophysiology, entomology and palaeozoology, especially on zootomy. His main works: *On the Morphology of the Carp, Internal Anatomy of the Black Finless Porpoise, Brain of the Tiger,* etc.

Cai Fangyin (1901–1964): Alias Mengqu. Native of Nanchang, Jiangxi Province. Architect and specialist in structural mechanics. Member of Academia Sinica Department of Technical Sciences. He graduated from MIT in USA. After returning from abroad in 1930, he became Professor at Northeastern College of Technology, Qinghua University; Dean of Technological College of Zhongzheng University; Professor and department head at Southwestern Union University. After liberation, Professor at Nanchang University of Jiangxi Province; Dean of College of Technology, First Deputy Director of the Committee of University Administrative Affairs; Chief Engineer and Vice-President of the Academy of Architectural Science of the Ministry of Architectural Engineering;

Member of the Third National Committee of the Chinese People's Political Consultative Conference. He spent all his time studying and teaching structural mechanics, achieving significant results in the design of steel structures. His main works were *General Theory of Structures* published in 3 volumes, *Theory and Design of Plates and Beams, Analysis of Rigid Frames with Various Sections* and its *Further Discussions, General Tables of Constants of Variable Beams, Modular Frames Hinged with Wedged Type Rods,* etc.

Chen Huanyong (1890—1971): Alias Wennong. Native of Xinhui, Guangdong Province. Plant systematist. Member of Academia Sinica Department of Biology. Before liberation, Professor at Jinling University, Southeastern University and Zhongshan University; Head and Director of Institute of Agriculture, Forestry and Botany, Zhongshan University; Director of Institute of Commercial Plants, Guangxi University. Since 1954, Director of the South China Institute of Botany, Academia Sinica. Taught and studied plant taxonomy all his life. In 1928 he founded the Institute of Agriculture, Forestry and Botany, Zhongshan University. He investigated, collected and studied the plants of the South China region, accumulating large amounts of precious plant specimens and data. In 1930 he started the botany-oriented journal *Zhongshan Specialized Publication*. In his last years, he was in charge of the compilation of *Flora of Hainan*, and, in association with Qian Chongshu, that of *Chinese Flora*. His main works were: *Chinese Commercial Trees, Cathaya Argyrophylla — An Endemic Coniferous Plant in China* and *New Genus and New Varieties of the Chinese Magnolia Family.*

Chen Jiangong (1893—1971): Native of Shaoxing, Zhejiang Province. Mathematician. Member of Academia Sinica Department of Mathematics, Physics and Chemistry. Before liberation, Professor at Zhejiang University. He wrote textbooks on mathematics in Chinese in the early part of the century. After liberation, he successively held the posts of Professor at Fudan University and Vice-President of Hangzhou University. He spent long years studying and teaching mathematics, solving a series of theoretical problems on the theory of functions, especially the theory of series of orthogonal functions, the theory of trigonometric series, the theory of schlicht functions and the theory of functional approximation. His main works included: *Theory of Trigonometric Series, Sum of Series of Orthogonal Functions* and *Theory of Functions of a Real Variable.*

Chen Rong (1888—1971): Alias Zongyi. Native of Anji, Zhejiang Province. Forestry specialist. Before liberation, Professor at Department of Forestry, Jinling University. After liberation, Director of Institute of Forestry Science, Chinese Academy of Forestry Science; Vice-President and then Acting President of the Chinese Society of Forestry. He taught and studied forestry all his life, making contributions to the establishment and development of Chinese arboriculture and silviculture. His works included: *Taxonomy of Chinese Trees, Summary of Silviculture, Special Topics on Silviculture, Plant Geography of Chinese Forests, Historical Data on Chinese Forests,* etc.

Chen Xintao (1904—1977): Native of Gutian, Fujian Province. Parasitologist. Before liberation, Professor at Lingnan University; Director of Laboratory for Hygiene, Jiangxi Province. After Liberation, Professor at South China Medical College and Zhongshan Medical College; Director of Guangdong Institute for the Prevention and Treatment of parasitic diseases and Guangdong Institute of Tropical Diseases. He joined the CPC in 1958. He made important contributions to the discovery and identification of schistosoma, especially the Chinese paragonimus, the eradication of schistosomiasis and the prevention and treatment of mites. Main works: *Medical Parasitology, Paragonimus in Yile Village, Review of Paragonimidae from the Study of Paragonimus Westermani,* etc. During his remaining years, he was chief editor of *Chinese Schistosoma.*

Chen Zhen (1894—1957): Aliases Xishan and Xiesan. Native of Yanshan, Jiangxi Province. Biologist and Member of Academia Sinica Department of Biology. Before liberation, Professor at Southeastern University, National Central University, Qinghua University and Beijing University. After liberation, Director of Institute of Zoology, Academia Sinica; President of the Chinese Society of Zoology. His study interests centred mainly on genetics, evolution and the variation of the goldfish. He was also interested in research on the social behavior of animals and the history of biology. His important works included: *Domestication and Variation of Goldfish, History of the Domestification of Goldfish and the Factors Affecting Variety Formation* and *The Influence of Ant Society upon Its Nesting.*

Cheng Maolan (1905—1978): Native of Boye, Hebei Province. Specialist of railway mechanical engineering. Member of Academia Sinica Department of Technical Sciences. He was engaged in railway reconstructional work from 1919. Before liberation, he was assigned as the Chief Inspector of railway engineering and Chief Engineer of the Middle-East Railway, Tianjin-Pukou Railway, Qingdao-Jinan Railway, Beijing-Shenyang Railway, Zhuzhou Railway etc. After 1947, he turned to education, and was appointed President of Jiaotong University in Shanghai; Professor at National Zhejiang University. After liberation, he was Representative at the First, Second and Third National People's Congress as well as Vice-President of Jiaotong University. He was a famous specialist and a scholar who pioneered railway mechanical engineering in China. He made contributions to railway construction and education in China.

Cheng Menxue (1902—1972): Native of Wuyuan Jiangxi Province. Specialist in traditional Chinese medicine. He learned medicine from Ding Ganren. He practised medicine and taught traditional Chinese medicine for many years. After liberation, Dean of Shanghai College of Traditional Chinese Medicine; President of the Shanghai Society of Traditional Chinese Medicine; Representative at the Second and Third National People's Congress; Member of the Scientific Committee of the Ministry of Public Health. He was a specialist in the theory of Evial Influences (Shanghan) and the theory of Wen-disease, and was esteemed for his rich clinical experience and devotion to teaching traditional Chinese medicine.

Cheng Xiaogang (1892—1977): Native of Yihuang, Jiangxi Province. Astrophysicist. In his early years, he studied in France while doing part-time jobs. He worked at the French Upper Porovans Observatory for a long time, focusing his study on star light spectra, which carried information of explosive variable stars (novae, recurrent novae and symbiotic stars); he studied the association of radiated rays with different physical conditions and by studying star light spectra, he could visualize the physical conditions and the transitions of these celestial bodies. Using the same method, he studied the luminescence phenomena of the atmosphere (aurora and night heavenly light) to determine the physical conditions of the altoatmosphere. He returned to China in 1957. He was in charge of the establishment of the Beijing Observatory, and made contributions to the development of modern astrophysics in China. He was Director of the Beijing Observatory, Academia Sinica; Representative at the Second and Third National People's Congress, and Member of the Fifth National Committee of the Chinese People's Political Consultative Conference.

Dai Fanglan (1893–1973): Alias Guanting. Native of Jiangling, Hebei Province. Mycologist. Member of Academia Sinica Department of Biology. Before liberation, Professor at Guangdong College of Agriculture, Southeastern University, Jinling University and Qinghua University. After liberation, Professor at Beijing Agricultural University; Director of the Institute of Applied Mycology, Academia Sinica; Director of the Institute of Micro-organisms, Academia Sinica. He joined the CPC in 1956. In his earlier years, he studied the diseases of rice and fruit trees and their prevention and control. He then studied the classification, forms and genetics of fungi (e.g. peronos peracae, erysiphales, uredinales). He made contributions to the establishment and development of mycology and plant pathology in China. His works included: *A Catalogue on Economic Plant Pathogeny in China, A General Collection of Fungi,* etc.

Dai Wensai (1911–1979): Native of Zhangzhou, Fujian Province. Astronomer, astrophysicist and cosmologist. As a young man he studied at Cambridge University, England. After returning from abroad in 1914, he became Senior Researcher at the Institute of Astronomy of the National Central Academy of Sciences; Professor at Yanjing University. After liberation, Professor and Head of Department of Astronomy at Beijing and Nanjing Universities; Vice-President of the Chinese Society of Astronomy; also a leader of the Astronomy Section of the State Scientific and Technological Commission. He joined the CPC at an advanced age. He was engaged in research on astrophysics and cosmogony, proposed his theory on the origin of the solar system, describing comprehensively, systematically and interconnectedly, the structure and dynamics of the solar system and the origins of various celestial bodies. He spent many years in education, and organized the writing of several kinds of textbooks. He stressed the importance of popularizing science, and wrote many articles on popular science. His main works were: *Astronomy of Stars, Evolution of Celestial Bodies, Evolution of the Solar System, An Analytical Study of the Light Scale of the Special Stars, Distributions of Mass and Angular Momentum of Galaxies,* etc. He was also chief author of *A Textbook of Astronomy, Methods of Astrophysics, Dictionary of Astronomy, An English-Chinese Glossary of Astronomy,* etc.

Deng Shuqun (1902–1970): Another name Zimu. Native of Fuzhou, Fujian Province. Biologist and mycologist. Member of Academia Sinica Department of Biology. Before liberation, Professor at Lingnan University, Jinling University and National Central University and Senior Research Scientist at National Central Academy of Sciences. After liberation, Vice-President of Shenyang College of Agriculture and Northeastern College of Agriculture; Vice-Director of Institute of Applied Mycology, Academia Sinica and Institute of Microbiology, Academia Sinica. He joined the CPC in 1956. In his early years of research, he studied forestry and plant pathology and then the classification of fungi and slime moulds. He was an authority on the classification of higher fungi. His major works included: *Chinese Higher Fungi, Chinese Fungi.*

Deng Zhiyi (1888–1957): Native of Dongguan, Guangdong Province. Agronomist and soil specialist. Before liberation, Professor at Guangdong University, Zhongshan University and concurrently, President of the Institute of Agriculture; Director of Guangdong Institute of Soil Surveying and Director of Institute of Soil, Zhongshan University. After liberation, Senior Research Scientist of Department of Soil and Agricultural Chemistry, North China Institute of Agricultural Sciences. He was engaged in agricultural education and soil research all his life. His works included: *Soil Science – A Textbook, General Aspects of Guangdong Soil and Division of Agricultural Utilization* and *On Soil Identification for Agricultural Production in the Chinese Ancient Era* and *Explorations on laws of Land Utilization.*

Ding Ying (1888–1964): Alias Zhuming. A native of Gaozhou, Guangdong. Agronomist, specializing in rice breeding and growing. Member of Academia Sinica Department of Biology. Before liberation, Professor at Agricultural College, Zhongshan University. After liberation, Dean of South China College of Agriculture, President of the Chinese Academy of Agricultural Sciences etc. Joined the Communist Party of China in 1956. Devoted all his life to the study of rice breeding and growing. Discovered wild rice in the Xiniuwei marsh in the eastern suburb of Guangzhou in 1926, and published the paper *Guangdong Wild Rice and the New Varieties Bred from It* in 1933, which demonstrated that our country was an origin of cultivated rice. Later, from an ecological viewpoint, he made a more systematic study of the origin and evolution of rice, its classification, regional division, and the techniques used for systematically selecting and cultivating farm varieties. Taking into account the specific conditions of rice growing, he selected and bred a batch of improved rice varieties, with the result that rice yield was increased and its quality improved. In his last years, he was chief writer of the book *Cultivation of Chinese Rice,* and in charge of the Laboratory of Rice Ecology. His works included: *The Ecotypes of Chinese Rice Varieties and Their Relation to the Development of Production, The Origin and Evolution of Chinese Cultivated Rice,* and *The Regional Division of Rice Areas of Sinica.*

Feng Lanzhou (1903–1972): Native of Linqu, Shandong Province. Parasitologist and medical entomologist; Member of Academia Sinica Department of Biology. Before liberation, Professor of parasitology at

PUMC (Peking Union Medical College); Professor and Head of Department of Parasitology, College of Medicine, Beijing University. After liberation, Professor and Head of Department of Parasitology, China Union Medical College; Director of Institute of Parasitology, Chinese Academy of Medical Sciences; concurrently Professor and Chief of Faculty of Parasitology, China Medical University. He spent all his life studying and teaching medical parasitology. He was the first one to discover the main species of mosquitoes responsible for transmission of malaria and filariasis in China. He investigated and tentatively systematized the distribution of Chinese mosquitoes. He also made a thorough study on the influence of the Chinese sand fly upon the spread of kala-azar. His research provided reliable theoretical basis for large-scale prevention and cure of the five major parasitic diseases in China, thus leading to their rapid control. In addition, he carefully studied the vermifuge action of pumpkin seeds and betel palm — traditional Chinese medicines — and made contributions to the study of traditional Chinese medicine. Main works: *Distribution of Filariasis and Its Mode of Spread in China, Study of Malaria and Its Dissemination in Xiamen, The Intermediate Host of Malayan Microfilariae in Zhejiang Province, China, Nomenclature of Mosquitoes in China, An Experimental Study of Kala-Azar in Hamsters Infected From Dogs by Sand Flies, Study on the Combined Use of Pumpkin Seeds and Betel Palm in the Treatment of Tape-Worm Infection, Further Studies of Malayan Filaria Transmitted by Chinese Anopheles*. He was also chief editor of *Collected Papers on the Description of Chinese Mosquitoes* and author of *Parasitology*.

Feng Zefang (1899—1959): Native of Yiwu County, Zhejiang Province. Agronomist and cotton specialist. Member of Academia Sinica Department of Biology. Before liberation, Professor at National Central University and concurrently, President of College of Agriculture. After liberation, Professor at Nanjing College of Agriculture; Senior Researcher of Institute of Cotton and concurrently, the President of the Chinese Academy of Agricultural Sciences. He was engaged in research on cotton throughout his career, specializing in the morphology, classification and genetics of Asian cotton and the genetics and cytology of the hybrid of Asian and American cotton. He actively advocated that the varieties of Stonevele and Delfos cotton should be grown in the Yellow River Valley and the Changjiang River Valley, thus making important contributions to developing cotton cultivation in China. His views that China was divided into five big cotton regions are widely accepted even today. His most important papers were published in a book entitled *Mr. Feng Zefang's Selected Papers on Cotton*. His other works were *Chinese Cotton, Foreign Cotton Suitable for Cultivation in China*.

Fu Lianzhang (1894—1968): Native of Changting, Fujian Province. Famous physician; President of the Chinese Medical Society. He engaged in revolutionary work in 1927, joining the Chinese Workers' and Peasants' Red Army in 1933; he participated in the Long March in 1934, and joined the CPC in 1938. He was Director of the Red Hospital of the Central Soviet Government; Director of the Central Hospital in Yanan and the Shanxi-Gansu-Ningxia Border Hospital; Vice-Minister of Hygiene of the

Central Revolutionary Military Commission; Vice-Minister of Public Health of the People's Republic of China; Vice-Minister of Hygiene of the General Logistics Department of the PLA; Member of the National Standing Committee of the Chinese People's Political Consultative Conference. He made important contributions to the cause of hygiene in China.

Gu Zhenchao (1920—1976): Native of Shanghai. Meteorologist and atmospheric physicist. He began studying in Sweden in 1947. He returned to China in 1950 to join in the socialist construction, and became a member of the Chinese Communist Party. He was successively Associate Senior Researcher, Senior Researcher and Head of Laboratory of the Institute of Geophysics, Academia Sinica; Director of Institute of Atmospheric Physics, Academia Sinica; Representative at the Third National People's Congress; Member of the Atmospheric Science Committee of the United Nations World Weather Organization etc. During the first years after liberation he worked hard to develop weather forecasting in China. He won the national prize in science for his article (in collaboration with Ye Duzheng) on *The Influence of the Xizang Plateau on the Atmospheric Circulation of Eastern Asia and the Weather of China*. Besides, he had a wide range of scientific research to his credit, covering numerical weather forecasts, physics of clouds, artificial weather control over small areas, physics of thunder, radar meteorology, and atmospheric discharge. His creative work contributed much to the establishment of these subjects. He wrote more than one hundred articles on meteorology and atmospheric physics.

Guo Moruo (1892—1978): Native of Leshan, Sichuan Province. The first President of the Academia Sinica. Member of Academia Sinica Department of Philosophy and Social Sciences. He went to Japan to pursue higher education in 1914, majoring in medicine. After returning from Japan, he turned to literature. He accepted Marxism-Leninism after 1924, and pioneered revolutionary literature. He joined the Nanchang uprising in 1927. While taking part in revolutionary activities, he remained a writer and scientist. He was President of the Academia Sinica for 29 years, from its establishment in 1949 (the year of liberation) to his death in 1978. He made great contributions to the development of Chinese science. Honorary membership was conferred on him by the Hungarian Academy of Sciences and the Bulgarian Academy of Sciences in 1953, membership of the Polish Academy of Sciences in 1954, and the Honorary membership of the Rumanian Academy of Sciences in 1975. In 1958, he was elected as member of the Soviet Union Academy of Sciences. He was a learned, talented, and highly productive writer, whose works covered literature, drama, history, archaeology, classical linguistics, including the *Guo Moruo Collected Writings*, etc.

He Zuolin (1900—1967): Native of Li County, Hebei Province. Geologist specializing in mineral rock. Member of Academia Sinica Department of Geoscience. Before liberation, Professor at Beijing University and Shandong University; Senior Research geologist of National Central Academy of Sciences. After liberation, Head of Department of Geology and Mineralogy and Dean of Shandong University; Senior

Researcher at Institute of Geology, Academia Sinica. He had a profound knowledge of crystallography, mineralogy, crystal optics and petrology. In his early years, he studied petrofabric and was the first to apply X-rays to petrofabric, and to introduce and apply Western research methods and techniques of optical mineralogy in China. Before liberation, he was the first to discover rare-earth metals and rare-earth minerals in the ore of Baiyunebo Baotou; he was also engaged in investigation and research on petrology, for example, research on granite diorite of Fangshan, Beijing. In his remaining years, he designed "The Box of Variable Temperature", and using a universal stage, made an enantiotropic measurement of mineral refractive index, and conducted a study of X-ray petrofabric. His chief works were: *Optical Mineralogy, Description of the Operation of Universal Stage, Application of Projection of Stereogram to Geological Science, Texture of Crystalline Body, A Guide to Transparent Mineral Identification in Thin Slices,* etc.

Hong Shilu (1894—1955): Native of Leqing, Zhejiang Province. Parasitologist. Before Liberation, Professor of pathology and parasitology in the Medical College of Beijing University; Head of Institute of Tropical Diseases in Hangzhou. After liberation, Director of the Zhejiang People's Experimental Institute of Hygiene; Chief of the Administration of Hygiene of Zhejiang Province; Dean of Zhejiang Medical College. He joined the CPC in 1954. He made contributions to research on parasitology and initiated the basic film colouring method and a quantitative method on Ankylostomiasis eggs, as well as to the morphology of Fasciolopis. His works were *General Discussions on Pathology, Monograph on Pathology, Ankylostomiasis and Trichostrongylus,* etc.

Hou Debang (1890—1974): Native of Minhou, Fujian Province. Chemist. Member of Academia Sinica Department of Technical Sciences. Before liberation, Chief Engineer and concurrently, Manager of Yongli Alkali Factory, Tanggu, and Yongli Ammonium Sulphate Factory, Nanjing; General Manager of Yongli Chemical Co. After liberation, Vice-Chairman of Association of Science and Technology of the People's Republic of China; Vice-Minister of Ministry of Chemical Industry; President of Chinese Society of Chemical Industry etc. He joined the CPC in 1957. In 1939, he was the first to suggest and adopt the continuous process of the combined alkali-manufacturing method and made a significant contribution to the pure alkali and nitrogen fertilizer industry. After liberation, he continued his work for the development of the Chinese chemical industry. Main works: *Manufacture of Alkali, Alkali Manufacturing,* and *Technology of Alkali Manufacturing.*

Hu Xiansu (1894—1968): Alias Buzeng. Native of Xinjian, Jiangxi Province, Biologist. Before liberation, Professor at Nanjing Higher Normal School, Southeastern University, Beijing University, Beijing Normal University; President of Zhongzheng University; examiner and academician of the National Central Academy of Sciences. With Bing Zhi (see above) he established the Institute of Biology of the Science Society of China and Jingsheng Biological Survey; and set up the Botanical Gardens of Lu Mountain, facilitating the study of animal and plant taxonomy in China. After liberation, Senior Researcher of Institute of Botany, Academia

Sinica. He studied phytotaxonomy, paleobotany and economic botany, published more than a hundred articles on the new genus and varieties of Strobodides, Sinojackia xylocarpa, Quince etc. and suggested the theory that angiosperm originated from a classification system of polygenesis. His chief works were *Chinese Flora* (with co-author Chen Huanyong), *Picture Collection of Chinese Pteridophyte* (with co-author Qin Renchang), *The Economic Botany, Handbook of Commercial Plants, Simple Classification of Plants,* etc.

Hu Zhengxiang (1896—1968): Native of Wuxi, Jiangsu Province Pathologist and educationist in medical science. He was engaged in pathological teaching and research work in the Peking Union Medical College (PUMC). After liberation, Professor of pathology; Chairman of the Chinese Society of Pathology; Vice-President of the Chinese Academy of Medical Science. He was extremely vigorous in his research and made contributions to the training of the specialized personnel of pathology and to the development of medical science of China. He wrote more than 40 articles on lymphocytic morphology, Kala-Azar, arteriosclerosis, liver disease and clinical oncological pathology etc. Chief author of *Pathology* (1951), with source materials mainly from China, which is widely read among medical circles.

Huang Minglong (1898—1979): Native of Yangzhou, Jiangsu Province. Organic chemist and member of Academia Sinica Department of Mathematics, Physics and Chemistry. He studied in Switzerland and Germany when he was young. He was Professor and Senior Researcher at Shanghai Tongde Medical College, Zhejiang Medical School, Institute of Chemistry of National Central Academy of Sciences, Southwestern Union University. He went abroad three times, working as Professor and Senior Researcher at Vitsburg University in Germany, Institute of Schering Pharmaceutical Factory in Germany, Institute of Biochemistry of the Middlesex Medical College in England, Harvard University in USA, and Mike Pharmaceutical Factory in USA. After returning from abroad in 1952, he was Senior Researcher, Chairman and Honorable Chairman of the Academic Committee of the Shanghai Institute of Organic Chemistry, Academia Sinica; Honorable editor of the international magazine *Quadron;* Deputy Head of the Birth Control Specialized Group of the State Scientific and Technical Commission; Director of the Chinese Society of Chemistry; Vice-President of the Chinese Society of Pharmacy; Representative of the Third National People's Congress and Member of the Second, Third and Fifth National Committee of the Chinese People's Political Consultative Conference. He studied plant chemistry in his earlier years, as well as the effective composition of Corydalis yanhusuo and the root of Chinese wild ginger (Asarum Sirboldii). Later, he studied the chemistry of steroids and discovered, with others, the reaction of "dienonic phenol" in steroids. While studying the stereo chemistry of santonin, he discovered the circular transformation of the four various stereoisomers of the deteriorated santonin under the influence of acids and alkalis. This new discovery had a theoretical significance in the field of natural organic chemistry at that time. The Huang Minglong reducing method, so named to remember his work in modifying the Kishner-Wolff reducing method, has been extensively used throughout the world. He was also the founder of the pharmaceutical industry for the production of steroidal hormone. He directed the study and production of Megestroli Acetas which was the first oral contraceptive in the world, and he made important contributions to birth control in China. He wrote nearly forty monographs and reviews, and another eighty papers in collaboration with his colleagues.

Jin Shuliang (1899—1964): Alias Donghua. Native of Xushui, Hebei Province. Metallurgist and specialist in iron smelting. Member of Academia Sinica Department of Technical Sciences. He graduated from Beiyang University. He managed the construction of the steelworks in Liuhegou, Dadukou, Weiyuan, etc. He went to Germany to study the steel industries there in 1937 and 1938. After liberation, he was Chief Engineer of the Benxi Steelworks; Dean of Northeastern College of Technology; Vice-President of the Northeastern Branch of Academia Sinica; Representative at the National People's Congress; Member of the National Standing Committee of the Chinese People's Political Consultative Conference. He joined the CPC in 1958. He was conferred the thesis Prize by the Chinese Engineers Society in 1943 for his improvement of the equipment for supplying material at the top of blast furnaces, and obtained several invention patents. After liberation, he took part in the reconstruction of the Anshan Steelworks and the Benxi Steelworks. Later he became President of Northeastern College of Technology, but often visited factories studying problems arising in production, for example, crystallization in the Benxi blast furnace, the "bag effect", the theory behind the formation of slag, and the smelting of V-Tizmagnetite ores in blast furnaces. He gave lessons to both college students and postgraduates and wrote *Modern Iron Smelting,* a textbook which documented iron production in China by blast furnaces.

Li Jitong (1897—1961): Native of Xinghua, Jiangsu Province. Botanist. Member of Academia Sinica Department of Biology. Before liberation, Professor at Jinling University, Nankai University, Qinghua University etc. After liberation, Professor at Beijing University and Inner Mongolia University. He studied forestry in his earlier years of research, and afterwards, he devoted himself to research on plant physiology, plant ecology and phytocoenology. He studied photosynthesis, the attractive forces of plant cells and plant hormones, and the cultivation of the ginkgo embryo in test tubes. He also conducted a survey of the vegetation in North China and Yunnan. He made contributions to the founding and development of plant physiology, plant ecology and phytocoenology in China. His main works were *Research on the Regeneration of the Physiological Apex, Development of the Ginkgo Embryo,* etc.

Li Siguang (1889—1971): Alias Zhongkui. Native of Huanggang, Hubei Province. Geologist. Founder of geological mechanics. Vice-President of the Chinese Academy of Sciences. Member of Academia Sinica Department of Geoscience. In his early years, he joined Tongmenghui (the Allied Party) and took part in the Xinhai (1911) Revolution. Afterwards, he was engaged in research and the teaching of palaeontology, glaciology and geological mechanics. After liberation, he was made

Minister for Geology, He was Director of the Institute of Palaeontology, Academia Sinica; Chairman of the Chinese Association of Science and Technology; Representative at the First, Second and Third National People's Congress; Vice-Chairman of the Second, Third and Fourth National Committee of the Chinese People's Political Consultative Conference; Member of the Ninth Central Committee of the CPC. One of his most important contributions to the theory of geology was the founding of geological mechanics. From the viewpoint of mechanics, he studied the phenomenon of crustal movement, explored the laws of crustal movement and mineral distribution, regarding the trace of every kind of structure as the product of earth stress activity, and established the "structural system" — a basic concept of geological mechanics. After analysing the characteristics of the geological structure in the eastern part of China, he forecast bright prospects for petroleum exploration in the three strips of settlement of the "Neocathayaian structural system", and made important contributions to petroleum exploitation in China. In the field of seismology and geology, he emphasized that observation of earth stress should be made on the basis of research on the activity of geological structures. This enlightening view opened up a new path to effective earthquake prediction. His works included: *Main Causes of Changes of the Earth's Surface Pattern, Fusulina in the Northern Part of China, Chinese Geology, Lushan in the Ice Age, Summary of Geological Mechanics, Earthquakes and Geology* and selected papers on Astronomy, Geology and Palaeontology.

Li Tao (1901–1959): Alias Yousong. Native of Fangshan (now attached to Beijing). Medical historian. He graduated from Beijing Professional School of Medicine (now Beijing Medical College). He was Professor at Beijing Medical College, and taught and studied the medical history. Major works include *Outline History of Medicine,* etc.

Liang Sicheng (1901–1972): Native of Xinhui, Guangdong Province. Architect. Member of Academia Sinica Department of Technical Sciences. Before liberation, he worked at the Chinese Academic Society of Architecture, studying ancient Chinese architecture, investigating, surveying and redrawing the ancient architectural works of important historical value, and sifting through ancient architectural documents. He was Professor and Head of the Architectural Department of Northeastern University and Qinghua University; Member of the Design Committee of the UN Building. After liberation, he was Chief of Department of Architecture of Qinghua University; Head of the Laboratory of Architectural History and Theory of the Chinese Academy of Architecture; Vice-President of the Chinese Society of Architecture; Representative at the First, Second and Third National People's Congress; Member of the Standing Committee of the Third National People's Congress; Member of the Standing Committee of the Chinese People's Political Consultative Conference. He joined the CPC in 1959. For many years he taught at universities and studied Chinese architectural history. He was one of the designers of the PRC emblem, and the Monument to the People's Heroes in Beijing. He made contributions to scientific research work on Chinese architecture. His chief works: *Examples of Architecture* *in the Qing Dynasty, History of Chinese Architecture,* and dozens of reports and papers on ancient architecture.

Liang Xi (1883—1958): Alias Shuwu. Native of Wuxing, Zhejiang Province. Forestry specialist. Member of Academia Sinica Department of Biology. He graduated from Zhejiang Military School and continued his study first in Japan, where he joined Tongmenghui (the Allied Party), and then in Germany. Before liberation, he was Professor of forestry at National Zhejiang University and National Central University. After Liberation, he was Chief Member of the Administrative Affairs Committee on Nanjing University; Vice-Chairman of the China Scientific and Technological Association; Minister of Forestry; Representative at the National People's Congress and Member of the Standing Committee of the Chinese People's Political Consultative Conference; Vice-Chairman of the Jiusan (September the Third) Society. He made contributions to the establishment of the subject of woods and chemistry of forestry production. His works: *Study of Woods,* etc.

Lin Ji (1897—1951): Alias Baiyuan. Native of Minhou, Fujian Province. Specialist in forensic medicine. He was a founder of modern forensic medicine in China. After graduating from Beijing Medical School, he continued his study of forensic medicine in Germany. After returning from abraod, he established the Institute of Forensic Medicine. He was successively Professor at Beijing University, National Central University and Nanjing University. He dedicated all his life to research on forensic medicine, personally dealing with doubtful cases and training coroners at different levels, and made contributions to the development of forensic medicine. His main works included: *Recent Developments in Identifying Methods in Forensic Medicine, Concise Forensic Medicine for Medical Doctors, Examples of Doubtful Cases Identified* and *Progress of Forensic Medicine in the Last Twenty Years.*

Liu Chengzhao (1900—1976): Native of Taian, Shandong Province. Zoologist specializing in amphibians and reptiles. Member of Academia Sinica Department of Biology. Before liberation, Professor at Yanjing University, Northeastern University, Soochow University, West China Union University etc. After liberation, Head of Department of Biology, Yanjing University; President of West China Union University and Sichuan Medical College. He joined the CPC in 1956. From 1929, he studied the problem of the classification of amphibians and reptiles, using the large number of specimens and data he had personally collected and accumulated. He had many original ideas, especially about Pelobatidae, and took great care and pains in the study of fauna in the classification of amphibians in the Hengduan Mountain Range. His main works were *Amphibia in West China* and *Chinese Tailless Amphibia.*

Liu Chongle (1901—1969): Alias Juemin. Native of Fuzhou, Fujian Province. Entomologist. Member of Academia Sinica Department of Biology. Before liberation, Professor at Qinghua University and Beijing Normal University. After liberation, Professor at Qinghua University and Beijing Agricultural University; Director of Institute of Insects, Beijing Agricultural University; Senior Research entomologist of Institute of Insects and Institute of Zoology, Academia Sinica; Vice-President of

Yunnan Branch of the Academia Sinica. He spent long years studying insect taxonomy, insect pathology and resource entomology and did much work to promote the development of these subjects and the training of specialized personnel in these fields. His important works included: *History of Chinese Economic Insects — Ladybugs Family.*

Liu Dunzhen (1897—1968): Alias Shineng. Native of Xinning, Hunan Province. Architectural historian. Member of Academia Sinica Department of Technical Sciences. Before liberation, he worked at the Chinese Academic Society of Architecture, studying Chinese ancient architecture, investigating, surveying and redrawing ancient architectural works of apparent historical value and studying ancient architectural documents. At the same time, he was also an educationist being a Professor, Head of the Department of Architecture, and Dean of College of Technology at the National Central University. After liberation, he successively held the posts of Professor and Head of Department of Architecture, Nanjing University and Nanjing College of Technology, Deputy Head of the Laboratory of Architecture, History and Theory, Academy of Architecture, and concurrently, Head of the Nanjing Branch of this Laboratory. He joined the CPC in 1956. Main works: *Synopsis of the Chinese Vernacular Dwellings, The Classic Gardens in Suzhou,* and dozens of reports and papers on ancient architecture. He was the chief editor of *An Outline of Chinese Architectural History* and the pictorial — *Architecture in Ten Years* (1959) presenting the architectural achievements of China after Liberation.

Liu Shene (1897—1975): Alias Shilin. Native of Mouping, Shandong Province. Botanist and forestry specialist. Before liberation, Senior Researcher and Director of Institute of Botany, Beijing Academy of Sciences; Professor at Beijing University, Zhongfa University, Zhongguo University, Furen University, Yunnan University and Northwestern College of Agriculture. After liberation, Deputy Director of Institute of Forestry and Soil, Academia Sinica. He collected large numbers of plant specimens in the Northwestern region, the Southwestern region, the Southeastern region, the Qinghai-Xizang Plateau and the Northeastern region of China, making a survey of flora and vegetation. His valuable work led to the establishment and development of the systematic study of botany and plant geography. He emphasized that scientific research should serve production, and also made contributions to protecting forest resources, preservation of the ecological equilibrium, and control of desertization. His major works were *Dynamic Geobotany, Plant Historical Geography* and he was chief editor of *Florastic Figures in the Northern Part of China, Florastic Figures of Woody Plants in Northeast China* and *Flora of Herbs in Northeast China.*

Liu Xianzhou (1890—1975): Native of Wan County, Hebei Province. Engineering educationist and mechanical engineering specialist. Member of Academia Sinica Department of Technical Sciences. In 1908, he joined Tongmenghui (the Allied Party). Before liberation, President of Beiyang University; taught at the Preparatory Class for Students Going to France. Attached to Yude Secondary School, Baoding, and at Hebei University; then, Professor at Northeastern University and Qinghua University. After liberation, first Vice-President of Qinghua University; President of the Chinese Society of Mechanical Engineering; President of the Chinese Society of Agricultural Mechanical Engineering; Representative at the First, Second, Third and Fourth National People's Congress. He joined the CPC in 1955. He was a famous educationist, spending all his life in teaching, being the earliest to teach engineering in China. At the same time, he studied the history of mechanical engineering. Since 1946, he had been actively engaged in the popularization of agricultural machinery and in research on the history of Chinese agricultural mechanical engineering, and made an important contribution in these fields. His works: *Principles of Mechanics, Heat Power Engineering* and *English-Chinese Dictionary of Mechanical Engineering* (the earliest specialized textbooks and dictionary ever compiled and written in China). Main works after Liberation included: *Invention History of Chinese Mechanical Engineering* and *Inventions of Chinese Agricultural Machinery.*

Lu Yuanlei (1894–1955): Native of Chuansha (now a county of Shanghai). Specialist in traditional Chinese medicine. Studied Literature first and then turned to learn traditional Chinese medicine from Yun Tieqiao. After liberation, Representative at the First National People's Congress; Chairman of Shanghai Society of Traditional Chinese Medicine. He worked for a long time as a practising doctor and teacher of traditional Chinese medicine. He attached great academic value to classical prescriptions and advocated the fusion of classical and modern knowledge. He wrote *Modern Explanation of the Theory of Evial Influence, Modern Explanation of the Outline of Jin Gui, A Collection of Medical Papers Written by Doctor Lu,* etc.

Luo Qingsheng (1898–1974): Native of Nanhai, Guangdong Province. Expert of veterinary medicine. Since 1924, Professor at Southeastern University; Professor and Dean of College of Agriculture, National Central University. After liberation, Professor and Deputy President of Nanjing College of Agriculture. In 1956, he joined the CPC. He dedicated all his life to the education of veterinary medicine and the prevention and cure of infectious diseases in domestic animals. The research programmes he directed on the prevention and cure of cattle pests, pig pests, duck pests and pig asthma were very successful. His publications: *Infectious Diseases of Livestock, Pig Diseases, Poultry Diseases, Study of Pig Asthma, Study of Duck Pests,* etc.

Luo Zongluo (1898–1978): Native of Huangyan, Zhejiang Province. Plant physiologist. Member of Academia Sinica Department of Biology. Before liberation, Professor at Zhongshan University, Jianan University, National Central University and Zhejiang University; Acting President of Taiwan University; Director of Institute of Botany, National Central Academy of Sciences. After liberation, Director of Shanghai Institute of Plant Physiology, Academia Sinica; President of Chinese Society of Plant Physiology. In the early years of his research career, he studied the effect of the concentration of hydrogen ions on the properties of the colloid of cell protoplast, and then researched on the absorption of ammonium nitrogen and nitrate nitrogen by plants and the effect of various metallic ions on the absorption of ammonium and nitrate ions. He started research on plant tissue culture and micro-elements, plant hormone etc. He also investigated water relationships, stress physiology and radiation physiology etc. In his last years, he devoted himself to the study of plant cytobiology. He stressed the importance of the applications of plant physiology and took part in several investigations of afforestation along the coast of the northern part of Jiangsu Province. He also studied the effect of drought and salinity on plant growth in the northwestern region and the chilling injury of rubber trees in the southern part of China. His papers were published in various specialized journals. He made important contributions to the establishment and development of plant physiology in China.

Ma Wenzhao (1886—1965): Native of Baoding, Hebei Province. Specialist in cytology and histology. Member of Academia Sinica Department of Biology. Before liberation, Professor and President of PUMC (Peking Union Medical College) and Medical College, Beijing University; Professor at Washington University, St. Louis, USA. After liberation, Professor of Beijing Medical College; Member of Scientific Committee, Ministry of Health; Member of National Committee of the Chinese People's Political Consultative Conference. He devoted more than forty years to research on the organelle of cells and lecithin, and did a great deal of work on the morphological changes of mitochondria and Golgi apparatus caused by diseases and changes of cell functions. Author of *Effects of Phosphorus Ester Upon Tissues*, etc.

Ma Rongzhi (1908—1976): Native of Ding County, Hebei Province. Pedologist and soil geographer. Before liberation, Acting Head of Soil Laboratory, Institute of Geological Surveys. After liberation, Director of Institute of Soil, Academia Sinica; Deputy Director of Comprehensive Survey Committee, Academia Sinica. Joined the CPC in 1956. Engaged in soil research all his life. The relatively systematic study he made of soil genesis, soil geographic distribution and the geology of the Quaternary Period, etc., promoted the development of soil science in China. His works included: *Law of Soil Geographic Distribution of China, Chinese Soil Division, Chinese Soil Maps* and *Maps of Chinese Soil Division*.

Pu Fuzhou (1888—1975): Native of Zitong, Sichuan Province. Physician of traditional Chinese Medicine. Vice-President of the Academy of Traditional Chinese Medicine; Member of the Standing Committee of the Chinese Medical Society; Member of the Central Committee of the Chinese Democratic Party of Peasants and Workers. He joined the CPC in 1962. He was engaged in traditional Chinese Medicine all his life and excelled at internal medicine, pediatrics and gynaecology. He had a rich and varied experience in the treatment of acute infectious diseases, such as epidemic meningitis B, pneumonia, and acute infectious hepatitis, and made contributions to popularizing traditional Chinese medicine. He wrote *Medical Records of Pu Fuzhou, Clinical Experience of Pu Fuzhou,* etc.

published in 1916 and 1917, were the earliest Chinese works in systematic botany and plant physiology. The article on *The Preliminary Observation of the Plants of Huang Mountain in Anhui Province* was one of the earliest works in geobotany and fauna. He made a thorough observation and study on the classification of Familia and genus of many Chinese forest plants and the flora and vegetation in certain areas. In his later years, he was co-author of *Chinese Flora* (in association with Chen Huanyong), *Chinese Forest Flora, Draft of the Chinese Vegetational Regionalisation,* etc.

Qian Chongshu (1883–1965): Alias Yunong. Native of Haining, Zhejiang Province. Botanist. Member of Academia Sinica Department of Biology. Before liberation, Professor at the First Nanjing College of Agriculture, Jinling University, Southeastern University, Qinghua University, Xiamen (Amoy) University, Sichuan University, Fudan University etc.; Director of Institute of Biology of the Science Society of China. After liberation, Director of Institute of Botany, Academia Sinica. His articles on *Two Asian Varieties of Binzhou Crowfoot* and *The Special Function of Ba, Sr, Ce for Special Effects on the System of Spirogyra*

Rao Yutai (1891–1968): Alias Shuren, Native of Linchuan, Jiangxi Province. Physicist. Member of Academia Sinica Department of Mathematics, Physics and Chemistry. In his youth, he studied in the USA. After returning to China, he worked as Head of the Department of Physics, Nankai University, Beijing University and Southwestern Union University; Dean of College of Sciences, Beijing University; Member of the Second and Third National Committee of the Chinese People's Political Consultative Conference; Member of the Standing Committee of the Fourth National Committee of CPPCC. From the twenties to the forties, he was engaged in research on atomic spectra and molecular spectra, first in the USA and then in Germany, and was one of the spectroscopists who studied Stark effects in their earlier periods. He achieved notable successes in the field of gaseous conduction and spectroscopy. He spent long years studying and teaching physics. He contributed much to the preparation of physics personnel in China.

Shen Kefei (1898–1972): Native of Sheng County, Zhejiang Province. Expert in surgery and Vice-President of the Chinese Medical Society. Before liberation, surgeon at PUMC; Chief of Surgery Section and Director of Central Hospital, Nanjing; Professor of Surgery at National Shanghai Medical College; Director of Zhongshan Hospital, Shanghai. After liberation, Vice-President of the Academy of Military Medical Sciences of the Chinese People's Liberation Army; Vice-President of Shanghai First Medical College; Member of the Standing Committee of the National Committee on Medical Sciences; Representative at the First, Second and Third National People's Congress. He zealously promoted medical education and made contributions to the training of surgeons and the development of surgery, especially neurosurgery, splenectomy, the treatment of late schistosomiasis, etc. in China. He was the chief editor of China's first *Textbook of Surgery,* and author of *Atlas of Abdominal Operations,* etc.

Shi Zhiren (1897–1972): Native of Leting, Hebei Province. Member of Academia Sinica Department of Technical Sciences. He graduated from Hongkong University and MIT, USA. Before liberation, Chief Engineer of National Railway Mechanical Factory; Manager of National Railway Mechanical Factory. After liberation, Counsellor at Advisors' Office, Ministry of Railways; Vice Minister of Ministry of Railways; Deputy Head of Section of Mechanical Engineering, National Scientific-Technical Commission; Head of Section of Railway, National Scientific-Technical

Commission; President of the Chinese Society of Mechanical Engineering; and Representative of the First, Second and Third National People's Congress.

Si Xingjian (1901—1964): Native of Zhuji, Zhejiang Province. Palaeobotanist. Member of Academia Sinica Department of Geoscience. Before liberation, Professor at Qinghua University, Beijing University, and Senior Researcher at Institute of Geology of National Central Academy of Sciences. After liberation, Director of Institute of Palaeontology and Institute of Geological Palaeontology, Academia Sinica. He studied plant fossils of the Late Mesozoic, Mesozoic and Cenozoic centuries, the classification and evolution of ancient plants, the classification and comparison of stratum layers as well as the geographical distribution of plants. He also directed attention to the study of stratum layers and made important contributions to palaeontology of China. His main works were *Plants in the Mesozoic Era of China, Plant Fossils of Coal Measures in Jixi, West Hubei; Plant Fossils in the Upper Devonian Period of China, The Plant Groups of the Mesozoic Era of the North Shanxi Stretching Layers,* etc.

Sun Benzhong (1897—1968): Native of Wujiang, Jiangsu Province. Chinese silkworm breeder. Before liberation, successively, Professor at National Central University and Zhejiang University; Head of Department of Sericulture, Central Laboratory of Agriculture. After Liberation, Senior Research Chinese silkworm breeder at the Sericultural Institute, Chinese Academy of Agricultural Sciences; President of the Chinese Society of Sericulture. His early research work in France concerning the cell physiology of the digestive organs of silkworms was creative and highly appreciated among academic circles. He succeeded in breeding a strain of bivoltine silkworm with yellow skin, "Zhongnong No. 29" and then, in breeding several other new strains, proposed fine breeding methods, such as double cross hybrid etc., thus making contributions to the breeding of Chinese silkworms.

Sun Yunzhu (1895—1979): Native of Gaoyou, Jiangsu Province. Palaeontologist and stratigrapher. Member of Academia Sinica Department of Geoscience. After liberation, Professor of Department of Geology, Beijing University; Head of Bureau of Education, Ministry of Geology; Vice-President of Academy of Geological Science; Representative of the Third National People's Congress; Member of the Second and Third National Committee of the Chinese People's Political Consultative Conference; Member of the Central Committee of Jiusan Society (September the Third Society). He spent years training Chinese geological personnel. From the forties to the sixties, he proposed the three principles of stratigraphic comparison, transgression in the Chinese Palaeozoic Era and biological zoning, and helped to establish the basic concept of stratigraphic palaeontological and comprehensive research. His book *Fossil Animals of the Cambrian Period in the Northern Part of China* published in 1924 was the first monograph on palaeontology in China.

Tan Xichou (1892—1952): Native of Wuqiao, Hebei Province. Geologist. He was engaged in geological surveys; discovered large numbers of well-preserved dinosaur and fish fossils in Mengyin of Shandong

Province, so that the period of local stratum layers was determined as the Cretaceous period of the Low Mesozoic Era, and the mistake of taking it as the Permo-carboniferous period as suggested by Leeschhoffen, a German, was corrected. He wrote *Outline of Geological History of Sichuan-Xikang, The Geology of Emei Mountain of Sichuan, The Geology of the Coalfields of Zichuan-Boshan in Shandong Province,* etc.

Tang Feifan (1897–1958): Native of Liling, Hunan Province. Microbiologist and one of the discoverers of the trachomatous virus. Member of Academia Sinica Department of Biology. Before liberation, Professor of bacteriology, National Shanghai Medical College. After liberation, Director of Beijing Institute of Biological Products. He was engaged in research on trachomatous pathogen for many years; in cooperation with ophthalmologist Zhang Xiaolou et al. of Tongren Hospital, Beijing, he succeeded in separating and culturing, for the first time, trachomatous pathogen by scraping materials of the conjunctive of trachomatous patients by means of chick embryo innoculation.

Tong Dizhou (1902–1979): Native of Ningbo , Zhejiang Province. Experimental embryologist; one of the main initiators of experimental embryology in China; Chief of Academia Sinica Department of Biology; Vice-President of the Academia Sinica. Before liberation, Professor at National Shandong University, National Central University, Tongji University of Fudan University; Senior Researcher at Institute of Psychology of National Central Academy of Sciences, Cambridge University in Britain and Yale University in USA. After liberation, Professor and Vice-President of Shandong University, Deputy Director of the Institute of Experimental Biology, Academia Sinica; Director of Institute of Oceanology, Academia Sinica and Institute of Zoology, Academia Sinica; Representative at the First, Second, Third, Fourth and Fifth National People's Congress and member of the Standing Committee of the Third, Fourth and Fifth National People's Congress; Vice-Chairman of the Fifth National Committee of the Chinese People's Political Consultative Conference. He joined the CPC in 1978. During his first years of research, he made the initial discoveries in the study of the ovum development of chordate animals, fish and amphibians. At the beginning of the fifties, he made a systematic study of the egg development of amphioxus, providing important evidence for determining the position of amphioxus in zoology and adding much to the theory of experimental embryology. In the late sixties, he formed some valuable original ideas concerning the relation between the nucleus and cytoplasm in the growth of each particular fish and fish eggs splitting and heredity of character. He also made contributions to the prevention and control of harmful oceanic organisms, the artificial cultivation of aquatic animals of commercial value, and new ways to develop the fishery industry. He wrote more than 70 papers and monographs.

Tu Changwang (1906–1962): Native of Hankou, Hubei Province. Meteorologist. Member of Academia Sinica Department of Geoscience. After liberation, Director of State Bureau of Meteorology; Member and Secretary of Secretariat of the First National Committee of the China Scientific and Technological Association; Vice-Chairman of the Fifth

Central Committee of the Jiusan (September the Third) Society. Later, he joined the Chinese Communist Party. He was one of the initiators of research work on long-term weather forecasts in China, and the first to suggest that the study of long-term weather forecasts in China be based on global weather changes, and to propose the study of the center of atmospheric movement, the wave movement of the atmosphere and the relation of ocean circulation to temperature and precipitation. Based on Zhu Kezhen's classification of climate, he further detailed the range of the weather districts in China and made important contributions to the development of Chinese meteorology.

Wang Dong (1906–1957): Alias Binjun. Native of Chongming (now a county of Shanghai). Specialist in science of prairie and livestock nutriology. Before liberation, Professor at Guizhou College of Agriculture and Engineering, Northwestern College of Agriculture and National Central University. After liberation, Professor at Nanjing University and Nanjing College of Agriculture. He spent all his life teaching and studying herbage, grassland management and livestock nutrition etc., and made contributions to the development of the Chinese livestock husbandry and grassland construction. Works: *Outlines of Herbage, Special Topics on Herbage, Science of Grassland Management, Livestock Nutriology,* etc.

Wang Jiaji (1897–1976): Alias Zhongji. Native of Fenqian (now a county of Shanghai). Zoologist. Member at Academia Sinica Department of Biology. Before liberation, Professor of National Central University; Senior Research Zoologist at Institute of Zoology, Science Society of China; Director of Institute of Zoology and Botany and Institute of Zoology, National Central Academy of Sciences. After liberation, Director of Institute of Aquatic Biology, Academia Sinica. In 1960 he joined the CPC. He dedicated all his life to the classification and ecological studies of protozoa and freshwater rotifers and made contributions to the establishment and development of Chinese protozoology. Author of *History of Chinese Freshwater Rotifers.* During his remaining years, he directed the compilation of *History of Chinese Zoology,* research on protozoa and rotifers in the comprehensive survey of Qinghai-Xizang Plateau, and also took care of the compilation of *Microfauna in Waste Water Treatment.*

Wang Jiayin (1911–1976): By name Yinzhi. Native of Yongnian, Hebei. Geologist. Before liberation, taught and did scientific research at Beijing University, Changsha Provisional University, Southwestern Union University and Institute of Geology among others. After liberation, Head of Faculty of Geochemistry, Beijing University. He made a thorough study of minerology and petrology, and had a profound knowledge of stress minerology. Important works: *Igneous Rock, Immersion Method – Identification of Transparent Minerals, Historical Data of Chinese Geology, Summary of Stress Minerals,* etc.

Wang Shou (1896–1972): Native of Qin County, Shanxi Province. Crop breeder and biological statistician. Before liberation, Professor at Jinling University; Dean of Northwestern College of Agriculture. After liberation, Head of Bureau of Food Production, Ministry of Agriculture; Director of Institute of Crop Genetics, Chinese Academy of Agricultural

Sciences; President of Shanxi College of Agriculture. The barley variety with high frost-resistance and rust-resistance bred by him in his earlier years of research was named "Wang's Barley" in the USA. The variety of soybean "332" bred by him has been widely planted in the Changjiang River Valley, and is called "Nanjing Soybean" abroad. In the sixties, improved varieties of soybean were bred which were suitable for cultivation in North China, and thus contributed to crop breeding in China. His major works included: *Statistical Methods of Practical Biology, Biological Statistics, Theory and Practice of Field Testing and Soybean.*

Weng Wenhao (1889—1971): Alias Yongni. Native of Ningbo, Zhejiang Province. Geologist. He studied in Belgium at the end of the Qing Dynasty. After returning to China in 1912, he worked in a geological survey set up by Ding Wenjiang et al. and then was director of the Geological Survey of the Beiyang Government and Acting President of Qinghua University. From 1936, in the Kuomingtang Government, he was the secretary-general of the Executive Yuan, Head of Resource Commission, Head of the Executive Yuan and Secretary-general of the House of the President. After returning from France in March, 1951, he became member of the National Committee of the Chinese People's Political Consultative Conference. He was the first one to suggest the existence of Yanshen Movement in China and its importance in the geological history of China. His works included: *A Brief Dictionary of Chinese Minerals, On Gansu Earthquakes, Earthquakes, Zhui Zhi Ji,* etc.

Wu Xian (1893—1959): Alias Taomin. Native of Fuzhou, Fujian Province. Biochemist Successively, Professor at and Dean of Department of Biochemistry, PUMC; Director of Institute of Nutrition, Central Academy of Experimental Hygiene and concurrently, President of Beijing Branch of Central Academy of Experimental Hygiene; Visiting Scholar and Senior Research Biochemist of Medical College, Columbia University, USA; Visiting Professor of Alabama Medical College; and President of the Chinese Society of Physiology. In the field of protein chemistry, he conducted systematic research on the variation of protein and as early as the thirties, proposed the theory of protein variation. In the field of immune chemistry, he determined the quantitative relation of combinations of antibody with antigen. In the field of blood chemistry, he established some analytical methods which are now still widely used. In the field of nutrilogy, he carried out a series of experiments, and systematically observed the influences of pure vegetative diets and mixed diets on the growth and health of animals over several generations. Author of *Table of Diet Components* (the earliest publication on this subject in China), *Outline of Nutrition,* and *Principle of Physical Biochemistry.*

Wu Youxun (1897—1977): Alias Zhengzhi. Native of Gaoan, Jiangxi Province. Physicist. Vice-President of the Academia Sinica. Member of Academia Sinica Department of Mathematics, Physics and Chemistry. Before liberation, Professor and Dean of Department of Physics, President of College of Sciences, Qinghua University and Southwestern Union University; President of National Central University; President of Chinese Society of Physics. After liberation, Vice-Chairman of the Association of Science and Technology of the People's Republic of China; Member of the

First and Second National Committee of the Chinese People's Political Consultative Conference and Standing member of the Third National Committee of CPPCC; Representative at the First and Second National People's Congress, and Member of the Standing Committee of the Third and Fourth National People's Congress. In the twenties, while doing scientific research at Chicago University, USA, he made an important contribution to the confirmation of the Compton effect. In the early thirties, he continued his study on the scattering of multiatomic gas by X-rays at Qinghua University. After Liberation, he was engaged in organisational activities, contributing much to the rapid development of science in China.

Xiao Longyou (1870–1960): Original name Fangjun; Alias Xiyuan. Native of Santai, Sichuan Province. Expert on traditional Chinese medicine. Member of Academia Sinica Department of Biology. He was given the title of Bagong (an academic degree in the Qing Dynasty) in 1897. He was an enthusiast of traditional Chinese medicine since his childhood. Not long after the founding of the Republic of China, he gave up a government post and began to practise medicine. In 1934, he founded Beijing College of Traditional Chinese Medicine together with Kong Bohua, the famous doctor of traditional Chinese medicine in Beijing. After liberation, he continued to practise medicine. He was successively Representative at the First and Second National People's Congress; Counsellor of the Academy of Traditional Chinese Medicine; Vice-Chairman of the Chinese Medical Society; Member of the Central Research Institute of Culture and History. He was familiar with the classical medical books and skilful in clinical practice. He emphasized that diagnosis and treatment should be based on an overall analysis of the illness and the patient's condition, advocated the diagnosis combining symptoms with signs, and encouraged the integration of traditional Chinese medicine with Western medicine. He contributed much to the development of traditional Chinese medicine in China. After liberation, he treated large numbers of patients, the medical records of whose cases have been bound into several thick books which are to be studied and systematized.

Xie Jiarong (1898–1966): Native of Shanghai. Geologist and specialist in mineral deposits. Member of Academia Sinica Department of Geoscience. Before liberation, Chief and Acting Director of the Beijing Geological Survey; Professor and Department Chief at Qinghua University and Beijing University; Chief of Bureau of Mineral Products Survey. After liberation, Chief Engineer of Ministry of Geology; Vice-Director of the Institute of Geological Mineral Deposits. He successively discovered the phosphorus mines at Fengdai in Anhui Province, the coal field at Bagong Mountain in Huainan, the aluminium and zinc mines at Qixia Mountain in Nanjing, as well as the copper mine at the Silver Plant in Gansu Province. His works covered: general laws governing mineralization, the orientation of discovering mineral deposits and the theory of mineralization of petroleum, natural gas, coal, metals, and non-metallic ores. He devoted his last years to writing *The Mineral Deposits of China*.

Xie Zhiguang (1899–1976): Native of Dongguan, Guangdong Province. Clinicoroentgenologist and one of the founders of radiology in China. He was Dean of the Medical College of Lingnan University; Honorary President of the Chinese Society of Radiology; Vice-President of the National Society of Cancer; Professor of radiology and Director of the Cancer Hospital attached to Zhongshan Medical College; Representative at the Third National People's Congress. He did outstanding work in clinicoroentgenology, radiophysics and radiobiology, and was a specialist in X-ray diagnosis of various systems of the human body. He completed the first comprehensive report on X-ray appearances of TB intestine and TB long bone among Chinese, and suggested the special angle for the posterior dislocation of the hip joint, which was called "Xie's position" when its high medical value was acknowledged at home and abroad.

Xin Shuzhi (1884–1977): Native of Linli, Hunan Province. He spent many years studying biology and agricultural science. He successively held the posts of Professor at Zhongshan University and National Central University; Director of National Publishing House of Compilation and Translation; Dean of Northwestern College of Agriculture; President of Lanzhou University. After liberation, he was Dean of Northwestern College of Agriculture; Vice-Chairman of the Chinese Society of Zoology. In 1947, he led the biological investigation team in Dayaoshan, Guangxi, discovering Yaoshan lizards, Shinisaurus crocodilurus and Shinisaurus crocodilidae. In his later years, he turned to research on ancient agricultural heritage of China. His works included: *Study of the History of Chinese Fruit Trees, New Interpretation of the Book Yugong,* and *Study of the History of Water and Soil Conservation in China.*

Xiong Qinglai (1893–1969): Native of Mile, Yunnan Province. Mathematician. Before liberation, Chief of Department of Mathematics of Southeastern University and Qinghua University; President of Yunnan University. After liberation, Senior Researcher at Institute of Mathematics, Academia Sinica; Member of National Committee of the Chinese People's Political Consultative Conference. For a long time he taught mathematics and wrote teaching materials on Higher Mathematical Analysis and Mechanics. He launched research on integral functions, meromorphic functions, algebroidal functions and normalized families. He wrote a book on *Meromorphic Functions and Algebroidal Functions – A Generalization of a Theorem of Nevanlinna.* Several outstanding mathematicians were his students. He established the mathematics department at Southeastern University and Qianghua University as well as *Acta Mathematica Sinica* and made several contributions to the development of mathematics and the growth of mathematical personnel in China.

Xu Baolu (1910–1970): Native of Hangzhou, Zhejiang Province. Mathematician. Member of Academia Sinica Department of Mathematics, Physics and Chemistry. Before liberation, Professor at Beijing University. After liberation, Professor of Department of Mathematics and Mechanics, Beijing University; Member of the National Committee of the Chinese People's Political Consultative Conference. Taught and studied mathematics for many years, and was distinguished for his unique contribution to mathematical statistics and the theory of probability, especially multivariate analysis, theory of limit distributions and experimental design. His work contributed much to the development of this branch of academic study in China. He also did significant work in matrix theory.

Yan Fuqing (1882–1970): Native of Shanghai. Medical educationist and specialist in public health; main founder and the first President of the Chinese Medical Society. He set up Xiangya (Yale) Medical College and Shanghai Medical College. Within 60 years (1910–1970) he was successively Professor, Vice-Dean, and Dean of Xiangya Medical College, PUMC, and Shanghai Medical College. After liberation, Representative at the First, Second and Third National People's Congress; Member of the National Committee of the Chinese People's Political Consultative Conference; Member of the Central Committee of Jiusan (September the Third) Society and Deputy Director of its Branch in Shanghai. He was involved in the prevention and cure of plague in 1911 at the Beijing-Hankou Railway, and in the investigation of the prevalence of ankylostomiasis, its prevention and cure in 1916 at Pingxiang Mine, and published papers at the same time. He was a meticulous researcher and made important contributions in training medical personnel in China.

Yang Weiyi (1897–1972): Alias Yizhi. Native of Shangrao, Jiangxi Province. Entomologist. Member of Academia Sinica Department of Biology. He studied the classification of insects of semipters in France, Britain and Germany. After returning to China, he began teaching at Jiangnan University. After liberation, he was Professor at and President of the Jiangxi College of Agriculture; Vice-President of the Jiangxi Branch of the Academia Sinica; Representative at the First, Second and Third National People's Congress. He joined the CPC in 1957. He devoted all his life to insect research and teaching. His views concerning insect classification and their geographical distribution was creative, and he had rich practical experience in the prevention and control of agricultural pests. He was the first one to suggest "Eliminating Snout Moth's Larva by means of Three Ploughing Methods" and the reforming measure of reserving the safflower seed for planting, which played an active part in the prevention and control of the rice pests and hence in increasing agricultural production. His works included: *Fauna of Chinese Economic Insects — the Stinkbug Family, Chinese Pests — the Rice-Stinkbug, Overall Prevention and Control of Rice Pests* and *A Report on the Investigation of Insects in Xinjiang.*

Yang Zhongjian (1897–1979): Alias Keqiang. Native of Hua County, Shanxi Province. Geologist Palaeontologist. Member of Academia Sinica Department of Geoscience. Before liberation, Professor at and President of Beijing University, Beijing Normal University, Chongging University and Northwestern University. After liberation, successively, Head of Laboratory of Vertebrate Palaeontology, Academia Sinica; Director of Institute of Vertebrate Palaeontology and Palaeoanthropology, Academia Sinica; and Head of Bureau of Compilation and Translation, Academia Sinica. He joined the CPC in 1956. He devoted all his life to the study of stratigraphic palaeontology, including stratigraphic palaeoanthropology and archaeology. He studied the Lufeng dinosaurian fauna and the reptilian fauna of the Permian and the Triassic Periods in Xinjiang and

Shanxi, and made an important contribution to the establishment and development of vertebrate palaeontology. *Fossils of Rodents in the Northern Part of China* published in 1927 was the earliest work in the field of Chinese vertebrate palaeontology. He was a scientist of great international prestige, being member of the Linnean Society of Britain, Honorary Member of the North American Society of Vertebrate Palaeontology and Honorary Member of the Moscow Association of Natural History. He also worked hard to popularize science. In his remaining years, he dedicated himself to the establishment and development of the Chinese Museums of Natural History and concurrently, was director of the Beijing Museum of Natural History. His main works were: *Fossils of Artiodactyla in Location 1 of Zhoukuodian, The Chinese Pseudosuchia and Fossil Dinosaur in Laiyang, Shandong Province,* etc.

Ye Liangfu (1894—1949): Alias Zuozhi. Native of Hangzhou, Zhejiang Province. Geologist. Taught and did survey research on geology all his life. Senior Researcher at Institute of Geology; Professor at Zhejiang University. Author of *History of Geology in the Western Hills of Beijing, Study of the Diorite in the Belt of Chinese Contact Iron Deposits* and co-author (with Yu Deyuan) of *History of Development of Igneous Rock Between Nanjing and Zhenjiang.*

Ye Qisun (1898—1977): Native of Shanghai. Physicist. Member of Academia Sinica Department of Mathematics, Physics and Chemistry. He graduated from the Graduate School, Harvard University in 1923. Before liberation, Professor at Southeastern University; Professor and Head of Department of Physics; Dean of College of Sciences, Qinghua University and President of the Chinese Society of Physics. After liberation, Head of the Committee of Administrative Affairs, Qinghua University; Professor at Beijing University; Representative of the First National Committee of the Chinese People's Political Consultative Conference and Representative of the First, Second and Third National People's Congress. His scientific contributions included the accurate measurement of Planck constant, the study of high-pressure magnetism and the study of the history of science. He spent many years training students of physics, thus giving great impetus to the development of physics in China.

Ye Zhupei (1902—1971): Native of Xiamen (Amoy), Fujian Province. Metallurgist. Member of Academia Sinica Department of Technical Science. In 1921, he went to USA to pursue higher education and returned in 1933. He was head of Chongqing Copper Refinery and Chongqing Sponge Iron works; General Manager of the Electro-chemical Smelting Factory. In 1944, he went to Europe and America to carry out an industrial survey and was assigned as the deputy head of a group at UNESCO. After returning from abroad in 1950, he was appointed as an adviser to the Ministry of Heavy Industry; Academic Secretary of Academia Sinica; Director of Institute of Chemical Engineering and Metallurgy, Academic Sinica; Member of the Second and Third National Committee of the Chinese People's Political Consultative Conference; Member of the Standing Committee of the Third National People's Congress. He was the first in China to propose, in the early fifties, that new techniques of "high blast pressure, temperature and humidity" be employed in the blast furnace. He made contributions to the comprehensive utilization and blast furnace smelting of Panzhihua magnetite ore containing vanadium and titanium, to developing the new technique of steel making with oxygen top-blown converters in China, and to research on smelting phosphorus fertilizer in shaft furnaces and the utilization of steel slag as phosphorus fertilizer. His works included: *On the Basic Problems in the Intensification of the Smelting Process in Blast Furnace, New Concept of the Iron-Smelting Process in the Blast Furnace, On the Problem of Steel Making by the Oxygen Top-Blown Converter Process in China.*

Yun Ziqiang (1899—1963): Alias Yun Daixian. Native of Wuchang, Hubei Province. Chemist. Vice-head of Academia Sinica Department of Mathematics, Physics and Chemistry. He graduated from the Department of Mathematics, Physics and Chemistry, Nanjing Higher Normal School in 1920, majoring in chemistry. Before liberation, he practised teaching at Nanjing Higher Normal School, Southeastern University, Second Normal School of Jirin and College of Pharmacology, Zhongfa University of Shanghai. In 1942, he went to the anti-Japanese democratic base area of North Jiangsu. From 1944 to 1947, he worked in Yanan, and was Vice-President of the Academy of Natural Sciences, Yanan University; Head of the Industrial Specialists School of Jin-Cha-Ji (Shanxi-Chahaer-Hebei); Vice-Dean of North China College of Technology etc. After Liberation, Chief of General Office, Academia Sinica; Vice-Head of Bureau of Compilation and Translation; Vice-President of Northeast China Branch, Academia Sinica. He helped much in the early organization and management of the chemical research establishments of Academia Sinica and of the Academia Sinica Departments.

Zhang Changshao (1909—1967): Native of Jiading (now attached to Shanghai). Pharmacologist. Professor of pharmacology at Shanghai First Medical College. Member of Academic Committee of Institute of Materia Medica, Academia Sinica; Head of Pharmaceutical Section, National Research Committee of Schistosomiasis. He devoted all his life to training large numbers of pharmacological personnel, and to research work in this field. His research included the antimalaria action of dichroine, pharmacology of antimony potassium tartrate, pharmacology of the autonomic nervous system, and the analgesic effect of morphine. His teaching and research work was a major contribution to pharmacology in China. His works were: *Modern Research on Traditional Chinese Drugs, Modern Pharmacology, General Pharmacology, Developments in Pharmacology,* etc.

Zhang Hongzhao (1877—1951): Aliases Yanqun, and Aicun. Native of Wuxing, Zhejiang Province. Geologist and one of the pioneers of Chinese geology. He established the first Institute of Geology for training Chinese geologists in 1913; and the first geological survey organization in China — Geological Survey — in 1916. He was also one of the sponsors for the Chinese Society of Geology and was its first President. He spent his old age writing and collating ancient Chinese works on geology. His works: *Shi-ya, Ancient Ores,* etc.

Zhang Jingyue (1895–1975): Alias Xianchai. Native of Wujin, Jiangsu Province. Botanist. Member of Academia Sinica Department of Biology. Before liberation, Professor at Southeastern University, National Central University, Beijing University, Southwestern Union University etc. After liberation, Professor at Beijing University. In 1922, he began studying and teaching plant morphological anatomy and made contributions to the establishment and development of botany, especially plant morphological anatomy. His works included *Origin and Differentiation of Rhizome Tissue of Fern, Study of Foreign Wood in Chinese Jurassic Period, Effect of Light Intensity upon the Growth of White Leaf Mustard* and *Resin Cell of Masson Pine (Pinus Massoniana)*.

Zhang Yun (1890–1977): Native of Pingyang, Zhejiang Province. Anatomist and medical educationist. Before liberation, Professor at and Dean of Medical School, Hebei University, Xiangya (Yale) Medical College of Hunan, Shanghai Medical College and Southeastern Medical College of Shanghai. After liberation, successively, Professor, Head of Department and Director of China Union Medical College, China Medical University, and Institute of Experimental Medicine, the Chinese Academy of Medical Sciences; Vice-President of the Chinese Academy of Medical Sciences; President of the Chinese Society of Anatomy. In 1911, he began teaching medicine and accumulated a rich teaching experience. He was chief editor of many medical textbooks, such as *Human Anatomy*. He also published some academic papers, and made contributions to the Chinese medical education and medical-scientific research.

Zhang Zhaoqian (1900–1972): Alias Guanchao. Native of Wenzhou, Zhejiang Province. Botanist specializing in plant taxonomy. Member of Academia Sinica Department of Biology. From 1933–1935, he studied at the British Royal Institute of Botany. Before liberation, Professor at the Agricultural schools of Zhejiang University and Guangxi University; Head of Department of Biology, Zhongzheng University; Technician at Beijing Institute of Jingsheng Biological Survey, and concurrently, Professor of Biology, Beijing University. After liberation, Senior Researcher and Deputy Director of Institute of Botany, Academia Sinica; Senior Researcher, Head of Laboratory of Plant Taxonomy, Deputy Director and Acting Director of the South China Institute of Botany, Academia Sinica. He joined the CPC in 1956. He was engaged in research on systematic botany all his life, studying the classification and distribution of compositae, podocarpaceae, violaceae, ranunculaceae etc. After liberation, he compiled and published *Flora of Hebei Province, The Key of Chinese Plant Families and Genus* and *The Figure and Sketch Description of the Main Chinese Plants*. He took part in the investigation of resource plants, studying Hainan Flora, the compositae and the flora in Shiwandashan, Guangxi. He wrote more than twenty papers, such as, the part on compositae in *Chinese Flora, Some Sp. nov. of the Chinese Compositae, The Sp. nov. of Senecio and Its Relative Genus*. In his remaining years, he was chief editor of *Flora of Hainan*. He was engaged in plant research for more than 40 years, and made contributions to plant taxonomy and regional flora in South Shina and the exploration and utilization of plant resources.

Zhao Chenggu (1885—1966): Alias Shimin. Native of Jiangyin, Jiangsu Province. Chemist. Member of Academia Sinica Department of Mathematics, Physics and Chemistry. Before liberation, Professor at Southeastern University and PUMC; Director of Institute of Materia Medica, Beijing Academy of Sciences. After liberation, Director of Institute of Materia Medica, Academia Sinica. One of the scientists who initiated the study of chemical constituents of Chinese herbs in China. After systematically analyzing the chemical constituents of more than thirty Chinese medicinal herbs, such as Chinese ephedra, Dichroa febrifuga, Corydalis yanhusuo, Radix Rauwolfiae, pseudo-ginseng, Stephania tetrandra, and core of lotus seed, he discovered many new alkaloids. In cooperation with pharmacological workers, he also studied the anticancer effect of alkaloid of Dichroa febrifuga and the analgesic effect of Corydalis B. He played a pioneering role in research on Chinese pharmocological chemistry and the preparation and training of personnel in this field.

Zhao Jiuzhang (1907—1968): Native of Wuxing, Zhejiang Province. Space physicist and meteorologist. Member of Academia Sinica Department of Geoscience. After liberation, Director of Institute of Geophysics, Academia Sinica; President of the Chinese Society of Meteorology; Representative at the Second National People's Congress; Member of the Third Standing Committee of the National People's Congress; Member of the Second National Committee of the Chinese People's Political Consultative Conference; Member of the Standing Committee of the Third National Committee of CPPCC; Member of the Standing Committee of the Fifth Central Committee of Jiusan (September the Third) Society. He pioneered research work on dynamic meteorology in China, investigated the question of the dynamics of trade winds and was the first one to discover the instability of the long wave of the westerlies. He made contributions to the development of geophysics and space physics in China, and to research on sea waves.

Zhou Ren (1892—1973): Alias Zijing. Native of Nanjing, Jiangsu Province. One of the pioneers in modern ferrous metallurgy and ceramics in China. Member of Academia Sinica Department of Technical Sciences. Before liberation, Dean of Jiaotong University, Shanghai; Director of Institute of Engineering Science, National Central Academy of Sciences. After liberation, Vice-President of East China Branch, Academia Sinica; Director of Institute of Metallurgy and Ceramics, Academia Sinica; President of Shanghai University of Science and Technology; President of Chinese Society of Metals; Representative of the First, Second and Third National People's Congress. He was the first one in China to develop successfully the steel-making process using the electric furnace. He achieved important results in research on nodular cast iron, for which he was given the National Scientific Prize of 1956. He was very successful in the smelting of Baotou Iron Ore in an experimental blast furnace. He also worked hard at the analysis and documentation of ancient Chinese porcelain; the papers he wrote concerning ancient porcelain have provided valuable data for the further development of ceramics in China. His main works: *Study of Ceramics in Jingde Town,* and *Behaviour of Fluorine in Blast Furnace Smelting.*

Zhu Kezhen (1890–1974): Native of Shaoxing, Zhejiang Province. Meteorologist; geographer; founder of modern meteorology in China; Vice-President of the Academia Sinica; Head of Academia Sinica Department of Geoscience. Successively, Vice-Chairman of the Chinese Association of Science and Technology; Head of Comprehensive Survey Committee, Academia Sinica; Honorary President of the Chinese Association of Meteorology; President of the Chinese Society of Geography; Member of the Standing Committee of the First, Second and Third National People's Congress. Joined the CPC in 1962. Famous for his contributions to Chinese climatogenesis, climatic characteristics, climatic division and climatic variation and also to bioclimatics and the history of natural sciences. He wrote many papers. At the same time, he paid attention to and took an active part in the popularization of science. He contributed much to the establishment and development of modern Chinese meteorology and geography.

Zhu Xi (1899–1962): Native of Linhai, Zhejiang Province. Experimental biologist. Member of Academia Sinica Department of Biology. Before liberation, Professor at Zhengshan University and Taiwan University; Senior Researcher of Beijing Academy of Sciences. After liberation, Director of Institute of Experimental Biology, Academia Sinica. He studied the egg-maturation fertilization and artificial parthenogenesis of animals and described the relation between the different states of ovary maturity and embryo development; he established the method of ovulation of toad ovary fragments in vitro and opened up a new way for researches into such problems as ovum maturity, fertilization and development. In research on the hybridization of mixed sperms of different strains in Bombyxmori, he discovered that the genetic character of the offspring could be affected by the supernumerary sperms of the different strains. He emphasized that scientific research should serve production, and made important contributions to introducing the Indian Erisilkworm and to artificial propagation of several kinds of economic fresh water bony fishes. Main works: *Biological Evolution,* etc.

Zhuang Changgong (1894–1962): Alias Pike. Native of Quanzhou, Fujian province. Chemist. Deputy Head of Academia Sinica Department of Mathematics, Physics and Chemistry. Before liberation, Head of Department of Chemistry, Northeastern University; Dean of College of Sciences, National Central University; Director of Institute of Chemistry, National Central Academy of Sciences. After liberation, Director of Institute of Organic Chemistry. He studied the structure of Ergosterol and the synthesis of steroid and polycyclic fatty compounds. Also worked on the structure of alkaloids, separated tetradrine and demethyltetradrine from the traditional Chinese medicine stephania tetrandra, and determined its structure. He played a pioneering role in investigating organic structure and synthesis by using micromethods, and also in training personnel.

SECTION V

PRIZES AND CERTIFICATES OF MERIT IN SCIENCE

Awards for Science in 1956

In 1956, there were 34 treatises which won prizes offered by the Academia Sinica in the fields of the natural sciences. Amongst them, 3 treatises were given first-grade awards, 5 the second-grade and 26 the third-grade. The authors who won prizes included members of the Academia Sinica, teachers of higher education, and technicians and engineers working in specialized fields. The announcement is reproduced below:

Announcement of the 1956 annual prizes for science (in the fields of the natural sciences)

The assessment of the treatises for prizes of the Academia Sinica in 1956 has been completed. The following is the list of the treatises that have won awards (arranged according to disciplines; treatises in the same discipline with the same grade of the award are arranged in order of the number of strokes in the Chinese characters of the authors' names).

Title of Treatise	Author	Author Affiliation	Grade of Prize
MATHEMATICS			
On the Theory of Functions of Several Complex Variables in the Classical Domains	Hua Luogeng	Institute of Mathematics, Academia Sinica	First Grade
On the Characteristic Classes and Pontrjagin Classes	Wu Wenjun	Institute of Mathematics, Academia Sinica	First Grade
Geometries of K-Spread Spaces and Generalized Metric Spaces, Theory of Curves in Projective Space	Su Buqing	Fudan University, Institute of Mathematics, Academia Sinica	Second Grade

Prizes and certificates

Title of Treatise	Author	Author Affiliation	Grade of Prize
MECHANICS			
Engineering Cybernetics	Qian Xuesen	Institute of Mechanics, Academia Sinica	First Grade
Problem of Large Deflection of Thin Elastic Circular Plate	Qian Weichang assisted by: Hu Haichang Ye Kaiyuan	Qinghua University, Institute of Mechanics, Academia Sinica	Second Grade
General Plastic Behaviour and Method of Solution for Plane Stress Problems with Axial Symmetry in Strain Hardening Range considering Finite Strain	Li Minhua (female)	Institute of Mechanics, Academia Sinica	Third Grade
Three-Dimensional Problems in the Theory of Elasticity of Transversely Isotropic Bodies	Hu Haichang	Institute of Mechanics, Academia Sinica	Third Grade
PHYSICS			
Investigation of Internal Friction and Mechanical Properties in Metals	Ge Tingsui et al.	Institute of Metals, Academia Sinica	Second Grade
Investigation of the Preparation Process of Nucleus Glue	He Zehui (female) assisted by: Lu Zuyin, Sun Hanchang	Institute of Physics, Academia Sinica	Third Grade
On the Theory of Lattices	Huang Kun	Beijing University	Third Grade
Manufacture of Halogen Counter and High Current Tubes, and Investigation of their Discharging Mechanisms	Dai Chuanzeng, Li Deping, assisted by: Xiang Zhilin, Tang Xiaowei, Li Zhongzhen (female)	Institute of Physics, Academia Sinica	Third Grade

Title of Treatise	Author	Author Affiliation	Grdae fo Prize
CHEMISTRY			
A Study of Fritillaria Alkaloids	Zhu Ziqing, assisted by: Lu Renrong, Huang Wenkui	Lanzhou University; Institute of Organic Chemistry, Academia Sinica	Third Grade
Studies on the Chemistry of Citrinin	Wang You, assisted by: Ding Hongxun, Tu Chuanzhong, Jia Chenwu	Institute of Organic Chemistry, Academia Sinica	Third Grade
Theory of Molecular Structure	Tan Aoqing	Northeastern People's University	Third Grade
Studies on the Determination of Molecular Weight of High Polymers	Qian Renyuan et al.	Institute of Chemistry, Academia Sinica; Institute of Applied Chemistry, Academia Sinica	Third Grade
ZOOLOGY			
On the Experimental Research of Indian Eri-Silk-worm	Zhu Xi, assisted by: Zhang Guo, Jiang Tianji, Wang Gaoshun et al.	Institute of Experimental Biology, Academia Sinica	Third Grade
BOTANY			
A New System for the Genus *Pedicularis*	Zhong Buqiu	Institute of Botany, Academia Sinica	Second Grade
Study on the Life Cycle of *Porphyra tenera Kjellm*	Zeng Chengkui, assisted by: Zhang Derui	Laboratory of Marine Biology, Academia Sinica	Third Grade
AGRONOMY			
Studies on the Lapinized Rinderpest Virus	Yuan Qingzhi, assisted by: Shen Rongxian, Ujiiye Hachira, Li Baoqi	Harbin Institute of Veterinary Science	Third Grade

Prizes and certificates

Title of Treatise	Author	Auhotr Affliation	Grade of Prize
MEDICAL SCIENCE			
The Transmission of *Wuchereria Malayi* in Nature by *Anopheles hyrcanus vasinensis*.	Feng Lanzhou	China Union Medical College; Institute of Entomology, Academia Sinica	Third Grade
GEOLOGY			
The Stratigraphy of Taizihe Valley, Liaoning	Wang Yu, Lu Yanhao, Yang Jingzhi, Mu Enzhi, Sheng Jinzhang	Nanjing Institute of Geopaleobiology, Academia Sinica	Third Grade
Paleogeographic Atlas of China	Liu Hongyun	Institute of Geology, Academia Sinica	Third Grade
New Investigations on Thermal Reactions and Control of Phase Transitions of Gibbsite and Kaolinite	Zhang Yuanlong	Institute of Geology, Academia Sinica	Third Grade
GEOPHYSICS			
The Influence of Tibetan Plateau on the General Air Circulation of East Asia and the Weather in China	Ye Duzheng, Gu Zhenchao	Institute of Geophysics, Academic Sinica	Third Grade
Some Problems in the Theory of Seismic Waves and Seismic Exploration	Fu Chengyi	Institute of Geophysics, Academia Sinica	Third Grade
DYNAMICS			
Studies on Gas Turbine	Wu Zhonghua	Qinghua University; Laboratory of Dynamics, Academia Sinica	Second Grade

Title of Treatise	Author	Auhotr Affliation	Grade of Prize
METALLURGY			
Research on Hydrogen in Steels	Li Xun et al.	Institute of Metals, Academia Sinica	Third Grade
Investigation of Low Alloy Steels for the Replacement of the USSR 40X	Wu Ziliang et al.	Institute of Metallurgy and Ceramics, Academia Sinica	Third Grade
Investigation of Nodular Cast Iron	Zhou Ren, Zhou Xingjian, Zhou Yuanxi, Li Lin (female) et al.	Institute of Metallurgy and Ceramics, Academia Sinica	Third Grade
On the Mechanism of Coherency of Austenitic Transformation	Ke Jun	Beijing Institute of Ferrous Metallurgy	Third Grade
PETROLEUM CHEMISTRY			
The Aromatization of Fischer-Tropsch Naphtha	Peng Shaoyi, Guo Xiexian, Chen Yingwu, Zhang Yuanqi et al.	Institute of Petroleum, Academia Sinica	Third Grade
Nitrided Fused Iron Catalysts for the Fluidized Bed Synthesis of Liquid Fuels	Lou Nanquan, Zhang Cunhao, Wang Shanyun, Lu Peizhang et al	Institute of Petroleum, Academia Sinica	Third Grade
ARCHITECTURE			
The Action of Direct Current in Soils and its Effect on the Physical and Mechanical Properties of the Soils	Wang Wenshao	Ministry of Water Conservancy, the People's Republic of China	Third Grade
Analysis of Rigid Frames with Non-Uniform-Section Members	Cai Fangyin	Ministry of Architectural Engineering, the People's Republic of China	Third Grade

Medals, monetary awards and certificates of merit will be given to the authors of the above research treatises by the Academia Sinica with 10,000 yuan for the first-grade award, 5,000 yuan for the second-grade and 2,000 yuan for the third-grade.

Prizes, certificates and medals for group efforts will be shared in accordance with the Temporary Prize Act of the Academia Sinica, Article 4, and a written evaluation of each treatise will be published in *Bulletin of Science.*

President Guo Moruo
January 24, 1957

Certificates Issued by the National Scientific Conference of 1978

826 advanced groups, 1192 advanced model researchers, 7657 units and individuals who were excellent in their scientific research projects, were awarded prizes at the National Scientific Conference. These advanced groups and model scientific researchers disregarded the heavy pressure exerted on them by the "Gang of Four" and made great contributions to the development of science and technology in our country. Highly respected by the Chinese people, they have duly won certificates of merits issued by our Party and State.

The following are the names of those who received the prizes on behalf of their provinces, municipalities, autonomous regions and departments in charge of the research work: Wang Tiemeng, Wang Jingwu, Liu Tianquan, Li Xilan, Yang Le, Min Yu, Shen Zhiquan, Chen Renfu, Jin Shanbao, Zhou Hao, Pang Rui, Zheng Guanghua, Zhao Naigang, Hu Zhong, Yao Jinzhong, Qian Lingxi, Huang Xiling, Peng Shilu, Tan Qinglin et al.

INDEX

A

abdominal ailments 132
acidification 64
acidifying agent 66, 67
acupuncture 5, 38, 128, 129, 227
acu-anaesthesia 128, 129
acu-analgesia 129
aftershocks 108
air-sea interaction 114
algorithm 61, 62
ambients 25
ammonoid 49
amylose agent 66, 67
anaesthesia 128, 135, 227
analgesia 128, 129
ancient acoustics 14
angiosperms 47
animal observing 96, 97
anomalies 107, 108
anthropology 41
architectural acoustics 17
aromatic sextet 74
artificial parthenogenesis 186
assemblages 47, 48, 49
asteroids 86
astrolable 85
astronomy 84
astrophysics 227
atomic bomb 200
atomic energy 162
aureamycin 36
Australopithecoid 43
automorphism 56

B

back-scattering 29
ballistic missiles 4
barotropical model 98
basal bone 42
benzologs 74
Betti number 57
biomacromolecules 33
biomembranes 37
blackbody 31
black holes 88
Borel directions 59
Bridgeman method 25
brightness 30
Brownian motion 140, 141
burns 133

C

Carboniferous 47, 49
   Early — 47
   Late — 47
carcinogenesis 37, 38
cardio-vascular diseases 132
carvone 81
cataracts 134
cathodoluminescence 29
celestial mechanics 86
cenophytes 47
cenozoic 48, 50
Cephalepoda 49
Changjiang bridge (Wuhan) 159, 174
chemical bonds 72
Chen's Theorems 55, 56
Chinese herbal drugs 135, 136
chloroplast 37
chromatogram 5
chromite ore 160
chronic diseases 132
cloud observing 96
coding 57
cohomology 58
cohomotopy 58
complex variables 56
conjugate 72
conodonts 50
control systems 62
Control Theory 62
Coriolis parameter 98
cornea transplant 161
Coronocephalus 49
cosmic science 4
cosmology 88, 89
cranium 42, 43
Cretaceous
   Early — 47
   Late — 47, 48, 49
Critical Path Method, CPM 55, 61
crust 109
crystal 24, 25
crystallography 34
crytomeria 210

D

daphne genkwa 137
decoctions 136, 137
Devonian 47, 48
diagnostic forecast 92, 93
diaphragms 22
dielectric 32

differential geometry   57, 58
dimers   65
diode   27
Diophantine   56
dipole   77
directional distributions   17
dislocations   28
dissociation   68, 81
DNA   35, 36, 38
DNP   35
doping   25, 29

E

earth crust   110
earthquake   100–111
earthquake engineering   110
earthquake percursors   106
earthquake prediction   106, 107, 108
eigenpolynomials   72, 76
eigenvectors   62
electroacoustics   22
electronic blood-corpuscle counter   196
electrophoresis   35
electroplating   27
endocrine system   130
endonucleases   38
energy levels   73, 75
enzymes   34–40
equi-tempered scale   15
Eskimoids   43
etching   27
Euclidean space   58
Euestherites bifurcatus   49
eukaryotic organisms   48
evolution   47
excitations   17, 32
extragalactic physics   4

F

fascicules   49
Fenchel's Theorem   57
FFT   22
flavin adenine dinucleotide (FAD)   34
Flos daturae   56
fluoroolefins   77, 82
Fluorophlogopite   75
fluorophore   35
forecast   91–99, 104, 105, 115
fossil   41–44, 48–50
fossilization   44
free electron model   72
free radicals   79, 81

functional analysis   59
fusulinids   46, 49, 50

G

GaAs crystals   25
galactic clusters   86
galaxies   88, 89
Gandisi Shan   156
gelation   65, 67
general theory of relativity   142
genetics   38
geomorphology   115
geotectonics   5
Gigantopithecus   43, 51
Glaciation Period   44
globoerythomycin   211
gluon   5
Goldbach's conjecture   4, 55, 56
gradient method   62
graph theory   72
graptolites   49
Great Hall   18
groups   56, 57
Grunwald's Theorem   57
gyromagnetometer   209

H

H-bomb   204, 205, 208
haemorrhoids   134
hail   92
Hall effect   28
halogen   76, 81
hemodynamic index   126
hepatoma   37, 38
herbal drugs   135, 136, 137
heterostructure   27
hippocampus   123
holograph   17
hominid   45
Homo erectus   41–45, 51
Homo sapiens   42, 44
homologous linearity   73, 75, 76
homologues   73, 74, 75
homology   58
homotopy   58
hormones   37
Hubble constant   89
human chorionic gonadotropin (HCG)   39
human fossils   41, 42, 44
hybrid orbital theory   72
hydration   80
hydrolases   35

hydrolysis 35
hydrophone 20
hyperon 5, 11

I

illumination 30, 31
information theory 121
inhomogeneity 31
insulin 201, 203
integrated circuits 26–29
interferometers 31, 84
invertebrate palaeontology 48
ionization 66
isolines 92
isomerization 73, 79
isomers 73, 77

J

Jacobian 98
Jacobian theorems 57
journals 55

K

kala-azar 6
kelp 5
kernels 59
Kerr effect 28
kidney 127, 128
kynurenine 36

L

Laplace sequence 57
lasers 27, 30–32
laver 5
lesions 129, 134
leucocytes 38
Lie algebra 56
Lie group 56, 57
Lingraph 16
Liujiang Man 44
logatoms 22
logic circuit 26
long-range forecast 94, 95
long-wave theory 94
Luoxue Mountains 10

M

magnetron 31
marine acoustics 115
marine biology 115
marine ecological studies 115
marine geology 115
marine ichthyology 115
marine science 112, 113, 116
Markovian process 55, 61
masers 31
mathematical logic 56
matrices 56
Mawangtui cadaver 39
Melanesoids 43
meromorphic function 55
mesozoic 48, 50
mesozoic era 184
metabolism 36
meteorology 91, 92
methylbenzene 157
microfaunas 50
microfossils 50
micropalaeontology 50
microscope (universal projective) 176
microwaves 31
mid-range forecast 93, 94
Milky Way 89
Mindel Glacial 44
Mindel-Riss Interglacial 44
mineral resources 46
Miocene
  Late — 43
mitochondrial 37
molecular orbitals 72, 76
Mongoloid 42, 43
monomers 65
MOS forecast 98
mosaic disease 39
Mount Everest 156
moxibustion 128
Mozi 15
mufflers 21
multiplicative groups 57
mutation 38
mycoplasma 39

N

nasion 42
nautiloid 49
nebular physics 87
nervous system 129, 130, 131
neurasthenia 120, 121
neurohormonal system 127
neutrino 5, 11
neutron stars 88
Nevanliuna Theory 55
Newtonian mechanics 147
nitrogenase 35

nodular cast iron 154
noise 20
noise control 20
nucleic acid 35, 36
nucleotide 35, 36
numerical weather forecast 97

O

observatory 84, 85
occipital 42
oceanographic chemistry 116
oceanographic physics 114
octave 15
odontae animals 184
olation 65, 70
operational research 61
optimum seeking method (OSM) 55, 61
Ordos Man 44
organic chemistry 71
orthopaedics 133
oxolation 65, 69

P

palaeoanthropology 41, 46
palaeobotany 47, 48
palaeontology 41, 46–52
Palaeozoic 48
Paleolithic
    Late — 44
    Middle — 44
palynology 48
Panama telescopic sight 152
partial differential equations 60
pathogens 130, 134
Pavlov's theory 117
Peckichara 48
Peking Man 41–44, 46, 158, 161
penicillin 154
pentatonic scale 15
peptides 34, 39
Permian 49
phosphorylation 37
photoelectric astrolabe 205
photolithography 26
photomagnetoelectric 28
piezoelectric crystals 19, 21
piezoquartz 19
planar process 26, 27
Planck's theory 31
planetarium 166, 174
Pleistocene
    Early — 43, 44, 51
    Late — 44, 51
    Middle — 43, 51
plenum chamber pressure 21
pleuronodoceras multinodosum 49
Pliocene
    Early — 43
polarisation 32, 39
polarity 79
polyhedrons 57
polymerization 31, 64–70
polymers 65, 68
polynomials 58, 59
polyterrafluoroethylene 200
pongid 45
precipitation area 93
premolar 42, 44
primitive group 45
procto-anal diseases 134
projective microscope 176
prostatitis 134
proton synchroton 219
protonation 64, 65, 69, 73
PSC 129, 130
psychodelic drugs 123
psychology 117–124
    child — 119, 120
    developmental — 123
    educational — 119, 123
    engineering — 121, 123
    experimental — 123
    general — 123
    labour — 119, 121
    medical — 119, 123
    physiological — 118, 121, 123
psychotechnology 119
psychotherapy 121
pteromalidae 189
pulsilogram 127
Pythagorean scale 15

Q

Q-process 55
Qi 101
Qing Hao Su 135
Qomolangma 156
quantum chemistry 72
Quaternary 48, 51
quasars 89
quinolinic acid 36

R

radicals 79

radioactive carbon 44
Raleigh distribution 17
Raman scattering 30
Ramapithecines 43
reflex 118, 119
refraction 86
regioselectivity 77, 78
resistivity 25, 28
resonance 15
rheology 131
riboflavin 37
Riemannian space 57, 58
ring 56, 57
Riss-Würm Interglacial 44
RNA 35, 38
Rochell salt 19

S

sand-siltation 116
satellites 86
saw-tooth wave 21
scanning microanalyser 216
schistomiasis 214
schizophrenia 121
Schlicht functions 59
seismic belts 109, 110
seismic wave 101, 102
seismic zones 109, 110
seismoscope 101
seismotectonic analysis 110
semiconductors 24—29, 172
set theory 56
Sicidium 48
silicic acid 64
silkworm 35, 36
singularity 62
soil 95
solar eclipse 176
solar physics 87
sound 15, 17, 19
soya beans 157
special theory of relativity 140, 142
spectroscopy 29
spline curve 55
statistical dynamical forecast 98
statistical probability forecast 97
statistics 61
Stegodon-Ailuropoda 44
stellar physics 87
stochastic process 61
streptomycin 36
subalgebras 56, 57
subharmonic waves 17

sunspot 87
surface acoustics wave (SAW) 19
symmetry rules 72
synoptic charts 91, 93, 96
synoptic dynamic model forecast 93

T

tectonic movements 110, 116
telefacsimile 208
teleprinter 164
temperature 69
Temple of Heaven 15, 17
tensor 52, 142
terminal effect 75
terramycin 166
theory of functions 58
theory of numbers 56
theory of probability 61
thermo-differential analyser 160
Three-Tone Stone 15, 17
tidal storms 114
tidal waves 114
tides 114
titanium 6
topology 58, 61
torus 43
Traditional Chinese Medicine (TCM) 125—140
Trans-Himalaya 159
transaminase 36
transient 28
Triassic
    Early — 47
    Late — 47
trichosanthin 34
trilobites 49
trimers 65
tropomyosin 34
tryptophon 36
turbine generator 200
Tyndall's acoustics 16

U

ultracentrifuge 35
ultrasonics 18, 19
ultraviolet catastrophe 140
unified field theory 142, 143
Upper Cave Man 43, 45
urinary diseases 134

V

vail 64

vanadium 6
vertebrate palaeontology 47
virus 36, 39
visceral organs 127
vocoder 21, 22

W

Wall of Echoes 15, 17
Wangtian tree 210
Waring's problem 56
waves 94, 114
weather models 94
westerly index 94
Wurm Glaciation 45

X

xylene 73

Y

Yang 101, 127
Yin 101
Yin-Yang 125, 128

Z

zenith tube 85
Zesterophyllum 47
zoning 109
zygomatic bone 43